Unconventional Raw Materials for Food Products

Unconventional Raw Materials for Food Products

Editors

Adriana Dabija
Georgiana Gabriela Codină
Lăcrămioara Rusu

MDPI

Basel • Beijing • Wuhan • Barcelona • Belgrade • Novi Sad • Cluj • Manchester

Editors
Adriana Dabija
Stefan cel Mare University of
Suceava
Suceava, Romania

Georgiana Gabriela Codină
Ștefan cel Mare University
Suceava, Romania

Lăcrămioara Rusu
Vasile Alecsandri University
of Bacau
Bacău, Romania

Editorial Office
MDPI
St. Alban-Anlage 66
4052 Basel, Switzerland

This is a reprint of articles from the Special Issue published online in the open access journal *Applied Sciences* (ISSN 2076-3417) (available at: https://www.mdpi.com/journal/applsci/special_issues/unconventional_raw_materials).

For citation purposes, cite each article independently as indicated on the article page online and as indicated below:

Lastname, A.A.; Lastname, B.B. Article Title. *Journal Name* **Year**, *Volume Number*, Page Range.

ISBN 978-3-0365-8554-3 (Hbk)
ISBN 978-3-0365-8555-0 (PDF)
doi.org/10.3390/books978-3-0365-8555-0

Contents

About the Editors

Adriana Dabija

Adriana Dabija is a Professor in the Department of Food Technologies, Food and Environmental Safety, Faculty of Food Engineering, Stefan cel Mare University of Suceava, Romania, and a PhD supervisor at the Doctoral School of Applied Sciences and Engineering, Ștefan cel Mare University, Suceava. She has experience in production, teaching and research in the field of food engineering (37 years), and her research activity is in the fields of Food Biotechnology, Microbiology of Food Technology and Quality Control in Food Industry, and Quality Management in Food Industry. She has published 54 ISI Thomson Reuters indexed papers, 177 articles in international data-based journals and proceedings, 86 papers in non-indexed scientific journals and proceedings, 3 patents, 1 Euro patent, and 31 patent proposals.

Georgiana Gabriela Codină

Georgiana Gabriela Codină (Professor Habilitate, PhD Eng.) joined the Ștefan cel Mare University, Faculty of Food Engineering (Suceava, Romania), in 2005, where she teaches different courses in the food science and technology field. She holds a PhD in Industrial Engineering from 2009 and became a PhD supervisor in Food Engineering in 2017, when she sustained her habilitation thesis. Dr. Codină has expertise in food rheology, food sensory analysis, bread making, beer industry, and food quality analysis through different rheological, textural and sensory methods, as well as in the design of experiments and data analysis. Her research activities focus on improving the technology of food products and the quality of different foods such as baked goods, beer, dairy products, etc. She was involved in more than 10 interdisciplinary research projects, has published more than 150 scientific papers, and has authored more than 35 patents under evaluation from which 6 were published.

Lăcrămioara Rusu

Lăcrămioara Rusu, Associate Professor Habilitate, PhD Eng., teaches at "Vasile Alecsandri" University of Bacau from Romania, Faculty of Engineering, Chemical and Food Engineering Department, in the fields of chemical engineering and food engineering (since 2002). She became a PhD supervisor in the environmental engineering field from 2022 when she sustained her habilitation thesis. She obtained her doctoral degree in 2001, in the field of chemical engineering, from "Gheorghe Asachi" Technical University of Iasi, Romania. Her research in recent years has focused on obtaining biocomposite materials with applications in food industry and environmental protection. Her scientific interests are centered around the valorization of bioresources, food industry by-products and waste. To this end, she takes an interdisciplinary approach to obtain biosorbents, biopesticides and functional foods by incorporating bioactive compounds. She has published more than 110 scientific papers from which 15 were in Q1 and Q2 areas, has been involved in 17 interdisciplinary research projects and holds 2 patents.

applied sciences

MDPI

Editorial

Studies on the Manufacturing of Food Products Using Unconventional Raw Materials

Adriana Dabija [1,*], Lăcrămioara Rusu [2] and Georgiana Gabriela Codină [1]

[1] Faculty of Food Engineering, Ştefan cel Mare University of Suceava, 13 Universitatii Street, 720229 Suceava, Romania; codina@fia.usv.ro
[2] Department of Chemical Engineering and Food, Vasile Alecsandri University of Bacău, 600115 Bacău, Romania; lacraistrati04@yahoo.com
* Correspondence: adriana.dabija@fia.usv.ro

Citation: Dabija, A.; Rusu, L.; Codină, G.G. Studies on the Manufacturing of Food Products Using Unconventional Raw Materials. *Appl. Sci.* **2023**, *13*, 7990. https://doi.org/10.3390/app13137990

Received: 1 July 2023
Revised: 5 July 2023
Accepted: 6 July 2023
Published: 7 July 2023

Today, companies blend innovation with tradition to create new products, as the food business is continually looking for fresh product ideas that reflect worldwide trends. The demand for functional foods, dietary foods, etc. as well as the depletion of conventional raw materials forces the adoption of novel raw materials in assortment diversification. Researchers are working to identify new sources of basic materials, such as those that can take the place of salt, sugar, and gluten from baked products, as well as animal proteins and lipids. For experts in the industry, the utilization of these unconventional raw materials in recipe creation presents a significant problem, both in terms of customer acceptance and applicable law.

Wang et al. [1] provided a summary of the research on the synergism of sweeteners that has been carried out so far, analyzed it, and offered comments on it. Additionally, although there has been significant progress in recent years in understanding the molecular mechanisms underlying the synergism of sweeteners, it is still hampered by a number of technological limitations. They then went over a number of hypotheses that have been put forth to explain the molecular mechanism underlying the synergism of sweeteners, particularly the interaction that results from the combination of sweeteners' effects on sweet taste receptors, and evaluated their scientific viability and shortcomings generally. Food and drink products such as candies, biscuits, and beverages have all been known to combine sweeteners. To improve the acceptability, functionality, and economics of the products, a combination of artificial intense sweeteners (such as sucralose, aspartame, and acesulfame-K) and natural bulk sweeteners (such as fructose, erythritol, and stevioside) can be used. These formulations include health and nutritional benefits in addition to reducing the amount of some sweetener components to comply with dosage restrictions (such as the current, well-liked sugar-free Coke) [1].

Another study illustrates the potential for greater capitalization of grape seeds from the seven grape types grown in Romania due to their unique qualities. Proanthocyanidins, which have an antioxidant potential 20 times higher than vitamin E and 50 times greater than vitamin C, are among the phenolic chemicals found mostly in grape seeds. Due to their antioxidant capacity, grape seeds have positive effects on cardiovascular disorders, anti-inflammatory, anti-cancer, and immune-boosting properties. They can currently play a significant role in finding new food as a source of bioactive compounds. Consumers are currently showing a lot of interest in items that are as natural as possible, which helps the body's immunity grow. Natural ingredients can be a significant alternative in the food industry [2].

Tița et al. [3] used three types of volatile oils, namely, volatile mint oil, volatile fennel oil, and volatile lavender oil to improve the quality of a kefir-type acid dairy product. In both conducted examinations, the finished product enhanced with volatile oils outperformed the control sample. Kefir samples that included volatile oils remained more sensory and texturally consistent during the duration of storage. All of these aspects demonstrated that the investigated items are accepted from a sensory point of view in addition to the positive effects they have on the health of the consumer. The investigated product follows current trends because it incorporates bioactive components that are advantageous for the consumer's health [3,4]. Another study sought to evaluate the antibacterial activity of volatile oils and the changes in the chemical composition of kefir enhanced with encapsulated volatile oils by enzymatic techniques. The fundamental benefit of employing encapsulation is that sodium alginate protects the volatile oils' beneficial constituents, allowing them to slowly diffuse into the kefir sample. Due to the volatile oils' antibacterial and antioxidant characteristics, products with high nutritional values were produced that are good for the consumer's health and have a longer shelf life [5].

An important source of non-conventional raw materials that can be successfully used to obtain food products is represented by secondary products resulting from various technological processes of the food industry. Spent grain is an important by-product that should be fortified to produce foods that are beneficial to health. A study used mathematical modeling and statistical optimization to provide a spelt pasta recipe that includes spent grain. Goods that have been enriched with spent grain are known as fortified foods. It is known that spent grain is a useful by-product rich in nutrients, including dietary fibers, minerals, vitamins, and lipids. A significant amount of bioactive substances with strong antioxidant properties, such as hydroxycinnamic acids, particularly ferulic and p-coumaric, are present in spent grain. According to the findings, 11.70% was the ideal determination for spelt pasta formulation that established a good balance between sensory and nutritional considerations. Products with a high fiber and protein content, antioxidant activity, and high polyphenol concentration include spelt pasta supplemented with spent grain. The achieved color of the pasta was acceptable, and the cooking losses were under 12%, which places them in the category of high-quality goods. These findings demonstrate that spent grain can be successfully included in a fortified pasta recipe to provide high-quality outcomes. Because they may enhance the products' nutritional value and have a lower glycemic index than pasta made with durum wheat white flour, spent grain flours can be employed in food compositions [6]. According to the results of another study, the inclusion of spent grain in the wafer formulation produced goods with a high level of acceptance, altered batter texture, enhanced all parameters, and only lowered adhesiveness. The samples that included spent grain had a lower pH and density. In comparison to the control sample, the fracturability of the goods with spent grain reduced as it was added, and the color darkened due to the spent grain's particular color. Adding spent grain boosts the amount of fiber and protein, antioxidant capacity, and baking loss because of the fibers it contains [7]. Another by-product, grape pomace, was used in obtaining flour by Gerardi et al. [8]. They studied the effects of time and temperature on the stability of bioactive molecules, color, and volatile compounds during the storage of grape pomace flour. They concluded that once pomace flour was stored for six months at four degrees Celsius, the activity of various health-relevant bioactive compounds remained constant, suggesting its potential use as a functional food ingredient [8].

Identification, measurement, and extraction of plant compounds with positive effects on human health that can be used in food have recently attracted increasing interest. New nutritional supplements or food products may be created as a result of the identification of these compounds. Through the application of contemporary extraction techniques, blackberries can offer even more to healthy foods and food ingredients. Anthocyanins may be found in agricultural and food processing waste from the blackberry industry. Blackberries can replace artificial food additives such as colorants and stabilizers. Blackberries are used as a traditional raw material to make jam, compote, as well as unfermented and fermented beverages. Additionally, adding blackberries increases the finished food's nutritional content and shelf life [9]. Due to the rising demand for the commercial synthesis of compounds with therapeutic properties, Crăciun and Gutt [10] established the ideal experimental settings for the extraction of trans-resveratrol in order to obtain a higher yield from the material of the vine prunings. Resveratrol has the potential to exert positive therapeutic effects on slowing down the aging process. It also has antioxidant properties that fight free radicals, anti-aging effects on the skin by inhibiting the enzymes that cause aging, antibacterial and antifungal properties, anti-inflammatory activity, anti-carcinogenic properties, cardioprotective properties, and benefits for diabetes symptoms.

The replacement of some conventional raw materials in the manufacturing processes of some fundamental foods is another area of research. For instance, Voinea et al. investigated the effects of substituting reduced sodium sea salt for sodium chloride in a bread recipe. It is well known that the World Health Organization (WHO) advises the food sector to lower the salt (sodium chloride, NaCl) content in foods to achieve a maximum salt intake of 5 g per day for people. One of the main sources of sodium in the average person's diet is bread and other bakery products. As a result, the majority of initiatives to cut back on sodium consumption concentrate on lowering the amount of sodium chloride in these items. With the increased level of sea salt with low sodium content added in wheat flour, the bakery goods produced with sea salt with low sodium content were of greater quality compared to the control sample, displaying better physical and textural qualities, a darker color, and being more well-liked by consumers [11]. Ziarno et al. [12] investigated the link between spreadability and other specified structural, physicochemical, and chemical factors by analyzing market samples of butter and butter replacements in terms of spreadability. The study's findings can be used to create new and enhanced butter and butter replacements that have comparable features for spreading without compromising other crucial qualities such as flavor, texture, and nutritional value. Since these fats vary not only in their calorific value and hence in their chemical composition, but also in their functional features, such as spreadability, it might be challenging for certain consumers to choose which fat to use in spreads.

Recent years have seen an increase in research towards the production of food items using previously ignored or underutilized basic materials. One of these is hemp, which is often processed for the textile industry due to its high fiber content, but in recent years, it has emerged as a novel raw material in food. Baldino et al. [13] researched the prospect of increasing the yield of oil extraction from seeds and enhancing the waste was examined using stable oil-in-water emulsions based on hemp oil and commercial hemp protein isolates. The created emulsion is stable, which can promote process development and the use of hemp seed extraction waste for the valorization of by-products and waste in order to produce complete food products with high nutritional content. For those with unique dietary demands, the product can alternatively be utilized as a foundation to obtain slightly structured products. Another study showed that adding hemp inflorescence to rice-based gluten-free bread can be beneficial. Significant alterations in the physical characteristics of the bread were brought on by the addition, which was utilized at concentrations between 1% and 4%. The bread's volume and the color of the crumb both considerably changed after 1% of the additive was added. Because hemp inflorescence has never been utilized to make bread and is a good raw material for health reasons, this study suggested using it to make gluten-free bread [14]. Ciocan et al. [15] suggested that the physicochemical and sensory characteristics of samples of wort and beer be evaluated in relation to the use of unmalted and malted buckwheat. Buckwheat is one of the pseudocereals that is most frequently used as a raw material in studies on the production of malts, gluten-free beers, and functional beers because it consistently produces excellent results in terms of productivity, enzymatic activity, and chemical composition of the finished product. The results obtained suggest that, in future studies, beer should only be made from unmalted buckwheat with enzyme addition, which is far more advantageous from an economic and technological aspect. This is due to the fact that by just adding enzyme preparations to the brewing process, buckwheat can be used to make beer without first being malted [15].

In relation to some of the chosen wort characteristics, Rydzak et al. [16] investigated the impact of vacuum impregnation on the steeping process and the modifications to the structure of barley grains. It can reduce the time it takes to moisten barley grain to 42% moisture content by around 6 h by vacuum-impregnating it under varied conditions.

The biggest benefit of vacuum impregnation in the malting of grain is that it makes it possible to drastically shorten the malt production cycle, which is now prolonged by the need to soak the grain before malting [16].

All of these studies have one thing in common: they all focus on finding new raw materials that can be used in food manufacturing recipes. These raw materials have been studied from a physico-chemical, sensory, and food safety perspective.

Author Contributions: A.D., L.R. and G.G.C. contributed equally to the preparation of the editorial. All authors have read and agreed to the published version of the manuscript.

Funding: This work was funded by Ministry of Research, Innovation and Digitalization within Program 1—Development of national research and development system, Subprogram 1.2—Institutional Performance—RDI excellence funding projects, under contract no. 10PFE/2021.

Conflicts of Interest: The authors declare no conflict of interest.

References

1. Wang, C.; Liu, Y.; Zhao, X.; Liu, B. Current Advances and Future Aspects of Sweetener Synergy: Properties, Evaluation Methods and Molecular Mechanisms. *Appl. Sci.* **2022**, *12*, 5096. [CrossRef]
2. Tița, O.; Lengyel, E.; Stegăruș, D.I.; Săvescu, P.; Ciubara, A.B.; Constantinescu, M.A.; Tița, M.A.; Rață, D.; Ciubara, A. Identification and Quantification of Valuable Compounds in Red Grape Seeds. *Appl. Sci.* **2021**, *11*, 5124. [CrossRef]
3. Tița, M.A.; Constantinescu, M.A.; Tița, O.; Mathe, E.; Tamošaitienė, L.; Bradauskienė, V. Food Products with High Antioxidant and Antimicrobial Activities and Their Sensory Appreciation. *Appl. Sci.* **2022**, *12*, 790. [CrossRef]
4. Tița, O.; Constantinescu, M.A.; Tița, M.A.; Opruța, T.I.; Dabija, A.; Georgescu, C. Valorization on the Antioxidant Potential of Volatile Oils of *Lavandula angustifolia* Mill., *Mentha piperita* L. and *Foeniculum vulgare* L. in the Production of Kefir. *Appl. Sci.* **2022**, *12*, 10287. [CrossRef]
5. Tița, M.A.; Constantinescu, M.A.; Opruța, T.I.; Bătuşaru, C.; Rusu, L.; Tița, O. Kefir Enriched with Encapsulated Volatile Oils: Investigation of Antimicrobial Activity and Chemical Composition. *Appl. Sci.* **2023**, *13*, 2993. [CrossRef]
6. Chetrariu, A.; Dabija, A. Valorisation of Spent Grain from Malt Whisky in the Spelt Pasta Formulation: Modelling and Optimization Study. *Appl. Sci.* **2022**, *12*, 1441. [CrossRef]
7. Chetrariu, A.; Dabija, A. Study of the Utilization of Spent Grain from Malt Whisky on the Quality of Wafers. *Appl. Sci.* **2022**, *12*, 7163. [CrossRef]
8. Gerardi, C.; Durante, M.; Tufariello, M.; Grieco, F.; Giovinazzo, G. Effects of Time and Temperature on Stability of Bioactive Molecules, Color and Volatile Compounds during Storage of Grape Pomace Flour. *Appl. Sci.* **2022**, *12*, 3956. [CrossRef]
9. Albert, C.; Codină, G.G.; Héjja, M.; András, C.D.; Chetrariu, A.; Dabija, A. Study of Antioxidant Activity of Garden Blackberries (*Rubus fruticosus* L.) Extracts Obtained with Different Extraction Solvents. *Appl. Sci.* **2022**, *12*, 4004. [CrossRef]
10. Crăciun, A.L.; Gutt, G. Optimization of Experimental Parameters in the Solvent Extraction of Trans-Resveratrol from Pruning Waste of Vitis vinifera, Fetească Neagră Variety. *Appl. Sci.* **2023**, *13*, 823. [CrossRef]
11. Voinea, A.; Stroe, S.-G.; Ropciuc, S.; Dabija, A.; Codină, G.G. The Effect of Sea Salt with Low Sodium Content on Dough Rheological Properties and Bread Quality. *Appl. Sci.* **2022**, *12*, 4344. [CrossRef]
12. Ziarno, M.; Derewiaka, D.; Florowska, A.; Szymańska, I. Comparison of the Spreadability of Butter and Butter Substitutes. *Appl. Sci.* **2023**, *13*, 2600. [CrossRef]
13. Baldino, N.; Carnevale, I.; Mileti, O.; Aiello, D.; Lupi, F.R.; Napoli, A.; Gabriele, D. Hemp Seed Oil Extraction and Stable Emulsion Formulation with Hemp Protein Isolates. *Appl. Sci.* **2022**, *12*, 11921. [CrossRef]
14. Pecyna, A.; Buczaj, A.; Różyło, R.; Kobus, Z. Physical and Antioxidant Properties of Innovative Gluten-Free Bread with the Addition of Hemp Inflorescence. *Appl. Sci.* **2023**, *13*, 4889. [CrossRef]
15. Ciocan, M.E.; Salamon, R.V.; Ambrus, Á.; Codină, G.G.; Chetrariu, A.; Dabija, A. Use of Unmalted and Malted Buckwheat in Brewing. *Appl. Sci.* **2023**, *13*, 2199. [CrossRef]
16. Rydzak, L.; Różyło, R.; Guz, T.; Natoniewski, M.; Vasiukov, K. The Use of Vacuum Impregnation of Barley Grain in the Production of Malt for Wort. *Appl. Sci.* **2023**, *13*, 1628. [CrossRef]

applied sciences

MDPI

Review

Current Advances and Future Aspects of Sweetener Synergy: Properties, Evaluation Methods and Molecular Mechanisms

Congrui Wang †, Yi Liu †, Xiangzhong Zhao and Bo Liu *

School of Food Science & Engineering, Qilu University of Technology (Shandong Academy of Sciences), Jinan 250353, China; wangcr07@163.com (C.W.); ly979365190@163.com (Y.L.); 13506416163@163.com (X.Z.)
* Correspondence: ertrdfgg@qlu.edu.cn; Tel.: +86-531-8963-1195
† These authors contributed equally to this work.

Abstract: Sweetener synergy is the phenomenon in which certain combinations of sweeteners work more effectively than the theoretical sum of the effects of each components. It provides benefits in reducing sweetener dosages and improving their sweetness. Many mixtures of sweeteners with synergistic effects have been reported up to now. Both artificial high-intensity sweeteners and natural sweeteners are popularly used in sweetener mixtures for synergism, although the former seem to display more potential to exhibit synergy than the latter. Furthermore, several evaluation methods to investigate sweetener synergy have been applied, which could lead to discrepancies in results. Moreover, structurally dissimilar sweeteners could cooperatively bind at the different sites in the sweet taste receptor T1R2/T1R3 to activate the receptor, and their hydration characters/packing characteristics in solvents could affect their interaction with the receptor, providing the preliminary explanations for the molecular basis of sweetener synergy. In this article, we firstly present a systematic review, analysis and comment on the properties, evaluation methods and molecular mechanisms of sweetener synergy. Secondly, challenges of sweetener synergy in both theory and practice and possible strategies to overcome these limitations are comprehensively discussed. Finally, future perspectives for this important performance in human sweet taste perception are proposed.

Keywords: sweetener; synergy; evaluation methods; sweet taste receptor; molecular basis; interaction; mixture

Citation: Wang, C.; Liu, Y.; Zhao, X.; Liu, B. Current Advances and Future Aspects of Sweetener Synergy: Properties, Evaluation Methods and Molecular Mechanisms. *Appl. Sci.* **2022**, *12*, 5096. https://doi.org/10.3390/app12105096

Academic Editors: Georgiana Gabriela Codină, Adriana Dabija and Lăcrămioara Rusu

Received: 28 March 2022
Accepted: 13 May 2022
Published: 18 May 2022

1. Introduction

Sweetener is an important food additive that provides a pleasant sense for humans and other animals. Some sweeteners such as sugars are important nutrient and energy sources [1,2]. According to the original producing methods of the sweeteners, they can be divided into natural sweeteners (e.g., sugars, sugar alcohols and sweet-tasting proteins) and artificial sweeteners (e.g., sucralose, saccharin and aspartame) [3,4]. With the development of diet and nutrition in human beings, high standards for sweeteners in food and beverages are needed [5].

A single sweetener often has the disadvantage of high calories (e.g., sucrose) that are a potential inducer for obesity, oral health problems, diabetes and hyperlipemia, or has an aftertaste and bitter taste (e.g., stevioside) [6,7]. Furthermore, artificial sweeteners have long been controversial for their safety, which requires their use only under strict control and below specific dose levels [8]. Therefore, different sweeteners are often used together to provide certain sensory properties and to take advantage of the cooperative effect that occurs in certain sweeteners combinations [9]. Based on the comparison of the sweet intensities of the mixed sweeteners with that of the sum of the unmixed components, the effect of sweetener mixtures can be categorized as addition (equal), synergism (greater) and suppression (less). Until now, many sweetener mixtures with synergistic effects have been reported [10–12].

Mixtures of sweeteners can be divided into binary and polynary combinations, according to the categories of sweeteners used. Generally, the polynary mixtures can lead to more complex and pleiotropic effects than the binary mixtures, accounting for their distinct properties in sweeteners' synergism [13]. In this paper, we summarize, analyze and comment on the findings for sweeteners' synergism reported up to now as well as their evaluation methods. Furthermore, insight into the molecular basis of sweeteners' synergism has seen substantial progress in recent years, although it is still limited due to several technological constraints [14]. We then describe several explanations reported so far for understanding the molecular mechanism of sweeteners' synergism, especially the consequent interaction between the sweeteners and sweet taste receptors due to the mixture of sweeteners, and analyze their scientific merit and deficiency in a broad context. Moreover, the future challenges and perspectives of sweeteners' synergism in performance improvement, development and application are proposed. This article provides meaningful and helpful guidelines for further research on sweetener synergy and promotion of its applications in the food industry.

2. Methods to Determine the Synergistic Effect of Sweetener Mixtures

Different methods have been applied in the evaluation of the synergistic effect of sweetener mixtures. Besides the long- and popularly used sensory evaluation, some novel methods have emerged in recent years, such as biosensors and electronic tongues [15,16]. However, most of the results for sweetener synergy were obtained by sensory evaluation. Basically, the discrepancy among these methods lies mainly in the comparison modes for the sweet intensity between the mixed sweeteners and the sum of the unmixed components. On the other hand, variable outcomes in the evaluations of specific sweetener mixtures could be obtained when different evaluation methods were adopted. We briefly describe the main points of these methods as follows.

2.1. Sensory Evaluation Method I

The first method is the most common and simple approach that has been popularly used [11,17]. In this method, which is called the simple additive model, the perceived sweetness of the mixture is directly compared to the sum of the sweetness of the unmixed components. If the sweetness of the mixture is equal to, greater or less than the summed sweetness of its components, then it is regarded as having additive, synergistic and suppressive effects, respectively. However, this method has been questioned by many researchers because it does not consider the effect of sweetener concentrations on the results; thus, contradictory findings in the research on sweeteners' synergism could often arise when different doses of the components were added in the sweetener mixtures [18]. Furthermore, it should be noted that in this method, the concentrations of used sweeteners were usually expressed as a % (weight of sweetener/weight of solvent) rather than a molarity (moles of sweetener/volume of solvent), which cannot address the fact that sweeteners have different molecular weights and thus are not scientifically rational for studying the mechanical interactions between the sweeteners as well as the sweeteners and receptors [17].

2.2. Sensory Evaluation Method II

Because the perceived sweetness of some sweeteners show a sigmoidal rather than linear relationship with their concentrations, so many researchers assessed the sweet intensity of self-mixtures (mixtures of each individual component with itself) to compare it with the sweet intensity of blend mixtures to determine the interactive effect of different sweeteners, which is essential to discriminate between superadditivity (which means that the self-mixture of a sweetener produces more sweetness than the summed sweetness of this sweetener in isolation) and synergism [19]. Based on this principle, the sweetness of sweetener mixtures is compared with the average of the sweetness ratings of the two pure components in the mixture (e.g., comparing the sweetness of a 3%A + 3%B mixture with the average sweetness of a 3%A + 3%A and 3%B + 3%B). This self-mixture model accounts for the concentration-dependent and non-linear psychophysical functions and considers the effect of interactions between sweeteners on their synergism [20,21]. Although this method has been popularly used, it was thought by some researchers to be too straightforward to elucidate the complex response toward the interactive sweet stimulus [22].

2.3. Sensory Evaluation Method III

A modified isobole analysis method was proposed by Sühnel in 1993, in which evaluating the effect of sweeteners' interactions was based on the comparison between the observed effect of the sweetener mixtures and what would be expected from each sweetener in isolation [23]. This method has been widely used to evaluate drug combinations with synergistic action [24]. Specifically, an equation $(cA/CA) + (cB/CB) = I$ is adopted in this approach, where cA and cB are the concentrations of sweeteners A and B in the mixture, respectively, and CA and CB are the concentrations of A and B that would individually produce the same sweet intensity as the mixture, respectively. A value of I less than 1 indicates synergy, more than 1 indicates suppression and around 1 indicates no interaction. This method has been suggested to have the advantage of demonstrating the empirical concentration–effect relationships, which is completely independent of the mechanism of interaction [22,25].

2.4. Other Evaluation Methods

Besides the traditional sensory evaluation methods, other methods have also been devised to assess the sweeteners' synergism. For instance, Fujiwara et al. carried out a cell-based Ca^{2+} mobilization assay that represented the elicited signal upon sweet-taste-receptor activation to investigate the synergistic effects of sweetener mixtures, and the results were in good agreement with that of sensory evaluations [26]. Moreover, the electronic tongue and bioelectronic tongue technologies, also called sensors and biosensors, respectively, have made significant advances in recent years, which show their great potential in evaluations of sweeteners' synergism [15,16,27,28]. For example, an in vitro binding assay with cloned human sweet taste receptors could well reflect the taste properties of sweeteners, which indicates that the taste-cell- and receptor-based biosensors can exhibit great advantages in obtaining taste-related qualitative/quantitative information of ingredients in food, thus enabling a more logical recognition of sweet substances in complex systems [28,29]. In this regard, with the improvement and development of the performance of these taste biosensors, their future practical application in evaluations of sweeteners' synergism could be greatly prospective [27,28].

3. Synergism in Mixtures of Sweeteners

3.1. Binary Mixtures of Sweeteners

Initially, research on sweeteners' synergy was mainly focused on high-intensity artificial sweeteners such as acesulfame-K, aspartame, cyclamate and sucralose. For example, early in 1969, Stone and Oliver reported that cyclamate was synergistic with sucrose and saccharin [30]. In 1989, Wells summarized the sweetener combinations exhibiting synergism, most of which were high potency sweeteners [31]. The most popular high-intensity sweeteners in these mixtures were acesulfame-K, cyclamate and saccharin. For instance, both acesulfame-K and cyclamate were synergistic with alitame, aspartame, stevioside and sucralose, respectively, while saccharin was synergistic with aspartame, cyclamate, neohesperidin dihydrochalcone (NHDC) and the sweet-tasting protein thaumatin. Based on these findings, it could be presumed that the high-intensity artificial sweeteners could have more potency to elicit synergy than natural sweeteners. Many other sweetener combinations showing synergism have also been reported [11,17]. However, there is no physicochemical or structural patterns or principles found in these mixtures of sweeteners that exhibit synergism.

Later, it was found that natural sugars and proteins could also elicit synergistic effects. In 1995, Schiffman et al. systematically examined the presence and degree of synergism among all binary mixtures of 14 sweeteners including 3 sugars (fructose, glucose and sucrose), 2 polyhydric alcohols (mannitol and sorbitol), 2 natural glycosides (rebaudioside-A and stevioside), 2 dipeptide derivatives (alitame and aspartame), 1 sweet-tasting protein (thaumatin), 1 sulfamate (sodium cyclamate), 2 amides (acesulfame-K and sodium saccharin) and NHDC [32]. Notably, the results of the synergism assessment were obviously concentration-dependent. Specifically, when two sweeteners with a sweetness equal to 3% sucrose were respectively added in the mixture, significant synergistic effects were found. However, only a few mixtures of sweeteners with a sweetness equal to 5% sucrose exhibited synergism, and no synergy was found for mixtures of sweeteners with a sweetness equal to 7% sucrose. Furthermore, two evaluation methods (Method I and II) were separately used in their investigation but led to some different conclusions, especially for mixtures of two sweeteners with a sweetness equal to 5% and 7% sucrose, respectively.

Reyes et al. advanced sweetener synergy research through an isobole analysis using Method III in 2019 [22]. They used 15 representative sweeteners at 3 different concentrations equal to 4%, 6% and 8% sucrose in sweet intensity, respectively, to investigate their synergism. However, no obvious effect of the concentrations on the results was found with this evaluation method. Furthermore, as expected, the highly synergistic mixtures ($I < 0.6$, see the equation in 2.3 Method III) included most high-intensity artificial sweeteners such as acesulfame-K and NHDC. For instance, the long-tested combinations acesulfame-K/aspartame and acesulfame-K/rebaudioside-A exhibited full synergistic effects, and NHDC showed significant synergism with acesulfame-K, sucralose, thaumatin and rebaudioside-A, respectively. The natural fructose showed partial synergy ($I = 0.6$–0.9) with acesulfame-K or rebaudioside-A. However, a few findings in their study, such as the mixture of acesulfame-K/rebaudioside-A, showed full synergy conflicts with the previously reported corresponding data that were revealed with Method I or II [22,32].

A lot of results about binary sweeteners' synergism have been reported up to now [11,17,33–39], which cannot be elaborated herein due to the page limit. In addition, alternative models were also proposed to account for the interaction between sweeteners in binary mixtures [40]. However, these models seem to be too mathematically complicated to be popularly used. Table 1 summarizes the previously reported most popular binary mixtures of sweeteners showing significant synergism, as well as their concentrations and the evaluation methods applied [20,22,32].

Table 1. The popular binary mixtures of sweeteners showing significant synergism reported up to now.

Sweetener 1	Sweetener 2	Concentration of 1	Concentration of 2	Evaluation Method	References
Acesulfame-K	Aspartame	Equal sweetness to 3% or 5% sucrose [1]	Same as sweetener 1	I	[32]
Acesulfame-K	Cyclamate	Equal sweetness to 3% or 5% sucrose	Same as sweetener 1	I	[32]
Alitame	NHDC [2]	Equal sweetness to 3% or 5% sucrose	Same as sweetener 1	I	[32]
Aspartame	Saccharin	Equal sweetness to 3% or 5% sucrose	Same as sweetener 1	I	[32]
Aspartame	Stevioside	Equal sweetness to 3% or 5% sucrose	Same as sweetener 1	I	[32]
NHDC	Fructose	Equal sweetness to 3% or 5% sucrose	Same as sweetener 1	I	[32]
Fructose	Sorbitol	Equal sweetness to 3% or 5% sucrose	Same as sweetener 1	I	[32]
NHDC	Glucose	Equal sweetness to 3% or 5% sucrose	Same as sweetener 1	I	[32]
Glucose	Sorbitol	Equal sweetness to 3% or 5% sucrose	Same as sweetener 1	I	[32]
Cyclamate	Stevioside	Equal sweetness to 3% or 5% sucrose	Same as sweetener 1	I	[32]
Saccharin	Rebaudioside-A	Equal sweetness to 3% or 5% sucrose	Same as sweetener 1	I	[32]
NHDC	Rebaudioside-A	Equal sweetness to 3% or 5% sucrose	Same as sweetener 1	I	[32]
NHDC	Stevioside	Equal sweetness to 3% or 5% sucrose	Same as sweetener 1	I	[32]
Cyclamate	Sucrose	5.6/12.5/22.6 mM [3]	0.1/0.25/0.5 M	II	[20]
Cyclamate	Aspartame	5.6/12.5/22.6 mM	0.7/2.7/4.9 mM	II	[20]
Cyclamate	Acesulfame-K	5.6/12.5/22.6 mM	1.0/4.0/6.0 mM	II	[20]
Cyclamate	Glucose	5.6/12.5/22.6 mM	0.39/0.88/1.69 M	II	[20]
Cyclamate	Xylitol	5.6/12.5/22.6 mM	0.2/0.6/1.0 M	II	[20]
Stevioside	Aspartame	0.25/0.62/1.1 mM	0.7/2.7/4.9 mM	II	[20]
Stevioside	Acesulfame-K	0.25/0.62/1.1 mM	1.0/4.0/6.0 mM	II	[20]
Saccharin	Sucrose	0.39/1.95/4.87 mM	0.1/0.25/0.5 M	II	[20]
Saccharin	Aspartame	0.39/1.95/4.87 mM	0.7/2.7/4.9 mM	II	[20]

Table 1. *Cont.*

Sweetener 1	Sweetener 2	Concentration of 1	Concentration of 2	Evaluation Method	References
Acesulfame-K	Sucrose	1.0/4.0/6.0 mM	0.1/0.25/0.5 M	II	[20]
Acesulfame-K	Glucose	1.0/4.0/6.0 mM	0.39/0.88/1.69 M	II	[20]
Acesulfame-K	Fructose	1.0/4.0/6.0 mM	0.15/0.39/0.74 M	II	[20]
Acesulfame-K	Xylitol	1.0/4.0/6.0 mM	0.2/0.6/1.0 M	II	[20]
Acesulfame-K	Aspartame	1.0/4.0/6.0 mM	0.7/2.7/4.9 mM	II	[20]
Acesulfame-K	Aspartame	Equal sweetness to 4%, 6% or 8% sucrose [4]	Same as sweetener 1	III	[22]
Acesulfame-K	NHDC	Equal sweetness to 4%, 6% or 8% sucrose	Same as sweetener 1	III	[22]
Acesulfame-K	Rebaudioside-A	Equal sweetness to 4%, 6% or 8% sucrose	Same as sweetener 1	III	[22]
NHDC	Sucralose	Equal sweetness to 4%, 6% or 8% sucrose	Same as sweetener 1	III	[22]
NHDC	Thaumatin	Equal sweetness to 4%, 6% or 8% sucrose	Same as sweetener 1	III	[22]
NHDC	Rebaudioside-A	Equal sweetness to 4%, 6% or 8% sucrose	Same as sweetener 1	III	[22]

[1] Each sweetener with specific concentrations in the mixtures displayed a sweetness intensity equal to that of 3% or 5% sucrose (*w*/*w*), respectively. [2] NHDC, neohesperidin dihydrochalcone. [3] Three molar concentrations divided by "/" were applied for each sweetener in the mixtures, respectively. [4] Each sweetener with specific concentrations in the mixtures displayed a sweetness intensity equal to that of 4%, 6% or 8% sucrose (*w*/*w*), respectively.

3.2. *Polynary Mixtures of Sweeteners*

The only detailed study on the synergism of polynary mixtures of sweeteners was performed by Schiffman et al. in 2000, who extended their binary mixture study to investigate the ternary mixture synergism with the same 14 sweeteners previously used [41]. Each sweetener in the mixture had a sweetness equal to 2% sucrose, and interactive effects were analyzed by comparing the sweetness of each ternary mixture with the average sweetness of the self-mixtures of its three constituent sweeteners (Method II). The results indicated that no obvious trend was found to predict the combinational effect in ternary mixtures of sweeteners, although the ternary combinations exhibited relatively less synergism than the binary combinations of their constituent sweeteners. Similar to the binary mixtures, the highest synergistic effect in ternary mixtures was found with three high-intensity sweeteners (alitame + NHDC + rebaudioside-A), whereas less synergistic effects in ternary mixtures were found with natural sugars (e.g., glucose and fructose). Considering the lower amount of information on the synergism of polynary mixtures, we proposed that further investigations should be performed to reveal their unique properties [35,41], such as studying the quaternary or polynary mixtures with multifarious concentrations, evaluation methods and sweeteners to be applied.

4. Mechanism for Synergism in Mixtures of Sweeteners

Why can combinations of two or more sweeteners exert synergism? It is well known that sweet taste perception is mediated by a class C G protein-coupled receptor (GPCR)—the heterodimer T1R2 (taste type 1 receptor 2)/T1R3 (taste type 1 receptor 3) [42,43]. Accordingly, this question certainly leads to elucidating the sweet taste elicitation process in which the sweeteners in mixtures cooperatively interact with and then activate the human sweet taste receptor T1R2/T1R3 [44–46]. However, due to the lack of structural information on the receptor, mechanical insights into the interplay between sweeteners and receptor are still elusive [47]. Nevertheless, molecular simulations and functional mutagenesis/chimera analysis can provide meaningful guidance for understanding the sweetener–receptor interaction as well as sweeteners' synergism [48–50].

In principle, different sweeteners have different weights, shapes and configurations, which definitely affect their specific interactions with the sweet taste receptor. In this regard, compared with the separate sweeteners, the mixing of sweeteners can result in a change or modification of their contact properties with the receptor, thus leading to activation of the receptor and eliciting their distinct sweetness [51]. Indeed, many previous studies have revealed that there are multiple binding sites in the sweet taste receptor for various sweeteners [47,48]. For example, the artificial high-intensity sweeteners aspartame, neotame, saccharin and acesulfame K interact with the VFTM (Venus flytrap module) of human T1R2, while natural sugars (e.g., glucose and sucrose) and sucralose bind to the

VFTMs of both T1R2 and T1R3 [52–54]. However, cyclamate and NHDC act on the TMD (transmembrane domain) of human T1R3 [55,56]. The sweet-tasting protein thaumatin was proposed to bind into the CRD (cysteine rich domain) of T1R3 [47,57]. We have demonstrated that the intense sweeteners perillartine and saccharin (at high concentrations above 3 mM) can bind to the TMD of human T1R2 (Figure 1) [58–60].

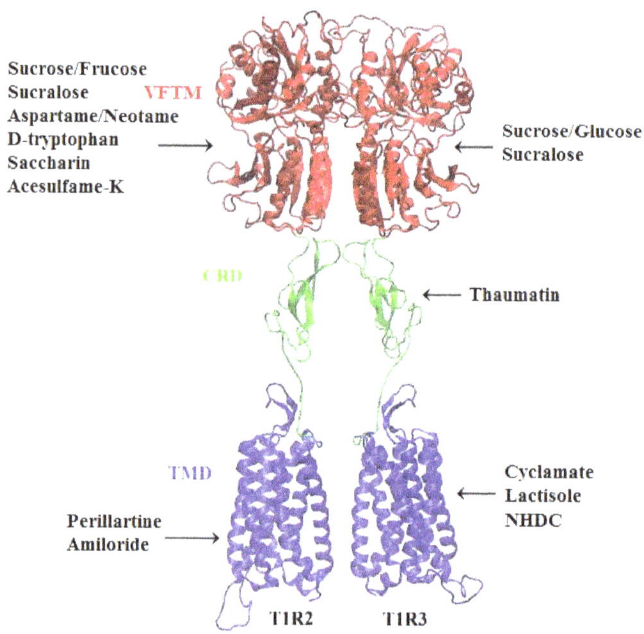

Figure 1. Graphic representation of the modeled structure of the human sweet taste receptor T1R2/T1R3 and its binding sites for different sweeteners. The extracellular VFTM (Venus flytrap module), CRD (cysteine rich domain) and TMD (transmembrane domain) are colored in red, light-green and blue, respectively. This figure was made with the PyMOL software. The binding sites of various sweeteners in the receptor are indicated by arrow lines.

According to these findings, it was postulated that the different sweeteners in the mixture could bind at different sites in the receptor to concertedly increase the efficiency of receptor activation, thus displaying synergism. This hypothesis has been supported by some sweetener combinations with synergistic effects that interact with different monomers/domains of human T1R2/T1R3, as demonstrated by a cell-based receptor function analysis [26]. Specifically, two high-intensity artificial sweeteners, NHDC and cyclamate, which have been identified as binding at the TMD of T1R3, significantly enhance the sweetness of various sweeteners including sucrose, aspartame, neotame, rebaudioside-A, saccharin, sucralose and thaumatin, which have been identified as interacting with the VFTM or CRD of T1R2/T1R3. Figure 1 shows the distinct binding sites of various sweeteners in the human sweet taste receptor. Based on these results, the authors suggest that NHDC and cyclamate act as positive sweet modulators to exert their synergistic effects with other sweeteners [50,61]. This finding is in line with the previous results that in mixtures, structurally dissimilar sweeteners interacting with different sites in the receptor have the potential to exhibit synergy, whereas structurally similar sweeteners interacting with parallel sites in the receptor have the potential to show suppression [22,32]. Furthermore, the relatively less effective synergism in polynary mixtures than binary mixtures could be partly explained by the distinct binding sites for various sweeteners, as the more sweeteners appear in the mixtures, the more possibilities for overlapping binding and contact with the receptor, resulting in an overall counteracted or alleviatived effect in sweeteners' synergism [41]. Moreover, a recent study has used molecular dynamics simulations to illuminate the potential structural basis of receptor activation by full synergistic combinations of sweeteners [14].

It was also proposed that synergy occurs when the components in the sweetener mixtures have identical hydration characters (e.g., hydrophobic or hydrophilic), which was explained by increases in the mobility of water molecules and reductions in the volume of hydrated sweetener molecules, but its link to the receptor activation and origin of sweetener synergy has not been elucidated [62]. Another study investigated the packing characteristics of blend sweetener molecules in water to account for their synergistic effect [63]. The results indicate that sweetener–sweetener, sweetener–solvent and solvent–solvent interactions have essential effects on the properties of binary sweeteners. The bulk sweeteners (sucrose and maltitol) appear to dominate the properties of the mixtures due to their molar excess over the intense sweeteners (acesulfame K, aspartame and cyclamate), while the intense sweeteners play an important role in modifying the structure of water in solution. The molecular volumes, hydrophobicity, ionic character and isentropic compressibility of the sweeteners in the binary mixtures are of obvious relevance to their interaction as well as the solvent's packing, thus affecting their accession to the sweet taste receptor and their sweet intensities [63]. Another interesting study pointed out that in some sweetener mixtures, across-adaption and synergism display opposite relationships, with the most across-adaption combinations showing the least synergism, and vice versa, suggesting a complex interplay between the sweetener components [64,65]. It is worthy to note that the sweeteners' synergism at the level of molecular interactions are sweetener-dependent, which should be analyzed carefully when specific sweetener combinations are considered or applied [63,65].

Other explanations of the molecular mechanisms of sweetener synergy have also been proposed. For instance, because some artificial sweeteners exhibiting synergism often have a bitter side taste, some researchers suggested that synergy may partly be due to the suppression of the bitter taste of one sweetener by another, whereas the bitterness of components could reduce their sweetness in isolation [66,67]. Furthermore, some researchers proposed that sweeteners' synergism may result from their simultaneous stimulation of multiple receptors, but no experimental evidence supports this hypothesis [68,69]. Moreover, it was suggested that different sweeteners could induce different downstream signal transduction cascades, leading to sweet taste signal convergence in a broadly tuned TRC (taste receptor cell) [70]. However, the mechanisms of sweetener synergy at the level of cellular signaling are still unknown, which should be explored in further investigations.

5. Factors Affecting the Results of Sweeteners' Synergism

As previously stated, there are usually obvious discrepancies among the results in evaluating the synergism of sweetener combinations [11,32]. A lot of factors may affect the evaluation results of the sweeteners' synergism. Firstly, the sweeteners used in these investigations may differ in source, purity and concentration, and the latter has been shown to be crucial for the resulted conclusions [19]. Furthermore, as described above, the evaluation methods have an important influence on the outcome, which has been shown in many previous investigations [22,32]. Based on their design principles, we propose that method III is the most logical approach for evaluating sweeteners' synergism. Moreover, the participants in the sensory tests were variable in age, nationality and health conditions, and different taste propensities in different populations have been reported [71]. All these factors could individually or cooperatively affect the results of synergism in mixtures of sweeteners, so researchers and manufacturers should pay attention to their conditions and background and use these data carefully.

6. Applications of Sweeteners' Synergism in Food and Beverages

Mixing of sweeteners has been widely used in food and beverage such as candy, chutty, biscuits and drinks [72]. Both natural bulk sweeteners (e.g., fructose, erythritol and stevioside) and artificial intense sweeteners (e.g., sucralose, aspartame and acesulfame-K) can be mixed to achieve optimization of the acceptability, functionality and economics of the products [73]. These formulations not only reduce the amount of some sweetener components to conform to their dosage restrictions, but also have healthy and nutritional advantages (e.g., the popular and welcomed sugar-free coke nowadays) [74]. It should be noteworthy herein that the aim of sweeteners' synergism applied in practice is not only to take advantage of the synergism of sweet intensities, but also to acquire a specific taste or flavor in certain sweetener mixtures. However, the latter topic is beyond the scope of the present paper and could be discussed elsewhere.

7. Challenges, Strategies and Outlooks

Since the phenomenon of sweeteners' synergism became of academic concern around 80 years ago, a great deal of data have been reported on this interesting topic. However, there seems to be confusion in some ways when analyzing and categorizing this information. This is largely due to the varied aims, conditions, methods or experimental designs in these different investigations [11,22,32]. In the future, it is necessary to set up relatively unified criterions effective both in research and practice for the sweeteners' synergism. For instance, specific dose ranges for certain sweeteners, generally acknowledged evaluation methods and standards requirements for various sweeteners, test panels and conditions should be normalized to better define the sweeteners' synergism. Furthermore, new evaluation methods (e.g., biomimetic sensors and biosensors) could be optimized to improve the accuracy of the results of sweeteners' synergism [15,16,27–29]. Moreover, emphasis should be taken on the functional performances of sweeteners' synergism (e.g., decreasing the usage of specific sweeteners, improving the sensory quality and reducing the calories of sweetener mixtures) to design the distinct and optimal combinations of chemically diverse sweeteners.

Despite several considerable proposals having been raised, the molecular mechanisms of sweeteners' synergism are still obscure, which limits the further development of the design and application of sweetener mixtures. The molecular basis for the onset of sweet perception that is dependent on the interaction between the sweetener (or mixture of sweeteners) and the sweet taste receptor is elusive due to the lack of structure of the receptor and its complexes with various sweeteners [75]. The recent breakthrough of AlphaFold in predicting GPCR structures could facilitate research on the mechanisms of interaction between the cooperative sweeteners and receptors, especially for molecular simulation studies [76]. Furthermore, with technological innovations such as cryo-electron microscopy, more and more structures of GPCRs have been solved in recent years, which could greatly promote the deciphering of the mechanisms of sweeteners' synergism at a structural basis [77,78]. Lastly, because sweeteners' synergism is a complex psychophysical process and involves many signal transduction pathways, elucidating the molecular mechanisms of sweeteners' synergism needs further multidisciplinary cooperation of scientists in many fields such as food, structural biology, neuroscience, physiology and biochemistry [79,80]. These efforts could jointly achieve the theoretical and practical development of sweeteners' synergism in the future, which will ultimately benefit the diet and health of human beings.

Author Contributions: Conceptualization, B.L.; software, C.W.; data curation, Y.L.; writing—original draft preparation, C.W.; writing—review and editing, B.L. and X.Z.; funding acquisition, B.L. and X.Z. All authors have read and agreed to the published version of the manuscript.

Funding: This research was funded by the National Natural Science Foundation of China (31970935), the Natural Science Foundation of Shandong Province (ZR2020KC035) and the Key Research & Development Project of Shandong Province, Major Science and Technology Innovation Project (2020CXGC010604).

Institutional Review Board Statement: Not applicable.

Informed Consent Statement: Not applicable.

Data Availability Statement: No new data were created or analyzed in this study. Data sharing is not applicable to this article.

Conflicts of Interest: The authors declare no conflict of interest.

References

1. Anderson, G.H. Sugars, sweetness, and food intake. *Am. J. Clin. Nutr.* **1995**, *62*, 195S–201S. [CrossRef] [PubMed]
2. Hess, J.; Latulippe, M.E.; Ayoob, K.; Slavin, J. The confusing world of dietary sugars: Definitions, intakes, food sources and international dietary recommendations. *Food Funct.* **2012**, *3*, 477–486. [CrossRef] [PubMed]
3. Sylvetsky, A.C.; Conway, E.M.; Malhotra, S.; Rother, K.I. Development of Sweet Taste Perception: Implications for Artificial Sweetener Use. *Endocr. Dev.* **2017**, *32*, 87–99. [CrossRef] [PubMed]
4. de Medeiros, A.C.; Filho, E.R.T.; Bolini, H.M.A. Impact of Natural and Artificial Sweeteners Compounds in the Sensory Profile and Preference Drivers Applied to Traditional, Lactose-Free, and Vegan Frozen Desserts of Chocolate Flavor. *J. Food Sci.* **2019**, *84*, 2973–2982. [CrossRef]
5. Trumbo, P.R.; Appleton, K.M.; de Graaf, K.; Hayes, J.E.; Baer, D.J.; Beauchamp, G.K.; Dwyer, J.T.; Fernstrom, J.D.; Klurfeld, D.M.; Mattes, R.D.; et al. Perspective: Measuring Sweetness in Foods, Beverages, and Diets: Toward Understanding the Role of Sweetness in Health. *Adv. Nutr.* **2021**, *12*, 343–354. [CrossRef]
6. Qi, X.; Tester, R.F. Lactose, Maltose, and Sucrose in Health and Disease. *Mol. Nutr. Food Res.* **2020**, *64*, 1901082. [CrossRef]
7. Antenucci, R.G.; Hayes, J.E. Nonnutritive sweeteners are not supernormal stimuli. *Int. J. Obes.* **2015**, *39*, 254–259. [CrossRef]
8. Mishra, A.; Ahmed, K.; Froghi, S.; Dasgupta, P. Systematic review of the relationship between artificial sweetener consumption and cancer in humans: Analysis of 599,741 participants. *Int. J. Clin. Pract.* **2015**, *69*, 1418–1426. [CrossRef]
9. Birch, G.G. Modulation of sweet taste. *Biofactors* **1999**, *9*, 73–80. [CrossRef]
10. Lawless, H.T. Theoretical note: Tests of synergy in sweetener mixtures. *Chem. Senses* **1998**, *23*, 447–451. [CrossRef]
11. Birch, G.G. Towards an improved understanding of sweetener synergy. *Trends Food Sci. Technol.* **1996**, *7*, 401–407. [CrossRef]
12. Bingham, A.I.; Birch, G.G.; de Graaf, C.; Behan, J.M.; Perring, K.D. Sensory Studies with Sucrose-Maltol Mixtures. *Chem. Senses* **1990**, *15*, 447–456. [CrossRef]
13. Frank, M.E.; Formaker, B.K.; Hettinger, T.P. Taste responses to mixtures: Analytic processing of quality. *Behav. Neurosci.* **2003**, *117*, 228–235. [CrossRef] [PubMed]
14. Jang, J.; Kim, S.K.; Guthrie, B.; Goddard, W.A., 3rd. Synergic Effects in the Activation of the Sweet Receptor GPCR Heterodimer for Various Sweeteners Predicted Using Molecular Metadynamics Simulations. *J. Agric. Food Chem.* **2021**, *69*, 12250–12261. [CrossRef] [PubMed]
15. Jeong, J.Y.; Cha, Y.K.; Ahn, S.R.; Shin, J.; Choi, Y.; Park, T.H.; Hong, S. Ultrasensitive Bioelectronic Tongue Based on the Venus Flytrap Domain of a Human Sweet Taste Receptor. *ACS Appl. Mater. Interfaces* **2022**, *14*, 2478–2487. [CrossRef] [PubMed]
16. Wasilewski, T.; Kamysz, W.; Gębicki, J. Bioelectronic tongue: Current status and perspectives. *Biosens. Bioelectron.* **2020**, *150*, 111923. [CrossRef]
17. Lanton, B. Recent developments in sweetener synergies. *Food Indust S. Afr.* **1988**, *41*, 23–25.
18. Tornout, P.V.; Pelgroms, J.; Meeren, J.V.D. Sweetness Evaluation of Mixtures of Fructose with Saccharin, Aspartame or Acesulfame K. *J. Food Sci.* **1985**, *50*, 469–472. [CrossRef]
19. Ayya, N.; Lawless, H.T. Quantitative and qualitative evaluation of high-intensity sweeteners and sweetener mixtures. *Chem. Senses* **1992**, *17*, 245–259. [CrossRef]
20. Frank, R.A.; Mize, S.J.S.; Carter, R. An assessment of binary mixture interactions for nine sweeteners. *Chem. Senses* **1989**, *14*, 621–632. [CrossRef]
21. Wolf, P.A.; Bridges, J.R.; Wicklund, R. Application of agonist-receptor modeling to the sweetness synergy between high fructose corn syrup and sucralose, and between high-potency sweeteners. *J. Food Sci.* **2010**, *75*, S95–S102. [CrossRef] [PubMed]
22. Reyes, M.M.; Gravina, S.A.; Hayes, J.E. Evaluation of Sweetener Synergy in Humans by Isobole Analyses. *Chem. Senses* **2019**, *44*, 571–582. [CrossRef] [PubMed]
23. Sühnel, J. Evaluation of interaction in olfactory and taste mixtures. *Chem. Senses* **1993**, *18*, 131–149. [CrossRef]
24. Tallarida, R.J. Drug combinations: Tests and analysis with isoboles. *Curr. Protoc. Pharmacol.* **2016**, *72*, 9–19. [CrossRef]
25. Fleming, E.E.; Ziegler, G.R.; Hayes, J.E. Investigating mixture interactions of astringent stimuli using the isobole approach. *Chem. Senses* **2016**, *41*, 601–610. [CrossRef] [PubMed]
26. Fujiwara, S.; Imada, T.; Nakagita, T.; Okada, S.; Nammoku, T.; Abe, K.; Misaka, T. Sweeteners interacting with the transmembrane domain of the human sweet-taste receptor induce sweet-synergisms in binary mixtures. *Food Chem.* **2012**, *130*, 561–568. [CrossRef]
27. Gupta, N.; Renugopalakrishnan, V.; Liepmann, D.; Paulmurugan, R.; Malhotra, B.D. Cell-based biosensors: Recent trends, challenges and future perspectives. *Biosens. Bioelectron.* **2019**, *141*, 111435. [CrossRef]
28. Zhang, N.; Wei, X.; Fan, Y.; Zhou, X.; Liu, Y. Recent advances in development of biosensors for taste-related analyses. *Trends Analyt. Chem.* **2020**, *129*, 115925. [CrossRef]
29. Bassoli, A.; Borgonovo, G.; Caremoli, F.; Mancuso, G. The taste of D- and L-amino acids: In vitro binding assays with cloned human bitter (TAS2Rs) and sweet (TAS1R2/TAS1R3) receptors. *Food Chem.* **2014**, *150*, 27–33. [CrossRef]

30. Stone, H.; Oliver, S.M. Measurement of the relative sweetness of selected sweeteners and sweetener mixtures. *J. Food Sci.* **1969**, *34*, 215–222. [CrossRef]
31. Wells, A.G. The use of intense sweeteners in soft drinks. In *Progress in Sweeteners*; Grenby, T.H., Ed.; Elsevier Applied Science: London, UK, 1989; pp. 169–214. ISBN 1-85106-270-7.
32. Schiffman, S.S.; Booth, B.J.; Carr, B.T.; Losee, M.L.; Sattely-Miller, E.A.; Graham, B.G. Investigation of synergism in binary mixtures of sweeteners. *Brain Res. Bull.* **1995**, *38*, 105–120. [CrossRef]
33. Kim, S.H.; Park, S.; Hong, J.H. Sweetness profiles of glycosyl rebaudioside A and binary mixtures with sugar alcohols in aqueous solution and a lemonade model system. *J. Sci. Food Agric.* **2022**, *102*, 2110–2119. [CrossRef] [PubMed]
34. Choi, J.H.; Chung, S.J. Sweetness potency and sweetness synergism of sweeteners in milk and coffee systems. *Food Res. Int.* **2015**, *74*, 168–176. [CrossRef] [PubMed]
35. Schiffman, S.S.; Sattely-Miller, E.A.; Graham, B.G.; Zervakis, J.; Butchko, H.H.; Stargel, W.W. Effect of repeated presentation on sweetness intensity of binary and ternary mixtures of sweeteners. *Chem. Senses* **2003**, *28*, 219–229. [CrossRef]
36. Bartoshuk, L.M.; Oeveland, C.T. Mixtures of substances with similar tastes: A test of a psychophysical model of taste mixture interactions. *Sens. Proc.* **1977**, *1*, 177–186.
37. Curtis, D.W.; Stevens, D.A.; Lawless, H.T. Peceived intensity of the taste of sugar mixtures and acid mixtures. *Chem. Senses* **1984**, *9*, 107–120. [CrossRef]
38. DeGraaf, C.; Frijters, J.E.R. Sweetness intensity of a binary sugar mixture lies between intensities of its components, when each is tasted alone and at the same total molarity as the mixture. *Chem. Senses* **1987**, *12*, 113–129. [CrossRef]
39. DeGraaf, C.; Frijters, J.E.R.; Van Trijp, H.C.M. Taste interaction between glucose and fructose assessed by functional measurement. *Percept. Psychophys.* **1987**, *41*, 383–392. [CrossRef]
40. Laffort, P.; Walsh, R.M.; Spillane, W.J. Application of the U and gamma' models in binary sweet taste mixtures. *Chem. Senses* **2002**, *27*, 511–520. [CrossRef]
41. Schiffman, S.S.; Sattely-Miller, E.A.; Graham, B.G.; Booth, B.J.; Gibes, K.M. Synergism among ternary mixtures of fourteen sweeteners. *Chem. Senses* **2000**, *25*, 131–140. [CrossRef]
42. Kim, U.K.; Breslin, P.A.; Reed, D.; Drayna, D. Genetics of human taste perception. *J. Dent. Res.* **2004**, *83*, 448–453. [CrossRef] [PubMed]
43. Nelson, G.; Hoon, M.A.; Chandrashekar, J.; Zhang, Y.; Ryba, N.J.; Zuker, C.S. Mammalian sweet taste receptors. *Cell* **2001**, *106*, 381–390. [CrossRef]
44. Kim, S.K.; Chen, Y.; Abrol, R.; Goddard, W.A., 3rd; Guthrie, B. Activation mechanism of the G protein-coupled sweet receptor heterodimer with sweeteners and allosteric agonists. *Proc. Natl. Acad. Sci. USA* **2017**, *114*, 2568–2573. [CrossRef] [PubMed]
45. Chéron, J.B.; Golebiowski, J.; Antonczak, S.; Fiorucci, S. The anatomy of mammalian sweet taste receptors. *Proteins.* **2017**, *85*, 332–341. [CrossRef] [PubMed]
46. DuBois, G.E. Molecular mechanism of sweetness sensation. *Physiol Behav.* **2016**, *164*, 453–463. [CrossRef]
47. Yang, L.; Cui, M.; Liu, B. Current Progress in Understanding the Structure and Function of Sweet Taste Receptor. *J. Mol. Neurosci.* **2021**, *71*, 234–244. [CrossRef] [PubMed]
48. Cui, M.; Jiang, P.; Maillet, E.; Max, M.; Margolskee, R.F.; Osman, R. The heterodimeric sweet taste receptor has multiple potential ligand binding sites. *Curr. Pharm. Des.* **2006**, *12*, 4591–4600. [CrossRef]
49. Liu, B.; Ha, M.; Meng, X.Y.; Kaur, T.; Khaleduzzaman, M.; Zhang, Z.; Jiang, P.; Li, X.; Cui, M. Molecular mechanism of species-dependent sweet taste toward artificial sweeteners. *J. Neurosci.* **2011**, *31*, 11070–11076. [CrossRef]
50. Zhang, F.; Klebansky, B.; Fine, R.M.; Liu, H.; Xu, H.; Servant, G.; Zoller, M.; Tachdjian, C.; Li, X. Molecular mechanism of the sweet taste enhancers. *Proc. Natl. Acad. Sci. USA* **2010**, *107*, 4752–4757. [CrossRef]
51. Zhao, X.; Wang, C.; Zheng, Y.; Liu, B. New Insight into the Structure-Activity Relationship of Sweet-Tasting Proteins: Protein Sector and Its Role for Sweet Properties. *Front. Nutr.* **2021**, *8*, 691368. [CrossRef]
52. Masuda, K.; Koizumi, A.; Nakajima, K.; Tanaka, T.; Abe, K.; Misaka, T.; Ishiguro, M. Characterization of the modes of binding between human sweet taste receptor and low-molecular-weight sweet compounds. *PLoS ONE* **2012**, *7*, e35380. [CrossRef]
53. Nie, Y.; Vigues, S.; Hobbs, J.R.; Conn, G.L.; Munger, S.D. Distinct contributions of T1R2 and T1R3 taste receptor subunits to the detection of sweet stimuli. *Curr. Biol.* **2005**, *15*, 1948–1952. [CrossRef]
54. Behrens, M.; Briand, L.; de March, C.A.; Matsunami, H.; Yamashita, A.; Meyerhof, W.; Weyand, S. Structure-Function Relationships of Olfactory and Taste Receptors. *Chem. Senses* **2018**, *43*, 81–87. [CrossRef] [PubMed]
55. Jiang, P.; Cui, M.; Zhao, B.; Snyder, L.A.; Benard, L.M.; Osman, R.; Max, M.; Margolskee, R.F. Identification of the cyclamate interaction site within the transmembrane domain of the human sweet taste receptor subunit T1R3. *J. Biol. Chem.* **2005**, *280*, 34296–34305. [CrossRef] [PubMed]
56. Winnig, M.; Bufe, B.; Kratochwil, N.A.; Slack, J.P.; Meyerhof, W. The binding site for neohesperidin dihydrochalcone at the human sweet taste receptor. *BMC Struct. Biol.* **2007**, *7*, 66. [CrossRef] [PubMed]
57. Masuda, T.; Taguchi, W.; Sano, A.; Ohta, K.; Kitabatake, N.; Tani, F. Five amino acid residues in cysteine-rich domain of human T1R3 were involved in the response for sweet-tasting protein, thaumatin. *Biochimie* **2013**, *95*, 1502–1505. [CrossRef] [PubMed]
58. Cai, C.; Jiang, H.; Li, L.; Liu, T.; Song, X.; Liu, B. Characterization of the sweet taste receptor Tas1r2 from an old world monkey species rhesus monkey and species-dependent activation of the monomeric receptor by an intense sweetener perillartine. *PLoS ONE* **2016**, *11*, e0160079. [CrossRef]

59. Zhao, X.; Liu, M.; Cui, M.; Liu, B. Multiple interaction modes between saccharin and sweet taste receptors determine a species-dependent response to saccharin. *FEBS Open Biol.* **2022**, *12*, 494–499. [CrossRef]
60. Zhao, M.; Xu, X.Q.; Meng, X.Y.; Liu, B. The Heptahelical Domain of the Sweet Taste Receptor T1R2 Is a New Allosteric Binding Site for the Sweet Taste Modulator Amiloride That Modulates Sweet Taste in a Species-Dependent Manner. *J. Mol. Neurosci.* **2018**, *66*, 207–213. [CrossRef]
61. Servant, G.; Tachdjian, C.; Tang, X.Q.; Werner, S.; Zhang, F.; Li, X.; Kamdar, P.; Petrovic, G.; Ditschun, T.; Java, A.; et al. Positive allosteric modulators of the human sweet taste receptor enhance sweet taste. *Proc. Natl. Acad. Sci. USA* **2010**, *107*, 4746–4751. [CrossRef]
62. Hutteau, F.; Mathlouthi, M.; Portmann, M.O.; Kilcast, D. Physicochemical and psychophysical characteristics of binary mixtures of bulk and intense sweeteners. *Food Chem.* **1998**, *63*, 9–16. [CrossRef]
63. Sneha, A.P.; Gordon, G.B.; Marie, O.P.; David, K. A study of the solution properties of selected binary mixtures of bulk and intense sweeteners in relation to their psychophysical characteristics. *Food Chem.* **1999**, *67*, 247–259. [CrossRef]
64. Lawless, H.T.; Stevens, D.A. Cross adaptation of sucrose and intensive sweeteners. *Chem. Senses* **1983**, *7*, 309–315. [CrossRef]
65. Froloff, N.; Lloret, E.; Martinez, J.M.; Faurion, A. Cross-adaptation and molecular modeling study of receptor mechanisms common to four taste stimuli in humans. *Chem. Senses* **1998**, *23*, 197–206. [CrossRef]
66. Lindley, M. Taste-ingredient interactions modulating sweetness. In *Optimising the Sweet Taste in Foods*; Spillane, W.J., Ed.; Woodhead: Cambridge, UK, 2006; ISBN 978-1-84569-965-9.
67. Lawless, H.T. Evidence for neural inhibition in bittersweet taste mixtures. *J. Comp. Physiol. Psychol.* **1979**, *93*, 538–547. [CrossRef]
68. Jakinovich, W., Jr.; Sugarman, D. Sugar taste reception in mammals. *Chem. Senses* **1988**, *13*, 13–31. [CrossRef]
69. McBride, R.L. Taste reception of binary sugar mixtures: Psychophysical comparison of two models. *Percept. Psychophys.* **1988**, *44*, 167–171. [CrossRef]
70. Margolskee, R.F. Molecular mechanisms of bitter and sweet taste transduction. *J. Biol. Chem.* **2002**, *277*, 1–4. [CrossRef]
71. Chamoun, E.; Mutch, D.M.; Allen-Vercoe, E.; Buchholz, A.C.; Duncan, A.M.; Spriet, L.L.; Haines, J.; Ma, D.W.L.; Guelph Family Health Study. A review of the associations between single nucleotide polymor-phisms in taste receptors, eating behaviours, and health. *Crit. Rev. Food Sci. Nutr.* **2018**, *58*, 194–207. [CrossRef]
72. Agu, H.O.; Onuoha, G.O.; Elijah, O.E.; Jideani, V.A. Consumer acceptability of acha and malted Bambara groundnut (BGN) biscuits sweetened with date palm. *Heliyon* **2020**, *6*, e05522. [CrossRef]
73. Wee, M.; Tan, V.; Forde, C. A Comparison of Psychophysical Dose-Response Behaviour across 16 Sweeteners. *Nutrients* **2018**, *10*, 1632. [CrossRef] [PubMed]
74. Scholey, A.B.; Kennedy, D.O. Cognitive and physiological effects of an "energy drink": An evaluation of the whole drink and of glucose, caffeine and herbal flavouring fractions. *Psychopharmacology* **2004**, *176*, 320–330. [CrossRef] [PubMed]
75. Miao, Y.; Ni, H.; Zhang, X.; Zhi, F.; Long, X.; Yang, X.; He, X.; Zhang, L. Investigating mechanism of sweetness intensity differences through dynamic analysis of sweetener-T1R2-membrane systems. *Food Chem.* **2022**, *374*, 131807. [CrossRef] [PubMed]
76. Varadi, M.; Anyango, S.; Deshpande, M.; Nair, S.; Natassia, C.; Yordanova, G.; Yuan, D.; Stroe, O.; Wood, G.; Laydon, A.; et al. AlphaFold Protein Structure Database: Massively expanding the structural coverage of protein-sequence space with high-accuracy models. *Nucleic Acids Res.* **2021**, *50*, D439–D444. [CrossRef] [PubMed]
77. García-Nafría, J.; Tate, C.G. Structure determination of GPCRs: Cryo-EM compared with X-ray crystallography. *Biochem Soc. Trans.* **2021**, *49*, 2345–2355. [CrossRef] [PubMed]
78. Lin, S.; Han, S.; Cai, X.; Tan, Q.; Zhou, K.; Wang, D.; Wang, X.; Du, J.; Yi, C.; Chu, X.; et al. Structures of G i-bound metabotropic glutamate receptors mGlu2 and mGlu4. *Nature* **2021**, *594*, 583–588. [CrossRef]
79. Servant, G.; Kenakin, T.; Zhang, L.; Williams, M.; Servant, N. The function and allosteric control of the human sweet taste receptor. *Adv. Pharmacol.* **2020**, *88*, 59–82. [CrossRef]
80. von Molitor, E.; Riedel, K.; Krohn, M.; Hafner, M.; Rudolf, R.; Cesetti, T. Sweet Taste Is Complex: Signaling Cascades and Circuits Involved in Sweet Sensation. *Front. Hum. Neurosci.* **2021**, *15*, 667709. [CrossRef]

applied sciences

MDPI

Article

Identification and Quantification of Valuable Compounds in Red Grape Seeds

Ovidiu Tița [1,*], Ecaterina Lengyel [1], Diana Ionela Stegăruș [2], Petre Săvescu [3], Alexandru Bogdan Ciubara [4], Maria Adelina Constantinescu [1], Mihaela Adriana Tița [1], Diana Rață [5] and Anamaria Ciubara [4]

1 Department of Agricultural Sciences and Food Engineering, Lucian Blaga University of Sibiu, Doctor Ion Rațiu 7, 550012 Sibiu, Romania; ecaterina.lengyel@ulbsibiu.ro (E.L.); adelina.constantinescu@ulbsibiu.ro (M.A.C.); mihaela.tita@ulbsibiu.ro (M.A.T.)
2 National Research and Development Institute for Cryogenics and Isotopic Technologies—ICSI Râmnicu Vâlcea, Uzinei 4, 240050 Râmnicu Vâlcea, Romania; diana.stegarus@icsi.ro
3 Faculty of Agronomy, University of Craiova, Libertatii 19, 200421 Craiova, Romania; petre.savescu@edu.ucv.ro
4 Faculty of Medicine and Pharmacy, Dunărea de Jos University of Galați, Alexandru Ioan Cuza 35, 800008 Galați, Romania; bogdan.ciubara@ugal.ro (A.B.C.); anamaria.ciubara@ugal.ro (A.C.)
5 SC ADEFARM Plus SRL, Str. Nicolae Bălcescu 9, 550159 Sibiu, Romania; rata.diana93@yahoo.com
* Correspondence: ovidiu.tita@ulbsibiu.ro; Tel.: +40-745-292-031

Abstract: Grape seeds are a by-product of the wine industry. They represent 38–52% of grape pomace and about 5% of the weight of grapes. The main objective of this study is to establish some important characteristics of grape seeds from red varieties cultivated in Romania. The analyzed grape varieties were Cabernet Sauvignon, Merlot, Pinot noir, Burgund Mare, Cadarcă, Syrah, Novac. The grape seeds were dried and ground and the following determinations were made: determination of total polyphenol content, antioxidant capacity, antiradical capacity and determination of phenolic compounds. The analyses were performed on the first day after obtaining the grape extract, on the 14th day and the 30th day. The obtained results demonstrate that all the analyzed samples have a high content of polyphenols and show antioxidant and antiradical capacity. The highest values were obtained on the first day after separation, drying, grinding and extraction of the grape seeds and began to decrease almost constantly in time, so that for 30 days from storage the values obtained could ensure good operating yields. The seeds from the Novac grape variety obtained the best results throughout the analysis period. In the case of the total polyphenol content, the average value of the three samples Novac was 394.57 mgGAE/g dry extract and the average value of antioxidant capacity was 284.35 mgAAE/g dry extract. The greatest antiradical capacity was presented by the seeds of the Syrah and Novac varieties. The average value of the three samples from the Syrah variety was 62.1%, and in the case of the Novac variety was 61.33%. The paper demonstrates the opportunity of superior capitalization of seeds from the seven grape varieties cultivated on the territory of Romania due to the characteristics it possesses. At present, there is a major interest of consumers in the most natural products, with a major contribution to increasing the body's immunity. The use of natural compounds in the food and pharmaceutical industry can be an important alternative.

Keywords: antioxidant capacity; antiradical capacity; natural products; biotechnology; polyphenols

Citation: Tița, O.; Lengyel, E.; Stegăruș, D.I.; Săvescu, P.; Ciubara, A.B.; Constantinescu, M.A.; Tița, M.A.; Rață, D.; Ciubara, A. Identification and Quantification of Valuable Compounds in Red Grape Seeds. *Appl. Sci.* **2021**, *11*, 5124. https://doi.org/10.3390/app11115124

Academic Editor: Adriana Dabija

Received: 30 April 2021
Accepted: 28 May 2021
Published: 31 May 2021

1. Introduction

Wine production is one of the most important agricultural activities, and grapes are a very valuable food product. The by-products of the wine industry are rich in a wide range of bioactive compounds. In recent years, grape by-products have been considered an alternative source for obtaining high value-added materials due to their antioxidant and antimicrobial activities. Grape seeds are a by-product of the wine industry. They represent 38–52% of grape pomace and about 5% of the weight of grapes. Grape seeds have a high content of fiber (40% w/w), protein (11% w/w), lipids (16% w/w), polyphenolic compounds (7% w/w), carbohydrates and minerals. Due to the benefits they bring to human health, the use of grape seed extracts has increased in recent years [1,2].

1.1. The Benefits of Grape Seeds

Grape seeds contain two main groups of compounds: polyphenols and oils [3]. Grape seeds contain significant concentrations of specific polyphenols such as proanthocyanidins [4], with an antioxidant role that helps lower blood pressure, protecting the heart, blood vessels and brain [5–8]. They form oligomeric proanthocyanidins complexes to protect against free radicals, acting positively in reducing allergies and together with phenolic acids and flavonoids combat oxidative stress [9–11]. It is involved in protecting healthy cells in the body by acting against toxins and stimulating nitric oxide. Studies to date confirm that polyphenols in grape seeds protect the heart and blood vessels leading to lower blood pressure and thrombotic complications, preventing the formation of clots [12,13]. Recent research claims that the use of grape seed extracts leads to an improved cognitive decline being recommended in the pathology of Alzheimer's disease and its prevention. These studies have shown that polyphenols in grape seeds reduce the damage induced by free radicals in the hippocampus by increasing the level of antioxidants, preventing the protein mutations responsible for this disease [14,15]. A beneficial role is also found in the antimicrobial effect of ointments based on grape seed extract, phenolic compounds acting in faster wound healing and infection prevention. Including internal organs can benefit from the beneficial action of grape seed components, such as resveratrol [16]. In this respect, studies carried out with polyphenolic extracts from grape seeds introduced as medicine have led to a decrease in liver enzymes, protecting the liver from the accumulation of fats (fatty liver), toxins, aflatoxins, heavy metals, chemicals or various medicines [17–20]. It was also noted a reduction in inflammation of the kidneys, with a sharp stimulation of renal functions and an increase in protective molecules. The beneficial compounds in these seeds have a positive role in stimulating insulin secretion, in rheumatoid arthritis, arthrosis, osteoarthritis, blood sugar, diabetes [21–24]. The antibacterial action of these polyphenol-rich grape seed extracts is manifested by inhibition and lysis of Gram-negative cells such as *Escherichia coli*, *Bacillus cereus*, *Staphylococcus aureus*, *Salmonella*, *Pseudomonas aeruginosa*, bacteria commonly found in human infectious pathology [25–30]. The US Food and Drug Administration (FDA) has awarded the grape seed extract the Certification Generally Recognized As Safe (GRAS) because it can be a potential nutritional supplement. High doses of grape extract (1–4 g/kg) have shown a protective and safe effect in various experiments on the metabolic effect in animals and humans [31]. In 2021, Zhao et.al demonstrated that grape seed extract improved the disturbance and metabolic function of intestinal flora induced by 2-Amino-1-methyl-6-phenylimidazo (4,5-b) pyridine (PhIP) in rats [32].

1.2. The Antioxidant and Antiradical Capacity of Grape Seeds

Most of the compounds with antioxidant effect that can be found in grapes are anthocyanins, catechin, gallic acid and resveratrol [33]. Resveratrol is a polyphenol found in grapes in greater quantities than in other fruits. This compound is found in vines in roots, seeds and stems, but the largest amount is found in the skin of grapes. The content of polyphenols with antioxidant action in wines is variable. It varies depending on the grape variety, the geographical location of the vineyard, the cultivation system, the climate, the type of soil, the harvest time or the oenological methods [34].

Catechin has a higher concentration of resveratrol in grapes. Malvidin 3,5-diglucoside is an anthocyanin compound found in grapes, which has a higher antioxidant activity than alpha-tocopherol. In red wine, anthocyanin compounds increase their antioxidant effect. The antioxidant effects of red wine are 50 times higher than that of white wine, so regular consumption of red wine reduces the risk of cardiovascular disease [34].

The polyphenolic content of wine, responsible for the antiradical activity, consists of two classes of components, flavonoids and non-flavonoids. They depend on the grape variety, the location of the vineyards, the cultivation system, the climate, the type of soil, the cultivation of the vines, the harvest time, the production and ageing process [35].

In a study conducted in 2016 by Karasu et al., five grape varieties were analyzed, and in all cases, the highest antiradical capacity was identified in seeds [36]. In 2016, Carbone and Fiordiponti analyzed a local variety of wine grown in the Lazio region and concluded that macerated wines showed the highest antiradical capacity using the DPPH (2,2-diphenyl-1-picrylhydrazyl) and ABTS (2,2′-Azinobis-(3-Ethylbenzthiazolin-6-Sulfonic Acid)) methods [37]. In 2009, Biagi et al. carried out a study in which they wanted to determine the antiradical capacity of Italian red wines. All types of wine analyzed showed high antiradical activity, always greater than 50% and a high antioxidant capacity, over 80%. Following the results obtained, it was concluded that the antiradical properties of the wine depend on the total polyphenolic content, especially on anthocyanins [35].

This research aims to highlight the importance of existing compounds in grape seeds. Most of the seeds left after processing the grapes are considered waste and remain unused. To carry out this study we analyzed seven varieties of grapes grown exclusively in Romania. We aim to highlight natural alternatives obtained from plants for use in the food and pharmaceutical industry.

2. Materials and Methods

2.1. Materials

Seven varieties of red grapes were selected for this study, namely Cabernet Sauvignon, Merlot, Pinot noir, Burgund Mare, Cadarcă, Syrah, Novac from the hilly area of Banat.

Cabernet Sauvignon is a red grape variety with the aroma of berries that have a thick and weather-resistant skinand the wine obtained has a high tannin content [38]. Following several studies and HPLC analysis, it was found that the Cabernet Sauvignon grape variety has a high content of resveratrol (52.3 and 49.6 mg/kgDW) [39].

Merlot is the second most planted variety in the world after Cabernet Sauvignon. The variety adapts to different microclimates, resistant to colder temperatures, in such conditions, it expresses itself better [40]. In the wine obtained from the Merlot grape variety, there is a high content of glycerol, ethyl acetate and succinic acid. These are giving the wine a specific taste and a well-pronounced flavor [41].

The grapes of the Pinot noir variety have thin skin and do not load the grape must. Pinot noir wines are always transparent. This variety likes cooler areas where it draws its freshness and subtle fruitiness that does not overwhelm other flavors [42]. Pinot noir is considered to be a difficult variety of product, being sensitive to the environment, with thin skin, has a high susceptibility to disease and lower concentrations of anthocyanin that often lead to reduced intensity of wine color [43].

The Burgund Mare variety has existed in Romania for over a century, but it has expanded considerably in culture after 1975 and offers high grape production. The wine does not have a specific personality, but it is balanced, with extractive and medium color intensity, reaching an alcoholic strength of 11–12 vol% [44].

"Cadarcă" or "Cadarcă de Miniș" is a traditional grape variety originating from the Miniș–Măderat vineyards, Arad area, where it was first prepared in 1744. The wine obtained is called from ancient times "bull's blood", a reputation that the locals keep even today, defining the red-ruby color, personality and strength. The grapes are medium-sized, cylindrical, sometimes winged, with the berries often placed in clusters and uneven in size. The bean is medium, spherical, slightly ovoid, with thin skin and unevenly colored in bluish-black. The variety is resistant to frost but has low resistance to drought, summer rains in sunny weather with a detrimental effect on plant development and medium resistance to grey rot of grapes [45].

Syrah is a red grape variety from France, from the Rohne Valley. Being a traditional French wine, Syrah expresses floral aromas in its youth, and over time, it develops aromas of black and white pepper, grassy notes, skin and smoke [46].

Novac is a grape variety with red grapes for red wines. It was obtained at SCDVV Drăgășani by sexual hybridization between the varieties Negru vârtos and Saperavi, being approved in 1987. It has medium-sized grains, weighing in the range of 168–315 g/100 grains, ovoid in shape, with black-bluish skin, covered with plum. The flesh is red, juicy, with a pleasant sweet and sour aroma and taste. The main climatic data from the test period were represented by the average temperature, with values between 11.4 and 12.5 °C [47].

Grapes were crushed by hand, the seeds separated by pulp and shell. The seeds were then washed under running water and then dried at room temperature. After drying, they were ground. The powder obtained was subjected to extraction with petroleum ether for 8 h. For each seed sample, 20 g of product was placed in an extraction cartridge and so on degreased, applying the Soxhlet method at a temperature of 50 °C. The samples were then homogenized with 50 mL solvent consisting of nine parts acetone and one part distilled water for the extraction of polyphenols for 12 h. Subsequently, they were centrifuged for 30 min at 3500 revolutions/min, filtered and evaporated in rotavapor—1:10. To obtain the most conclusive results, three samples of each variety selected for the study were used (1, 2, 3). The extracts obtained were stored at a temperature between 0 and 4 °C in dark glass containers.

2.2. Procedure Methods

The analyses were performed after extraction, in three periods, on the first day after their drying, on the 14th day and on the 30th day of storage of the grape seed extract. The purpose of making the determinations in the three periods was to demonstrate the preservation of the characteristics of the grape seed extract and the possibility of preserving it. To eliminate as many errors as possible, such as those related to equipment, human errors, temperature or humidity, we decided to analyze three samples from each grape extract on each day of analysis.

2.2.1. Determination of the Total Polyphenol Content of the Grape Seeds

The determination of the total polyphenol content was carried out by the Folin–Ciocâlteu method. A UV–VIS Cecil 1200 spectrophotometer was used, measuring at the wavelength of 665 nm. The results obtained were compared with the standard curve and expressed in mg gallic acid equivalent—GAE/g dry extract [48].

2.2.2. Determination of the Antioxidant Capacity of the Grape Seeds

The antioxidant capacity was determined by the method of reduction with phosphomolibdenum (Prieto et al. 1999), the reading of the samples being carried at the wavelength length of 695 nm with the spectrophotometer Cecil 1200. The results are expressed in mg ascorbic acid—AAE/g dry extract. In total, 0.1 mL of sample was homogenized in an Eppendorf tube with 1 mL of 0.6 M sulfuric acid, 28 mm sodium phosphate and 4 mm ammonium molybdate. The coated tubes were incubated for 90 min at 95 °C, cooled, then the absorbance was read at a wavelength of 695 nm on the Cecil 1200 spectrophotometer in the presence of the control sample. The control sample consists of the same reagents used and the procedure minus the extract. The results are expressed in mg ascorbic acid (AAE/g dry extract), the calibration line was performed in a range of 50 mg/g–350 mgAAE/g dry extract [49].

2.2.3. Determination of the Antiradical Capacity of the Grape Seeds

The antiradical capacity was determined by the Brand–Williams et al. method and Yalcin et al. method which involves homogenizing the grape seed extract with 2,2-diphenyl-1-picrilhydrazil in the presence of methanol, keeping in the dark for 30 min and then reading the absorption at 516 nm with the Cecil 1200 spectrophotometer. A blind sample lacking the seed extract was used as the standard sample. The results are expressed by applying the standard formula which implies [50,51]:

$$Ar\,[\%] = 100 \times (1 - Ap{:}Ae) \quad (1)$$

where *Ap* represents the absorption of the sample; *Ae* represents the absorption of the standard sample.

2.2.4. Determination of the Phenolic Compounds of the Grape Seeds

The determination of phenolic compounds was carried out by the HPLC (Knauer) a method involving the direct injection of the samples and the separation was carried out on a monolithic column C18. The chromatograph was equipped with a UV/DAD detector, and the solvent had a constant flow rate of 1.2 mL/min at a temperature of 22 °C. Solvent A was composed of distilled water and glacial acetic acid 0.15%, and solvent B of acetonitrile and glacial acetic acid 0.15%. The extraction steps were established according to the protocol as follows: gradient B: 0–4 min 9–12%, 4–10 min 12–15%, 10–20 min 25–70%, 20–30 min 100%. The amount injected was 1 µL and the reading was performed at the wavelengths of the spectra of the identified phenolic compounds: at 225 nm vanilic acid; at 280 nm: gallic acid, catechin, epicatismin, epicatchin gallate, P-hydroxybenzoic acid, M-hydroxybenzoic acid, syringic acid; at 305 nm P-cumaric acid, resveratrol; at 330 nm caffeic acid and chlorogenic acid; at 360 nm rutin and quercetin. Calibration curves were designed on compound concentrations starting from: 0.1, 0.5, 1, 5, 10, 15, 25, 50, 100, and 150 mg/L, all standards being of chromatographic purity—Sigma Aldrich [52].

2.2.5. Statistical Analysis

Results are expressed by mean ± standard deviation (SD) for each group. Graphical representation and statistical processing were performed using the Minitab 14 statistical softwareand $p < 0.05$ was considered significant.

3. Results

3.1. Determination of the Total Polyphenol Content of the Grape Seeds

Figure 1A shows the total polyphenols content in the seeds on the first day of the following grape varieties: Cabernet Sauvignon, Merlot, Pinot noir, Burgund mare, Cadarcă, Syrah, Novac. The highest total polyphenol content was recorded in the case of Novac seeds, the average value of the three samples being 394.57 mgGAE/g dry extract. A high content of polyphenols was also recorded in the case of Syrah seeds. The average value of the three samples is 384.36 mgGAE/g dry extract. Lower values of the total polyphenol content were recorded for the varieties Cadarcă (344.42 mgGAE/g dry extract), Cabernet Sauvignon (340.49 mgGAE/g dry extract), Pinot noir (334.38 mgGAE/g dry extract) and Burgund mare (326.64 mgGAE/g dry extract). Among the analyzed varieties, the lowest level of the total polyphenols content was registered in the case of seeds from the Merlot variety, being registered an average value of 230.45 mgGAE/g dry extract.

Figure 1B shows the total polyphenol content determined from the seeds of the seven grape varieties on the 14th day. Compared to the first day of analysis, we find a slight decrease in the total polyphenol content for all seven grape varieties. On this day of analysis, the Novac variety has the highest polyphenol content, the average value of the three samples being 390.83 mgGAE/g dry extract. The lowest polyphenol content on this day of analysis was recorded in the Merlot variety, the average value of the three samples being 218.42 mgGAE/g dry extract.

Figure 1C shows the total polyphenol content determined from the seeds of the seven grape varieties on the 30th day. The lowest values of polyphenol content of the seven grape varieties were recorded on this day of analysis. Even on this day, the Novac variety recorded the highest polyphenol content, the average value of the three samples being 387.36 mgGAE/g dry extract. The lowest polyphenol content on this day of analysis was recorded in the Merlot variety, the average value of the three samples being 211.21 mgGAE/g dry extract.

3.2. Determination of the Antioxidant Capacity of the Grape Seeds

Figure 2A shows the antioxidant capacity of grape seeds on the first day from the seven varieties studied: Cabernet Sauvignon, Merlot, Pinot noir, Burgund mare, Cadarcă, Syrah, Novac. The highest antioxidant capacity was registered in the case of Novac seeds. The average value of the three samples is 284.35 mgAAE/g dry extract. The Pinot noir seeds reported a fairly high antioxidant capacity, the average value of the three samples being 241.07 mgAAE/g dry extract. The seeds of the Syrah, Cadarcă and Cabernet Sauvignon varieties have antioxidant capacities between 220.58 and 237.52 mgAAE/g dry extract. The lowest antioxidant activities were reported in the Burgund mare and Merlot varieties. The average value of the three samples in the case of the Burgund mare variety is 214.38 mgAAE/g dry extract, and in the case of the Merlot variety, it is 133.91 mgAAE/g dry extract.

Figure 2B shows the antioxidant capacity determined from the seeds of the seven grape varieties on the 14th day. Compared to the first day, there is a slight decrease in the antioxidant capacity of the seven grape varieties. On this day, the highest antioxidant capacity was recorded for the Novac variety, the average value of the three samples being 281.27 mgAAE/g dry extract. The lowest value was obtained in the case of the Merlot variety, the average value of the three samples being 130.37 mgAAE/g dry extract.

Figure 2C shows the antioxidant capacity determined from the seeds of the seven grape varieties on the 30th day. On this day, the lowest values of antioxidant capacity were registered. Additionally, on this day, in the case of the Novac variety the highest values were obtained, the average value of the three samples being 277.04 mgAAE/g dry extract. The Merlot variety recorded the lowest values on this day, the average value of the three samples being 127.46 mgAAE/g dry extract.

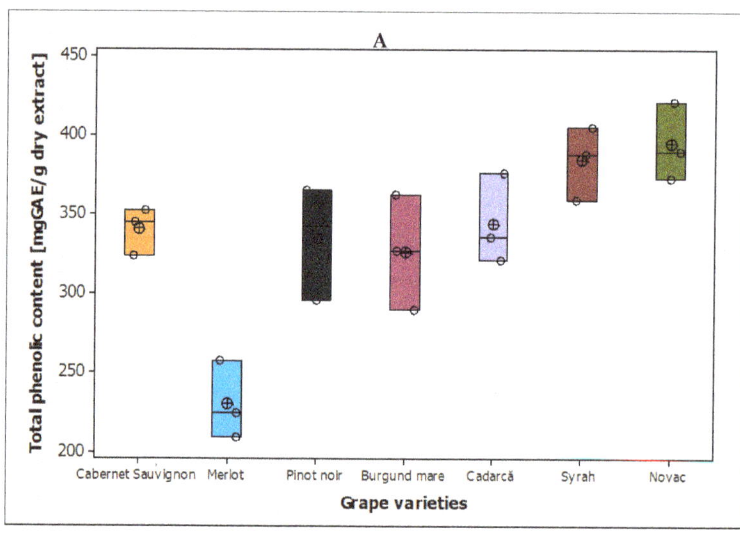

Figure 1. *Cont.*

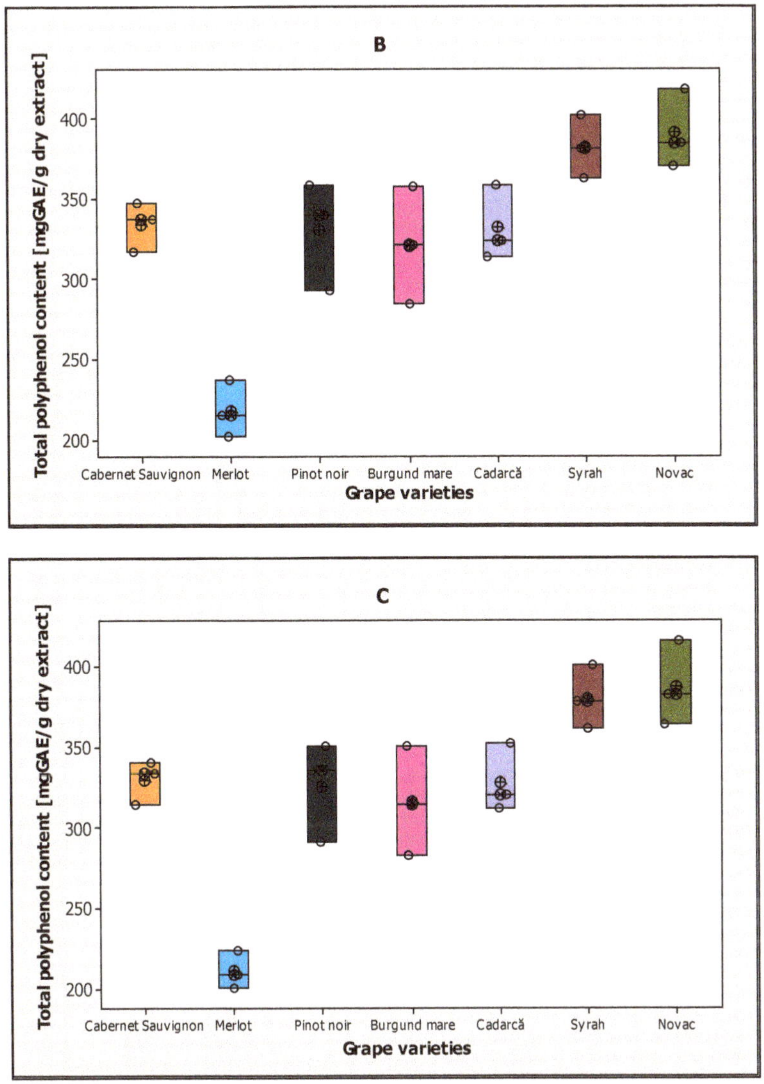

Figure 1. (**A**) The total polyphenol content determined from the seeds of the seven grape varieties on the first day; (**B**) the total polyphenol content determined from the seeds of the seven grape varieties on the 14th day; (**C**) the total polyphenol content determined from the seeds of the seven grape varieties on the 30th day.

3.3. Determination of the Antiradical Capacity of the Grape Seeds

The antiradical capacity of extracts obtained from the red grape seed is an important indicator in assessing their quality for their widespread use in various contexts.

Figure 3A shows the antiradical capacity of grape seeds on the first day from the seven varieties studied: Cabernet Sauvignon, Merlot, Pinot noir, Burgund mare, Cadarcă, Syrah, Novac. The greatest antiradical capacity is presented by the seeds of the Syrah and Novac varieties. The average value of the three samples from the Syrah variety is 62.1%, and in the case of the Novac variety, it is 61.33%. The antiradical capacities of Pinor noir and Cadarcă varieties are quite close, the average values being 47.45% and 45.36%. The lowest values were obtained for the Cabernet Sauvignon, Merlot and Burgund Mare varieties. The average values of their antiradical capacities are 33.75%, 33.55% and 33.3%.

Figure 3B shows the antiradical capacity determined from the seeds of the seven grape varieties on the 14th day. Compared to the first, there is a slight decrease in antiradical capacity. The highest values of antiradical capacity were recorded for Syrah and Novac varieties. The average value of the three Syrah samples was 59.84% and the average value of the three Novac samples was 59.64%. The lowest value of the antiradical capacity was registered in the case of the Burgund Mare variety, the average value of the three samples being 31.61%.

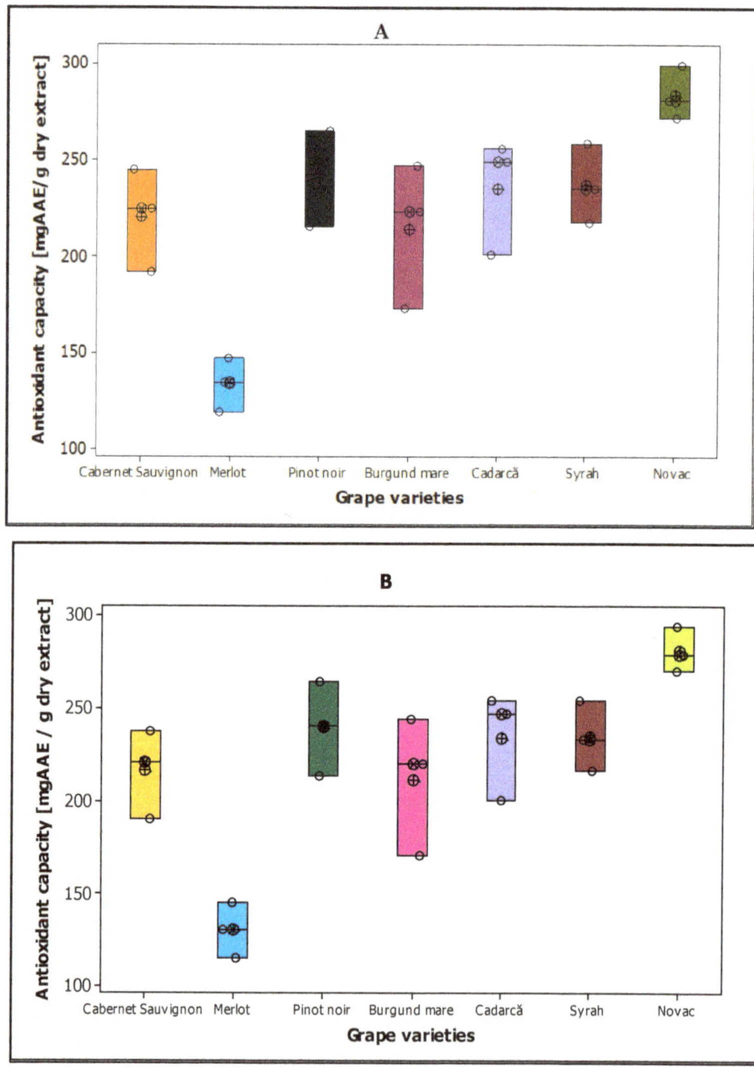

Figure 2. *Cont.*

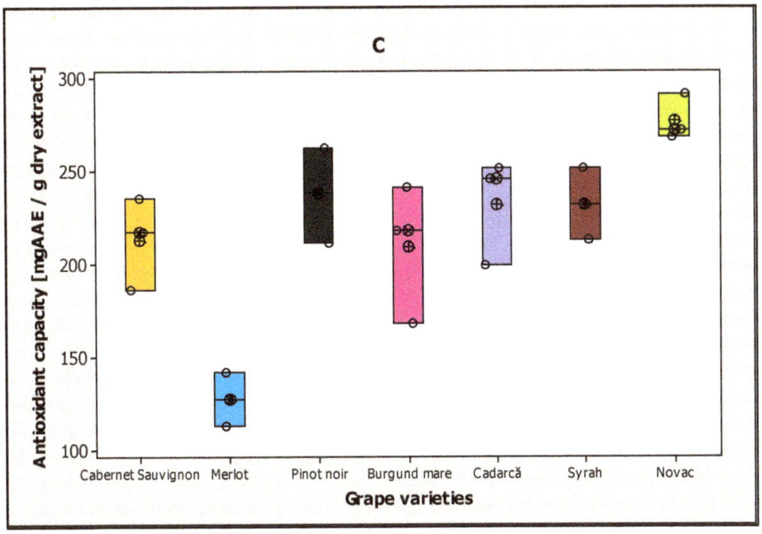

Figure 2. (**A**) Antioxidant capacity determined from the seeds of the seven grape varieties on the first day; (**B**) antioxidant capacity determined from the seeds of the seven grape varieties on the 14th day; (**C**) antioxidant capacity determined from the seeds of the seven grape varieties on the 30th day.

Figure 3C shows the antiradical capacity determined from the seeds of the seven grape varieties on the 30th day. The lowest values were recorded on this day of analysis. The highest values were recorded for Syrah and Novac varieties. The average value of the three Syrah samples was 58.38% and the average value of the three Novac samples was 58.33%. The lowest values were obtained in the case of the Burgund Mare variety, the average value of the three samples being 29.92%.

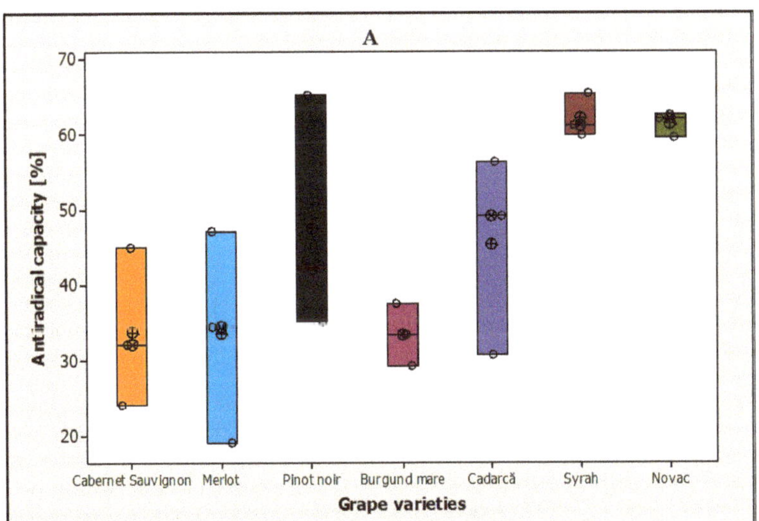

Figure 3. *Cont.*

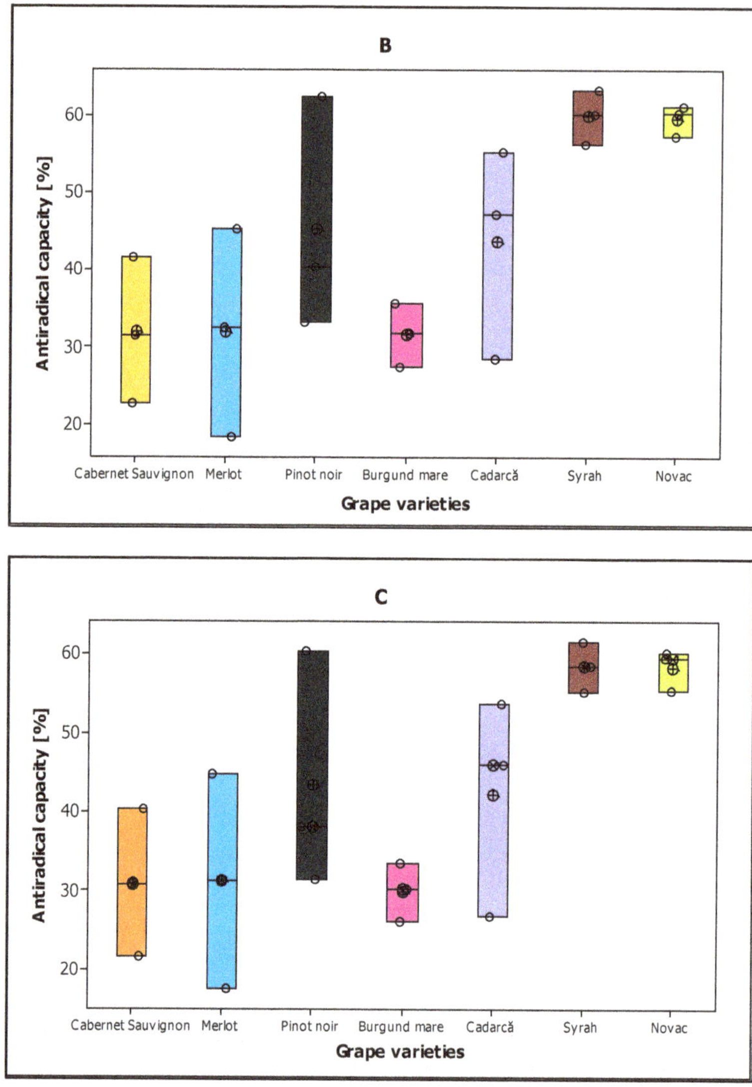

Figure 3. (**A**) Antiradical capacity determined from the seeds of the seven grape varieties on the first day; (**B**) antiradical capacity determined from the seeds of the seven grape varieties on the 14th day; (**C**) antiradical capacity determined from the seeds of the seven grape varieties on the 30th day.

3.4. Determination of the Phenolic Compounds of the Grape Seeds

Table 1 shows the phenolic compounds identified in the seven grape varieties studied: Cabernet Sauvignon, Merlot, Pinot noir, Burgund mare, Cadarcă, Syrah, Novac. The phenolic compounds identified and quantified by the HPLC (high performance liquid chromatography) method have values that vary greatly depending on the nature of the compound but also on the variety from which the seeds come. It is noted that the most significant values are syringic acid, being between 121.22 and 136.66 mg/L. The richest varieties in syringic acid are Syrah and Novac, followed by Burgund Mare and Cadarcă. The lowest values are observed in the case of Cabernet Sauvignon and Merlot extracts. Gallic acid and vanillic acid are found in amounts of tens of mg/L, the most significant values being present in the Novac variety—39.22 and 20.91 mg/L. Catechin has values between 6.05 and 9.01 mg/L and chlorogenic acid has values between 4.82 and 8.02 mg/L. M-hydroxybenzoic acid is about 2–3 times more significant than p-hydroxybenzoic acid, with values ranging from 1.2 to 5.62 mg/L. Epicatechin and p-coumaric acid are around five with oscillations of 1–2 units up or down. Epicatechin gallate varies from 1.84 to 2.56 mg/L, the most significant values being in the case of seed extract from the Pinot noir variety. Caffeic acid is present with values starting from 0.94 mg/L and reaching a maximum of 1.56 mg/L, specific to the Novac variety. Subunit values are observed in the case of ferulic acid with a maximum of 0.96 mg/L for Pinot noir, quercetin with a maximum of 0.96 mg/L for Syrah and 0.41 mg/L for Cadarcă. Resveratrol is found in the seeds of all varieties with values between 1.91 and 2.92 mg/L for Cabernet Sauvignon, 1.41 and 2.23 mg/L for Merlot, 1.93 and 2.37 mg/L for Pinot noir, 1.66 and 1.88 mg/L for Burgund Mare, 1.71 and 2.46 mg/L for Cadarcă, 2.12 and 2.34 mg/L for Syrah and 2.16 and 2.38 mg/L for Novac variety.

Table 1. Identification and quantification of valuable phenolic compounds in the seeds of the seven grape varieties.

Compound	Composition (mg/L ppm (Parts per Million))						
	Cabernet Sauvignon	**Merlot**	**Pinot noir**	**Burgund Mare**	**Cadarcă**	**Syrah**	**Novac**
Catechin	8.17 ± 0.01	7.28 ± 0.01	6.98 ± 0.01	7.15 ± 0.01	8.12 ± 0.01	8.46 ± 0.01	8.78 ± 0.01
Epicatismin	5.11 ± 0.01	5.12 ± 0.01	5.16 ± 0.01	4.82 ± 0.01	4.43 ± 0.01	5.98 ± 0.01	6.17 ± 0.01
Epicatechin gallate	2.15 ± 0.01	2.18 ± 0.01	2.26 ± 0.01	1.95 ± 0.01	2.24 ± 0.01	2.02 ± 0.01	2.46 ± 0.01
Gallic acid	26.34 ± 0.05	0.71 ± 0.01	31.46 ± 0.05	31.02 ± 0.05	32.46 ± 0.05	33.80 ± 0.05	35.46 ± 0.05
P-hydroxybenzoic acid	1.85 ± 0.05	1.23 ± 0.05	1.77 ± 0.05	2.04 ± 0.05	2.09 ± 0.05	2.22 ± 0.05	2.30 ± 0.05
Vanilic acid	18.02 ± 0.01	15.57 ± 0.01	16.15 ± 0.01	13.24 ± 0.01	18.65 ± 0.01	18.45 ± 0.01	20.08 ± 0.01
Syringic acid	122.87 ± 0.25	130.13 ± 0.25	129.40 ± 0.25	129.81 ± 0.25	133.20 ± 0.25	134.26 ± 0.25	134.12 ± 0.25
M-hydroxybenzoic acid	4.58 ± 0.01	4.60 ± 0.01	4.20 ± 0.01	3.69 ± 0.01	3.44 ± 0.01	5.32 ± 0.01	5.26 ± 0.01
Caffeic acid	1.01 ± 0.01	1.08 ± 0.01	0.96 ± 0.01	1.19 ± 0.01	1.27 ± 0.01	1.33 ± 0.01	1.35 ± 0.01
Ferulic acid	0.59 ± 0.01	0.47 ± 0.01	0.90 ± 0.01	0.61 ± 0.01	0.70 ± 0.01	0.62 ± 0.01	0.54 ± 0.01
Chlorogenic acid	7.67 ± 0.01	6.95 ± 0.01	8.24 ± 0.01	5.32 ± 0.01	6.54 ± 0.01	6.67 ± 0.01	7.90 ± 0.01
P-coumaric acid	5.22 ± 0.01	3.50 ± 0.01	4.49 ± 0.01	4.68 ± 0.01	4.45 ± 0.01	5.27 ± 0.01	6.01 ± 0.01
Resveratrol	2.27 ± 0.01	1.22 ± 0.01	0.84 ± 0.01	1.74 ± 0.01	2.01 ± 0.01	2.28 ± 0.01	2.41 ± 0.01
Rutin	0.15 ± 0.01	0.41 ± 0.01	0.21 ± 0.01	0.36 ± 0.01	0.37 ± 0.01	0.36 ± 0.01	0.31 ± 0.01
Quercetin	0.82 ± 0.01	32.43 ± 0.05	2.14 ± 0.01	0.47 ± 0.01	5.84 ± 0.01	2.33 ± 0.01	1.06 ± 0.01

4. Discussion

All four ascertainments, the determination of the total content of polyphenols, the determination of antioxidant and antiradical activity, as well as the determination of phenolic compounds were performed on all seven grape varieties. The grapes came from plantations in Romania, more precisely from the Banat area. The analyses were performed after extraction, in three periods, on the first day after their drying, on the 14th day and on the 30th day of storage of the grape seed extract. To eliminate as many errors as possible, such as those related to equipment, human errors, temperature or humidity, we decided to analyze three samples from each grape extract on each day of analysis.

According to the obtained results, in the case of determining the total content of polyphenols, antioxidant and antiradical capacity, a slight decrease of these contents was found in the three days of analysis. This decrease was slight in all seven grape samples. The highest values were obtained in the case of the Novac variety. This is also demonstrated by the analysis of phenolic compounds because this variety has high values for all identified components. Merlot seeds have the lowest values in all three determinations, analysis of total polyphenol compounds, antioxidant and antiradical analysis.

Using the HPLC method, we identified several phenolic compounds contained in the analyzed seeds. In all seven samples, the following proportion has the highest proportion: syringic acid, gallic acid and vanilic acid. Syringic acid is an excellent compound to be used as a therapeutic agent and possesses antioxidant, antimicrobial, anti-inflammatory and antiendotoxic capabilities [53]. Gallic acid has various properties, including antifungal, antimicrobial, and anticancer capabilities [54]. Vanillic acid reduced collagen accumulation and hydroxyproline content [55]. According to the results obtained from this determination, we notice that the seeds of the Novac variety contain the highest amounts of syringic acid, gallic acid and vanillic acid, and the seeds of the Merlot variety have a low content of vanillic acid and gallic acid. These values justify the results obtained in the case of determining the antioxidant and antiradical capacity. In the case of both determinations, the best results were obtained in the case of seeds of the Novac variety, and the lowest in the case of seeds of the Merlot variety. In addition to the three compounds, HPLC analysis showed others that possessed antioxidant properties, such as catechin, caffeic acid, ferulic acid, coumaric acid, resveratrol and quercetin, and antiradical properties, such as chlorogenic acid and rutin [56–63]. Extraction yields differ depending on the quality of the solvent and the working conditions. The obtained results attest to the fact that grape seeds contain significant amounts of polyphenols with strong antioxidant activity.

Antioxidants are widely used as food additives to prevent food degradation. Antioxidants also play an important role in preventing a variety of lifestyle-related diseases and ageing, as they are closely linked to active oxygen and lipid peroxidation. Finding viable solutions for the realization of basic products in food, with the widest possible destination, which in addition to a longer life cycle to ensure at the same time a healthy lifestyle, is of utmost importance. Making foods that increase the body's immunity and bring several benefits to the consumer is an effective alternative to ensuring physical and mental health. In addition, it can be a sustainable and efficient activity [64].

5. Conclusions

Grape seeds are an important source of valuable compounds for human health. Grape seeds mainly contain phenolic compounds, such as proanthocyanidins with an antioxidant capacity 20 times higher than vitamin E and 50 times higher than vitamin C. Due to their antioxidant capacity, grape seeds have antiallergic, anti-inflammatory, anticancer, immune boosting action, as well as beneficial effects in cardiovascular diseases. They are a source of bioactive substances and can currently be an important solution in obtaining new medicines of plant origin.

Several very valuable products have been identified qualitatively and quantitatively such as catechin, epicatismin, epicatechin gallate, gallic acid, P-hydroxybenzoic acid, vanilic acid, syringic acid, M-hydroxybenzoic acid, caffeic acid, ferulic acid, chlorogenic acid, P-coumaric acid, resveratrol, rutin and quercetin with special importance demonstrated for antioxidant and antiradical activity.

The analyses performed at 30 days showed a reduction in the content of useful substances contained in grape seeds, a reduction due to transformations that take place during the storage period of the seeds under analysis, oxidation and degradation. Under the required storage conditions, it is clear that the extraction yields for the active compounds justify the interest. It is more than certain that more adequate special storage conditions—protection against oxidative processes or seed storage humidity in the range of 10–12%—can significantly extend the storage period of yields, on the extraction of useful products on a longer period.

The superior recovery of these by-products and the obtaining of high value added bioproducts for use in the food, pharmaceutical and cosmetics industries are becoming an increasingly important practice. Bioproducts will have a complex biochemical composition and high antioxidant potential, is intended for the prevention and diet therapy of diseases caused by oxidative stress. The orientation towards the use of natural products with bioactive compounds must be a priority for the processors in the food and pharmaceutical industry both from a sustainable and economic point of view. Grape seeds fit perfectly into these product categories and represent an important alternative.

Author Contributions: Conceptualization, O.T., E.L., D.I.S. and M.A.C.; methodology, P.S., M.A.T. and D.R.; software, M.A.C.; validation, O.T., E.L., P.S. and M.A.T.; formal analysis, D.I.S. and D.R.; investigation, E.L., M.A.T., A.B.C. and A.C.; resources, P.S. and M.A.C.; data curation, O.T.; writing—original draft preparation, D.I.S., P.S. and D.R.; writing—review and editing, E.L., M.A.C. and M.A.T.; visualization, O.T. and A.B.C.; supervision, O.T.; project administration, O.T.; funding acquisition, O.T., A.B.C. and A.C. All authors have read and agreed to the published version of the manuscript.

Funding: This research was funded by Lucian Blaga University of Sibiu and Hasso Plattner Foundation, grant number LBUS-IRG-2020-06.

Institutional Review Board Statement: Not applicable.

Informed Consent Statement: Not applicable.

Data Availability Statement: Not applicable.

Acknowledgments: We would like to express our sincere gratitude to the Research Center in Biotechnology and Food Engineering (C.C.B.I.A.), Lucian Blaga University of Sibiu and Hasso Plattner Foundation for the entire support granted throughout the research period. We also appreciate the editor and the anonymous reviewers for their constructive comments and insightful suggestions on the manuscript.

Conflicts of Interest: The authors declare no conflict of interest.

References

1. Silva, A.; Silva, V.; Igrejas, G.; Gaivão, I.; Aires, A.; Klibi, N.; Dapkevicius, M.E.; Valentão, P.; Falco, V.; Poeta, P. Valorization of winemaking by-products as a novel source of antibacterial properties: New strategies to fight antibiotic resistance. *Molecules* **2021**, *26*, 2331. [CrossRef] [PubMed]
2. Arvanitoyannis, I.S.; Ladas, D.; Mavromatis, A. Wine Waste management: Treatment methods and potential uses of treated waste. In *Waste Management for the Food Industries*; Arvanitoyannis, I.S., Ed.; Academic Press: Cambridge, MA, USA, 2008; pp. 413–452.
3. Xiuzhen, H.; Tao, S.; Hongxiang, L. Dietary polyphenols and their biological significance. *Int. J. Mol. Sci.* **2007**, *8*, 950–988.
4. Sano, A. Safety assessment of 4-week oral intake of proanthocyanidin-rich grape seed extract in healthy subjects. *Food Chem. Toxicol.* **2017**, *108*, 519–523. [CrossRef] [PubMed]
5. Jayaprakasha, G.; Tamil, S.; Sakariah, K.K. Antibacterial and antioxidant activities of grape (Vitis vinifera) seed extracts. *Food Res. Int.* **2003**, *36*, 117–122. [CrossRef]
6. Feng, Y.; Liu, Y.-M.; Leblanc, M.H.; Bhatt, A.J.; Rhodes, P.G. Grape seed extract given three hours after injury suppresses lipid peroxidation and reduces hypoxic-ischemic brain injury in neonatal rats. *Pediatr Res.* **2007**, *61*, 295–300. [CrossRef]
7. Feringa, H.H.; Laskey, D.; Dickson, J.E.; Coleman, C. The effect of grape seed extract on cardiovascular risk markers: A meta-analysis of randomized controlled trials. *J. Am. Diet. Assoc.* **2011**, *111*, 1173–1181. [CrossRef]
8. Jin, H.; Liu, M.; Zhang, X.; Pan, J.; Han, J.; Wang, Y.; Lei, H.; Ding, Y.; Yuan, Y. Grape seed procyanidin extract attenuates hypoxic pulmonary hypertension by inhibiting oxidative stress and pulmonary arterial smooth muscle cells proliferation. *J. Nutr. Biochem.* **2016**, *36*, 81–88. [CrossRef]
9. Fine, A.M. Oligomeric proanthocyanidin complexes: History, structure, and phytopharmaceutical applications. *Altern. Med. Rev.* **2000**, *5*, 144–151.
10. Song, Q.; Shi, Z.; Bi, W.; Liu, R.; Zhang, C.; Wang, K.; Dang, X. Beneficial effect of grape seed proanthocyanidin extract in rabbits with steroid-induced osteonecrosis via protecting against oxidative stress and apoptosis. *J. Orthop. Sci.* **2015**, *20*, 196–204. [CrossRef]
11. Yamakoshi, J.; Sano, A.; Tokutake, S.; Saito, M.; Kikuchi, M.; Kubota, Y.; Kawachi, Y.; Otsuka, F. Oral intake of proanthocyanidin-rich extract from grape seeds improves chloasma. *F. Phytother Res.* **2004**, *18*, 895–899. [CrossRef] [PubMed]
12. Bijak, M.; Sut, A.; Kosiorek, A.; Saluk-Bijak, J.; Golanski, J. Dual anticoagulant/antiplatelet activity of polyphenolic grape seeds extract. *Nutrients* **2019**, *11*, 93. [CrossRef] [PubMed]
13. Clifton, P.M. Effect of grape seed extract and quercetin on cardiovascular and endothelial parameters in high-risk subjects. *J. Biomed. Biotechnol.* **2004**, *5*, 272–278. [CrossRef] [PubMed]
14. WWang, Y.-J.; Thomas, P.; Zhong, J.-H.; Bi, F.-F.; Kosaraju, S.; Pollard, A.; Fenech, M.; Zhou, X.-F. Consumption of grape seed extract prevents amyloid-β deposition and attenuates inflammation in brain of an alzheimer's disease mouse. *Neurotox. Res.* **2009**, *15*, 3–14. [CrossRef]
15. Ferruzzi, M.G.; Lobo, J.K.; Janle, E.M.; Cooper, B.; Simon, J.E.; Wu, Q.L.; Welch, C.; Ho, L.; Weaver, C.; Pasinetti, G.M. Bioavailability of gallic acid and catechins from grape seed polyphenol extract is improved by repeated dosing in rats: Implications for treatment in Alzheimer's disease. *J. Alzheimer's Dis.* **2009**, *18*, 113–124. [CrossRef] [PubMed]
16. Soleymani, S.; Iranpanah, A.; Najafi, F.; Belwal, T.; Ramola, S.; Abbasabadi, Z.; Momtaz, S.; Farzaei, M.H. Implications of grape extract and its nanoformulated bioactive agent resveratrol against skin disorders. *Arch. Dermatol. Res.* **2019**, *311*, 577–588. [CrossRef]

17. Argani, H.; Ghorbanihaghjo, A.; Vatankhahan, H.; Rashtchizadeh, N.; Raeisi, S.; Ilghami, H. The effect of red grape seed extract on serum paraoxonase activity in patients with mild to moderate hyperlipidemia. *Sao Paulo Med. J.* **2016**, *134*, 234–239. [CrossRef]
18. Asbaghi, O.; Nazarian, B.; Reiner, Ž.; Amirani, E.; Kolahdooz, F.; Chamani, M.; Asemi, Z. The effects of grape seed extract on glycemic control, serum lipoproteins, inflammation, and body weight: A systematic review and meta-analysis of randomized controlled trials. *Phytother. Res.* **2020**, *34*, 239–253. [CrossRef]
19. Rajput, S.A.; Sun, L.; Zhang, N.; Khalil, M.M.; Ling, Z.; Chong, L.; Wang, S.; Rajput, I.R.; Bloch, D.M.; Khan, F.A.; et al. Grape seed proanthocyanidin extract alleviates aflatoxinb1-induced immunotoxicity and oxidative stress via modulation of NF-κB and Nrf2 signaling pathways in broilers. *Toxins* **2019**, *11*, 23. [CrossRef] [PubMed]
20. Terra, X.; Valls, J.; Vitrac, X.; Mérrillon, J.-M.; Arola, L.; Ardèvol, A.; Bladé, C.; Fernández-Larrea, J.; Pujadas, G.; Salvado, J.; et al. Grape-seed procyanidins act as antiinflammatory agents in endotoxin-stimulated RAW 264.7 macrophages by inhibiting NFkB signaling pathway. *J. Agric. Food Chem.* **2007**, *55*, 4357–4365. [CrossRef]
21. El-Awdan, S.A.; Abdel Jaleel, G.A.; Saleh, D.O. Grape seed extract attenuates hyperglycaemia-induced in rats by streptozotocin. *Bull. Fac. Pharm. Cairo Univ.* **2013**, *51*, 203–209. [CrossRef]
22. Kar, P.; Laight, D.; Rooprai, H.K.; Shaw, K.M.; Cummings, M. Effects of grape seed extract in type 2 diabetic subjects at high cardiovascular risk: A double blind randomized placebo-controlled trial examining metabolic markers, vascular tone, inflammation, oxidative stress and insulin sensitivity. *Diabet Med.* **2009**, *26*, 526–531. [CrossRef] [PubMed]
23. Turki, K.; Charradi, K.; Boukhalfa, H.; Belhaj, M.; Limam, F.; Aouani, E. Grape seed powder improves renal failure of chronic kidney disease patients. *EXCLI J.* **2016**, *15*, 424–433.
24. Visser, J.; Van Staden, P.J.; Soma, P.; Buys, A.V.; Pretorius, E. The stabilizing effect of an oligomeric proanthocyanidin on red blood cell membrane structure of poorly controlled Type II diabetes. *Nutr. Diabetes.* **2017**, *7*, 275. [CrossRef] [PubMed]
25. Leone, A.; Longo, C.; Gerardi, C.; Trosko, J.E. Pro-apoptotic effect of grape seed extract on MCF-7 involves transient increase of gap junction intercellular communication and Cx43 up-regulation: A mechanism of chemoprevention. *Int. J. Mol. Sci.* **2019**, *20*, 3244. [CrossRef]
26. Hamza, A.A.; Heeba, G.H.; Elwy, H.M.; Murali, C.; El-Awady, R.; Amin, A. Molecular characterization of the grape seeds extract's effect against chemically induced liver cancer: In vivo and in vitro analyses. *Sci. Rep.* **2018**, *8*, 1270. [CrossRef]
27. Al-Habib, A.; Al-Saleh, E.; Safer, A.-M.; Afzal, M. Bactericidal effect of grape seed extract on methicillin-resistant Staphylococcus aureus (MRSA). *J. Toxicol. Sci.* **2010**, *35*, 357–364. [CrossRef]
28. Bagchi, D.; Swaroop, A.; Preuss, H.G.; Bagchi, M. Free radical scavenging, antioxidant, and cancer chemoprevention by grape seed proanthocyanin: An overview. *Mutat. Res.* **2014**, *768*, 69–73. [CrossRef] [PubMed]
29. Mao, J.T.; Xue, B.; Smoake, J.; Lu, Q.-Y.; Park, H.; Henning, S.M.; Burns, W.; Bernabei, A.; Elashoff, D.; Serio, K.J.; et al. MicroRNA-19a/b mediates grape seed procyanidin extract-induced anti-neoplastic effects against lung cancer. *J. Nutr. Biochem.* **2016**, *34*, 118–125. [CrossRef] [PubMed]
30. Uchiyama, H.; Uehara, K.; Nagashima, T.; Nakata, A.; Sato, K.; Mihara, Y.; Komatsu, K.I.; Takanari, J.; Shimizu, S.; Wakame, K. Global Liver gene expression analysis on a murine metabolic syndrome model treated by low-molecular-weight lychee fruit polyphenol [Oligonol(R)]. *Anticancer Res.* **2016**, *36*, 3705–3713. [PubMed]
31. Kadri, S.; El Ayed, M.; Kadri, A.; Limam, F.; Aouani, E.; Mokni, M. Protective effect of grape seed extract and orlistat co-treatment against stroke: Effect on oxidative stress and energy failure. *Biomed. Pharmacother.* **2021**, *136*, 1–13. [CrossRef] [PubMed]
32. Zhao, X.; Wu, Y.; Liu, H.; Hu, N.; Zhang, Y.; Wang, S. Grape seed extract ameliorates PhIP-induced colonic injury by modulating gut microbiota, lipid metabolism, and NF-κB signaling pathway in rats. *J. Funct. Foods* **2021**, *78*, 1–12. [CrossRef]
33. Carle, R.; Claus, A.; Kammerer, D.; Schieber, A. Polyphenol screening of pomace from red and white grape varieties (Vitis vinifera L.) by HPLC-DAD-MS/MS. *J. Agric. Food Chem.* **2004**, *52*, 4360–4367.
34. Xu, Z. Important antioxidant phytochemicals in agricultural food products. In *Analysis of Antioxidant-Rich Phytochemicals*, 1st ed.; Xu, Z., Howard, L.R., Eds.; Wiley-Blackwell: Hoboken, NJ, USA, 2012; pp. 1–24.
35. Biagi, M.; Miraldi, E.; Figura, N.; Giachetti, D. Antiradical Activity and in vitro Inhibition of Helicobacter pylori by Italian Red Wines. *Nat. Prod. Commun.* **2009**, *4*, 255–260. [CrossRef]
36. Karasu, S.; Başlar, M.; Karaman, S.; Kiliçli, M.; Abdullah, A.U.; Yaman, H.; Sağdiç, O. Characterization of some bioactive compounds and physicochemical properties of grape varieties grown in turkey: Thermal degradation kinetics of anthocyanin. *Turk. J. Agric. For.* **2016**, *40*, 177–185. [CrossRef]
37. Carbone, K.; Fiordiponti, L. Colour evaluation, bioactive compound content, phenolic acid profiles and in vitro biological activity of Passerina del Frusinate white wines: Influence of pre-fermentative skin contact times. *Molecules* **2016**, *21*, 960. [CrossRef]
38. Cabernet Sauvignon. Available online: https://www.rewine.ro/blog/cabernet-sauvignon/ (accessed on 3 November 2020).
39. Tobar, M.; Fiore, N.; Pérez-Donoso, A.G.; León, R.; Rosales, I.M.; Gambardella, M. Divergent molecular and growth responses of young "Cabernet Sauvignon" (Vitis vinifera) plants to simple and mixed infections with Grapevine rupestris stem pitting-associated virus. *Hort. Res.* **2020**, *7*, 1–14. [CrossRef] [PubMed]
40. De ce Merlot Este Vinul Prieten cu Toate Gusturile? Available online: https://vinlavin.ro/vin-merlot-pentru-gusturi-pretentioase/ (accessed on 3 November 2020).
41. HHu, B.; Gao, J.; Xu, S.; Zhu, J.; Fan, X.; Zhou, X. Quality evaluation of different varieties of dry red wine based on nuclear magnetic resonance metabolomics. *J. Appl. Biol. Chem.* **2020**, *63*, 1–8.

42. Fii "Snob" Vara Asta. Degusta Pinot Noir! Available online: https://www.rewine.ro/blog/fi-snob-vara-asta-degusta-pinot-noir/ (accessed on 3 November 2020).
43. Martin, D.; Grab, F.; Grose, C.; Stuart, L.; Scofield, C.; McLachlan, A.; Rutan, T. Vintage by vine interactions most strongly influence Pinot noir grape composition in New Zealand. *OENO One* **2020**, *54*, 881–902. [CrossRef]
44. Blaufraenkisch ... Soiuri de Struguri. Available online: http://vinpenet.blogspot.com/2013/02/despre-blaufraenkisch-2010-lacerta.html (accessed on 3 November 2020).
45. Balla Géza a Reinventat Cadarca și Nu s-a Oprit Aici. Available online: https://vinul.ro/balla-geza-a-reinventat-cadarca-si-nu-s-a-oprit-aici.html (accessed on 4 November 2020).
46. Syrah sau Shiraz. Available online: https://www.rewine.ro/blog/syrah-sau-shiraz/ (accessed on 4 November 2020).
47. Novac, Soi de Perspectivă pentru Vinuri Roșii. Available online: https://www.lumeasatului.ro/articole-revista/agrotehnica/6169-novac-soi-de-perspectiva-pentru-vinuri-rosii (accessed on 4 November 2020).
48. Lamuela-Raventós, R.M. Folin–Ciocalteu method for the measurement of total phenolic content and antioxidant capacity. In *Measurement of Antioxidant Activity & Capacity: Recent Trends and Applications*, 1st ed.; Apak, R., Capanoglu, E., Shahidi, F., Eds.; Wiley: Hoboken, NJ, USA, 2018; pp. 107–114.
49. Prieto, P.; Pineda, M.; Aguilar, M. Spectrophotometric quantitation of antioxidant capacity through the formation of a phosphomolybdenum complex: Specific application to the determination of vitamin, E. *Anal. Biochem.* **1999**, *269*, 337–341. [CrossRef]
50. Brand-Williams, W.; Cuvelier, M.E.; Berset, C. Use of a free radical method to evaluate antioxidant activity. *LWT Food Sci. Technol.* **1995**, *28*, 25–30. [CrossRef]
51. Yalcin, H.; Kavuncuoglu, H.; Ekici, L.; Sagdic, O. Determination of fatty acid composition, volatile components, physico-chemical and bioactive properties of grape (Vitis Vinifera) seed and seed oil. *J. Food Process. Preserv.* **2016**, *41*, 1–7. [CrossRef]
52. Salas, P.G.; Soto, A.M.; Carretero, A.S.; Gutierrez, A.F. Phenolic-compound extraction systems for fruit and vegetable samples. *Molecules* **2010**, *15*, 8813–8826. [CrossRef] [PubMed]
53. Srinivasulu, C.; Ramgopal, M.; Ramanjaneyulu, G.; Anuradha, C.; Kumar, C.S. Syringic acid (SA)–A review of its occurrence, biosynthesis, pharmacological and industrial importance. *Biomed. Pharmacother.* **2018**, *108*, 547–557. [CrossRef] [PubMed]
54. Rosas, E.C.; Correa, L.B.; Henriques, M.D.G. Anti-inflammatory properties of Schinus terebinthifolius and its use in arthritic conditions. In *Bioactive Food as Dietary Interventions for Arthritis and Related Inflammatory Diseases*, 2nd ed.; Watson, R.R., Preedy, V.R., Eds.; Academic Press: Cambridge, MA, USA, 2019; pp. 489–505.
55. Janel, N.; Noll, C. Polyphenols in chronic diseases and their mechanisms of action. In *Polyphenols in Human Health and Disease*; Watson, R.R., Preedy, V.R., Zibadi, S., Eds.; Academic Press: Cambridge, MA, USA, 2014; pp. 1401–1419.
56. Dias, T.R.; Alves, M.G.; Silva, B.M.; Oliveira, P.F. Nutritional factors and male reproduction. In *Encyclopedia of Reproduction, Second Edition*; Skinner, M., Ed.; Academic Press: Cambridge, MA, USA, 2018; pp. 458–464.
57. Kim, J.; Lee, K.W. Coffee and its active compounds are neuroprotective. In *Coffee in Health and Disease Prevention*; Preedy, V.R., Ed.; Academic Press: Cambridge, MA, USA, 2015; pp. 423–427.
58. Calabrese, V.; Mancuso, C.; De Marco, C.; Stella, A.M.G.; Butterfield, D.A. Nitric oxide and cellular stress response in brain aging and neurodegenerative disorders. In *Oxidative Stress and Neurodegenerative Disorders*; Qureshi, G.A., Parvez, S.H., Eds.; Elsevier Science: Amsterdam, The Netherlands, 2007; pp. 115–134.
59. Abramovič, H. Antioxidant properties of hydroxycinnamic acid derivatives: A focus on biochemistry, physicochemical parameters, reactive species, and biomolecular interactions. In *Coffee in Health and Disease Prevention*; Preedy, V.R., Ed.; Academic Press: Cambridge, MA, USA, 2015; pp. 843–852.
60. Risuleo, G. Resveratrol: Multiple activities on the biological functionality of the cell. In *Nutraceuticals Efficacy, Safety and Toxicity*; Gupta, R.C., Ed.; Academic Press: Cambridge, MA, USA, 2016; pp. 453–464.
61. Ay, M.; Charli, A.; Jin, H.; Anantharam, V.; Kanthasamy, A.; Kanthasamy, A.G. *Quercetin. Nutraceuticals Efficacy, Safety and Toxicity*; Gupta, R.C., Ed.; Academic Press: Cambridge, MA, USA, 2016; pp. 447–452.
62. Zuo, J.; Tang, W.; Xu, Y. Anti-hepatitis B virus activity of chlorogenic acid and its related compounds. In *Coffee in Health and Disease Prevention*; Preedy, V.R., Ed.; Academic Press: Cambridge, MA, USA, 2015; pp. 607–613.
63. Patel, K.; Patel, D.K. The beneficial role of rutin, a naturally occurring flavonoid in health promotion and disease prevention: A systematic review and update. In *Bioactive Food as Dietary Interventions for Arthritis and Related Inflammatory Diseases*, 2nd ed., Watson, R.R., Preedy, V.R., Eds.; Academic Press: Cambridge, MA, USA, 2019; pp. 457–579.
64. Bătușaru, C.M. Sustainability of the small business environment in Romania in the context of increasing economic competitiveness. *Manag. Sustain. Dev.* **2019**, *11*, 37–41.

applied sciences

MDPI

Article

Food Products with High Antioxidant and Antimicrobial Activities and Their Sensory Appreciation

Mihaela Adriana Tița [1,*], Maria Adelina Constantinescu [1], Ovidiu Tița [1], Endre Mathe [2], Loreta Tamošaitienė [3] and Vijolė Bradauskienė [3,4]

1 Department of Agricultural Sciences and Food Engineering, Lucian Blaga University of Sibiu, Doctor Ion Rațiu No. 7, 550012 Sibiu, Romania; adelina.constantinescu@ulbsibiu.ro (M.A.C.); ovidiu.tita@ulbsibiu.ro (O.T.)
2 Institute of Nutrition, Faculty of Agricultural and Food Sciences and Environmental Management, University of Debrecen, Böszörményi út No. 138, 4032 Debrecen, Hungary; endre.mathe64@gmail.com
3 Department of Food Technology and Nutrition, Klaipeda State University of Applied Sciences, Bijūnų g. No. 10, 91223 Klaipeda, Lithuania; loreta.tamo14@gmail.com (L.T.); v.bradauskiene@kvk.lt (V.B.)
4 Food Institute, Kaunas University of Technology, Radvilenu Road No. 19C-413, 44239 Kaunas, Lithuania
* Correspondence: mihaela.tita@ulbsibiu.ro; Tel.: +40-751-057-321

Abstract: (1) Background: The demand for healthy and nutritious food is growing worldwide. Fermented dairy products are highly valued by consumers for their health benefits. Kefir is a fermented dairy product that brings many benefits to the consumer due to its antioxidant, anticancer, antidiabetic, antihypertensive and antimicrobial properties. Extracts from various plants in the form of volatile oils have a beneficial efct on consumer health. Following the research, their antioxidant and antimicrobial activities were demonstrated. (2) Methods: In the present study, the main purpose was to obtain a fermented dairy product with a high nutritional value; therefore, kefir, enriched with three types of volatile oils, namely, volatile mint oil, volatile fennel oil and volatile lavender oil, was made. The kefir samples obtained were sensory and texturally analyzed. The beneficial effect on health must also be studied in terms of the acceptability of these products by consumers from a sensory point of view. A non-numerical method based on several multi-personal approval criteria was used to interpret the results obtained in the sensory analysis. In the textural analysis, the consistency, cohesiveness and firmness of the kefir samples were analyzed. (3) Results: The samples enriched with volatile oils obtained superior results compared to the control sample in both conducted examinations. Kefir samples with volatile oils retained their sensory and textural characteristics for a longer time during storage. (4) Conclusions: The volatile oils added to kefir positively influenced the sensory and textural characteristics of the finished product.

Keywords: kefir; volatile oils; sensory analysis; textural analysis; bioactive compounds

Citation: Tița, M.A.; Constantinescu, M.A.; Tița, O.; Mathe, E.; Tamošaitienė, L.; Bradauskienė, V. Food Products with High Antioxidant and Antimicrobial Activities and Their Sensory Appreciation. *Appl. Sci.* **2022**, *12*, 790. https://doi.org/10.3390/app12020790

Academic Editors: Codină Georgiana Gabriela, Adriana Dabija and Lăcrămioara Rusu

Received: 9 December 2021
Accepted: 7 January 2022
Published: 13 January 2022

1. Introduction

Fermented dairy products are highly valued by consumers for their health benefits. They can be consumed from an early age [1]. The nutritional characteristics of fermented dairy products are determined by the nutrients present in the milk, those from other ingredients and those resulting as metabolites generated by the fermentation of lactic acid bacteria. Lactic fermentation causes changes in the composition, consisting of the production of lactic acid from lactose and the formation of peptides and amino acids from proteins and fatty acids from lipids [2].

The demand for healthy and nutritious foods is growing worldwide. It is desirable to make food products that, in addition to satisfying the need for food, also bring numerous nutritional contributions to the health of the consumer [3]. Due to the intensification of food production, many changes take place in terms of the quality and safety of these products [4]. Creating products capable of meeting food and health needs is of great interest due to the increase in economic competitiveness [5]. Food processors must produce foods that prevent consumers from becoming sick and help to improve their physical and mental health [3].

A major stress factor for people is the current situation caused by the COVID-19 pandemic, as well as the strict but necessary measures that have been taken to prevent the spread of the virus. Doctors recommend adopting a healthy lifestyle so that the transition is easier over this period and to mitigate the effects caused by the pandemic stress. Without adopting a healthy lifestyle, or avoiding sedentarism as much as possible, people can develop anxiety, depression or cardiovascular and mental diseases [6].

After several studies, it was found that volatile oils make an extraordinary contribution to the health of consumers. These oils are extracted either from the aerial part of the plant or from its root [7]. Volatile oils are increasingly being used in the pharmaceutical and food industries due to their antimicrobial and antioxidant properties. The rising interest in natural substances has led to the growing interest in finding new applications for these substances [8].

The volatile oils extracted from certain plants have antimicrobial action. This action is visible on the pathogenic bacteria *Listeria monocytogenes*, *Listeria innocua* and *Salmonella typhimurium*. Volatile oils also have antioxidant activity. Free radicals cause the oxidation of biomolecules, including proteins, amino acids and DNA, and produce molecular changes related to aging, arteriosclerosis and cancer [9].

Fennel (*Foeniculum vulgare* L.) belongs to the *Umbelliferae* family and is an aromatic plant considered one of the oldest cultivated medicinal plants in the world [10]. In Romania, fennel is grown in the Moldova, Dobrogea, Timiș and Oltenia areas. The most important fennel compounds are trans-anethole (63.30%), pinene (11.11%) and fencone (8.32%) [11]. Fennel volatile oil has antioxidant and antimicrobial capacities [12]. Regarding antimicrobial activity, in 2013, Thompson et al. conducted a study investigating its antimicrobial capacity against *Escherichia coli* bacteria. The results showed that this type of volatile oil has high antimicrobial activity [13]. In 2018, Salma et al. conducted another study using the disc diffusion method, which is also called the Kirby–Bauer method. The result of this method demonstrated high antimicrobial activity [14]. In the study conducted by Pande and Preetha in 2017, the antioxidant activity of fennel essential oil was measured using the DPPH (2,2-diphenylpicryl hydrazyl) method. This study showed that fennel volatile oil has a high antioxidant activity [15].

Lavender (*Lavandula angustifolia*) is part of the *Lamiaceae* family and is an aromatic plant used in folk medicine to relieve stress and anxiety [16]. Lavender volatile oil is a complex blend of mono alcohols and sesquiterpenoids, esters, oxides and ketones [17]. The main components of lavender volatile oil are the monoterpenoids linalool, linalyl acetate, 1,8-cineole, β-ocimen, terpinen-4-ol and camphor [18]. Lavender essential oil is recognized for being used in the treatment of anxiety, migraines, stress, irritability, exhaustion, depression, headaches, digestion, colds, flatulence, insomnia, loss of appetite, upset stomach, liver disease, nervousness and aroma [19]. In 2017, Zhao et al. showed that lavender volatile oil is effective in inhibiting the tumor growth of human carcinogenic xenografts in mice. Linalool mainly contributed to this effect [20]. Regarding antimicrobial activity, in 2015, Kunicka-Styczyńska et al. investigated this against *Staphylococcus aureus*, *Escherichia coli*, *Candida* spp. and *Aspergillus niger*. The result obtained was a positive one; this type of oil showed high antimicrobial activity [21]. In the 2017 study by Andrys et al., the antioxidant activity of volatile lavender oil was investigated. The DPPH method was used according to the procedures described by Kumaran and Karunakaran (2007) and Wojdyło et al. (2007). The reduction in the DPPH radical was determined spectrophotometrically by measuring the absorbance at 517 nm. Finally, it has been shown that volatile lavender oil has a high antioxidant capacity [22].

Mint (*Mentha piperita* L.) is part of the *Lamiaceae* family and is a perennial plant with a characteristic taste and smell [23]. The main components of volatile mint oil are menthol, mentofuran, isomentone, caryophyllene, eucalyptol, linalool, limonene, carvone, pulegone and α-terpinol [24]. Numerous specialized studies have been performed over the years to demonstrate the antimicrobial and antioxidant activities of volatile mint oil. In 2017, Ramos et al. conducted a study demonstrating the antimicrobial and antioxidant activities of volatile mint oil. The antimicrobial activity of this oil was tested against two strains of bacteria: *Staphylococcus aureus* and *Escherichia coli*. The test results indicated high antimicrobial activity against these bacteria. Regarding the antioxidant activity, the 2,2-diphenyl-1-picrylhydrazyl method was used. The reading was performed on a spectrophotometer at an absorbance of 517 nm, and it was shown that volatile mint oil has a high antioxidant activity [25]. In Singh's 2015 study, the antioxidant activity of volatile mint oil was also studied using the DPPH method. The reading was also performed on a spectrophotometer at an absorbance of 517 nm, and the result was identical to that of the previously presented study [26].

Kefir is a dairy product that is obtained from kefir grains that contain a specific combination of bacteria and yeast [27]. It is obtained by adding a culture called "kefir grains" to milk, thus producing a creamy texture, sour taste and low effervescence [28]. The microbial composition of kefir varies depending on the type and composition of the milk, the culture medium, the fermentation period and the temperature, as well as the storage conditions. *Lactobacillus*, *Lactococcus*, *Streptococcus* and *Leuconostoc* are the most common bacteria and *Saccharomyces*, *Kluyveromyces* and *Candida* are the most common yeasts in kefir [27]. Kefir has various health benefits due to its antioxidant, anticancer, antidiabetic, antihypertensive and antimicrobial properties [29].

In the present study, the main objective was to obtain a fermented dairy product with a high nutritional value, and for this purpose, kefir enriched with three types of volatile oils, namely, volatile mint oil, volatile fennel oil and volatile lavender oil, was made. The kefir samples obtained were sensory and texturally analyzed. The beneficial effect on health must also be studied in terms of the acceptability of these products by consumers from a sensory point of view.

2. Materials and Methods

2.1. Extraction of Volatile Oils

Mint and lavender, the dried and crushed aerial parts (grass), and fennel seeds were used to extract and dose the volatile oils. The volatile oil was extracted by entrainment with water vapor using the Neo-Clevenger apparatus modified by Moritz (method according to the Romanian Pharmacopoeia edition X) [30]. The volatile oils obtained had a characteristic odor and were pale yellow to greenish yellow (for mint). The plants were harvested from plantations in Sibiu, Romania.

2.2. Encapsulation of Volatile Oils in Sodium Alginate

Volatile oils extracted from various vegetable products require special storage conditions and are very sensitive to the action of environmental factors. Because of this, it was decided to encapsulate these oils in sodium alginate. Alginate is a natural polysaccharide extracted from brown algae [31]. To make the capsules with volatile oils, three varieties of oils were used: volatile mint oil, volatile fennel oil and volatile lavender oil. From each assortment of volatile oil, 30 µL and over 10 mL of 2% sodium alginate solution were added. The capsules obtained had a characteristic odor of each type of volatile oil used and an opalescent white color. The size of each capsule was about 240 µm, and they had a gelatinous structure.

2.3. Obtaining Samples of Kefir from Cow's Milk with the Addition of Volatile Oils Encapsulated in Sodium Alginate

Raw milk was pasteurized for 25 min at a temperature of 85–90 °C. Cow's milk came from a farm in Sibiu, Vulpăr village. After cooling to 20 °C, pre-mixed powder milk was added with a quantity of warm milk and a starter culture. A mix of LYOFAST cultures, MS 059 DT, manufacturer SACCO, was used. Starter culture composition was *Lactococcus lactis subsp. lactis, Lactococcus lactis subsp. cremoris, Lactococcus lactis subsp. lactis biovar diacetylactis* and *Leuconostoc*. The amount of powder milk used was 150 g for 2 L of milk, and the amount of starter culture was 0.15 g for the same amount of milk. The seed milk was poured into plastic containers with a capacity of 250 g, and then the capsules with volatile oils were added. Then, 1 g of the capsules was added to 100 g of the kefir sample. The thermostat was set in two stages; thermostat 1 was set at 18 °C for 10 h, and thermostat 2 was set at 10 °C for 8 h. The plastic containers with kefir samples were stored in a refrigerator at a temperature of 4 °C and covered with cling film. In the end, four types of kefir were obtained: cow's milk kefir with the addition of encapsulated volatile mint oil, cow's milk kefir with the addition of encapsulated lavender volatile oil, cow's milk kefir with the addition of volatile fennel oil encapsulated and a control sample (cow's milk kefir in which no volatile oil was added).

2.4. Sensory Analysis

The tasting was carried out by a team of seven amateur tasters who regularly consume kefir, and the selected periods were the first day, on the 10th day and on the 20th day of storage. To perform the sensory analysis, a non-numerical method was used based on several multi-person approval criteria described by Fadhil and collaborators in 2017 and 2020 [32].

The assessed characteristics of each kefir assortment were consistency, color, viscosity, taste and odor. Table 1 shows the evaluation scale used to gather the tasters' opinions, and Table 2 shows the level of importance of the criteria based on the scale.

Table 1. Linguistic assessment scale [32].

Scale	Description	Abbreviation
1	Like very much	LV
2	Like moderately	LM
3	Like slightly	LS
4	Neither like nor dislike	NT
5	Dislike slightly	DS
6	Dislike moderately	DM
7	Dislike very much	DV

Table 2. Criteria importance level [32].

Scale	Description	Abbreviation
1	Very high	LV
2	High	LM
3	Neither like nor dislike	NT
4	Low	DM
5	Very low	DV

After establishing the evaluation scale and the level of importance of the criteria, a matrix of evaluation criteria was formulated based on the opinion of the evaluators and the chosen alternatives. By using Formula (1), the denial importance level of criteria was determined.

$$\text{Neg}\ (W_k) = (W_{q-k+1}) \quad (1)$$

where:
Neg (W_k) = negation of criteria k;
k = index;
q = scale amount.

For the approval process based on criteria, Formula (2) was used.

$$V_{ij} = \min\ [\text{Neg}\ (W_{ak}) \vee V_{ij}\ (a_k)] \quad (2)$$

where:
V_{ij} = alternative i by person j;
V_{ij} (a_k) = alternative i by person j on criteria k;
k = 1, 2, . . . , m.

Formula (3) was used to determine the value weights.

$$Q_k = \text{Int}\,[1 + (k \cdot (q - 1)/r)] \tag{3}$$

where:
Q_k = score k;
Int = integer;
R = number of assessors.

Formula (4) was used to determine the tasting process of the tasters.

$$Vi = f(Vi)\ \max\ [Qj \wedge bj] \tag{4}$$

where:
Vi = total score for alternative i;
Qj = score j;
j = 1, 2, . . . , m;
bj = order from the biggest alternative score i from alternative score j [32].

2.5. *Texture Profile Analysis*

The analysis of the texture profile of the kefir samples was performed at room temperature (25 °C) using the texture analyzer TA.XT*puls*C equipped with specialized software for texture analysis. The calibration of the force was performed with the help of a weight of 160 g [33]. A disc extruder was used to extrude the product up and around the edge of the disc. The effort to do this was measured, and the results indicate viscosity [34]. Firmness is defined as the peak force obtained during the first compression cycle. Cohesion is defined as the working area of the negative force, representing the work required to remove the compression probe. Consistency is defined as the energy required to chew solid food [33].

3. Results

3.1. *Sensory Analysis*

3.1.1. Determining Alternatives

In the initial stage, the calculation of the negation of the importance level of the criteria was performed using Formula (1), so the negation of the value of the criteria weight was obtained based on each index k. The results of the calculation of the negation of the criteria are presented in Table 3.

Table 3. Negation of criteria importance level [32].

Criteria Importance Level	The Negation of Criteria Importance Level
Criteria 1 = Very high	Criteria 1 = Very low
Criteria 2 = High	Criteria 2 = Low
Criteria 3 = Neither like nor dislike	Criteria 3 = Neither like nor dislike
Criteria 4 = Low	Criteria 4 = High
Criteria 5 = Very low	Criteria 5 = Very high

The opinions of the tasters obtained from the distribution of the questionnaire are presented in Table 4.

Table 4. Criteria of assessment by each person of all alternatives.

Person	Alternative	Criteria														
		Consistency			Viscosity			Color			Taste			Smell		
		Day 1	Day 10	Day 20	Day 1	Day 10	Day 20	Day 1	Day 10	Day 20	Day 1	Day 10	Day 20	Day 1	Day 10	Day 20
D1	A1	LM	LV	LM	LS	LV	LM	LM	LV	LM	LS	LM	LM	LM	LV	LM
	A2	LS	LM	LM	LS	LM	LS	LM	LM	LS	LM	LV	LM	LM	LV	LM
	A3	LM	LM	LM	LM	LM	LM	LM	LV	LM	LM	LV	LM	LM	LV	LM
	A4	LM	LM	LS	LS	LM	NT	LM	LM	NT	LS	LS	NT	LS	LS	NT
D2	A1	LM	LV	LM	LM	LV	LM	LV	LV	LM	LM	LV	LM	LM	LV	LM
	A2	LM	LM	LM	LM	LM	LS	LM	LV	LM	LV	LV	LM	LV	LV	LM
	A3	LM	LV	LM	LM	LV	LM	LV	LV	LM	LV	LM	LS	LV	LV	LM
	A4	LM	LM	LM	LS	LM	LS	LM	LM	NT	LS	LS	NT	LM	LM	NT
D3	A1	LM	LV	LM	LM	LV	LM	LM	LM	LS	LV	LM	LS	LV	LV	LS
	A2	LM	LM	LS	LV	LM	LS	LV	LV	LM	LV	LV	LS	LV	LV	LS
	A3	LM	LM	LM	LM	LV	LM	LV	LV	LM	LM	LV	LM	LM	LV	LM
	A4	LM	LM	NT	LM	LS	NT	LM	LM	NT	LM	LS	NT	LM	LS	NT
D4	A1	LS	LM	LS	LS	LM	LS	LV	LV	LS	LM	LV	LS	LM	LV	LS
	A2	LS	LM	LS	LS	LM	LM	LM	LM	LM	LS	LM	LS	LS	LM	LS
	A3	LM	LM	LM	LM	LV	LM	LM	LV	LM	LM	LM	LM	LM	LV	LM
	A4	LS	LM	NT	LS	LM	NT	LS	LM	NT	LS	LS	DS	LM	LS	DS
D5	A1	LM	LM	LS	LM	LV	LS	LV	LV	LM	LM	LV	LS	LV	LV	LS
	A2	LM	LV	LM	LS	LM	LM	LM	LM	LM	LM	LM	LS	LM	LM	LS
	A3	LS	LM	LS	LS	LV	LM	LM	LM	LM	LM	LM	LM	LM	LV	LM
	A4	LS	LM	NT	LS	LM	NT	LS	LS	NT	LS	LS	NT	LS	LS	NT
D6	A1	LM	LM	LS	LM	LV	LS	LV	LV	LM	LV	LV	LM	LM	LV	LS
	A2	LM	LM	LS	LS	LM	LS	LM	LM	LS	LS	LM	LS	LM	LM	LS
	A3	LM	LM	LM	LM	LM	LS	LM	LM	LS	LM	LM	LS	LM	LM	LM
	A4	LS	LS	NT	LS	LM	DS	LS	LM	DS	LS	LS	DS	LM	LM	DS
D7	A1	LM	LV	LM	LV	LV	LM	LM	LM	LS	LM	LV	LS	LM	LV	LS
	A2	LM	LM	LM	LM	LM	LM	LS	LM	LS	LM	LV	LM	LM	LV	LM
	A3	LM	LM	LM	LS	LM	LS	LS	LM	LS	LM	LM	LS	LM	LV	LS
	A4	LS	LM	DS	LS	LM	DS	LS	LM	DS	LM	LM	NT	LM	LM	NT

A1 = Cow's milk kefir enriched with encapsulated lavender volatile oil; A2 = Cow's milk kefir enriched with encapsulated mint volatile oil; A3 = Cow's milk kefir enriched with encapsulated fennel volatile oil; A4 = Control sample.

3.1.2. Determining the Criteria

According to the tester's opinion, the approval criteria for each alternative were calculated using Formula (2), so, for each alternative, the following results were obtained:

1. Day 1 of storage

   The results of the approval criteria for alternative 1 are =LS, LM, LM, LS, LM, LM and LM.

   The results of the approval criteria for alternative 2 are =LS, LM, LM, LS, LS, LS and LS.

   The results of the approval criteria for alternative 3 are =LM, LM, LM, LM, LS, LM and LS.

   The results of the approval criteria for alternative 4 are =LS, LS, LM, LS, LS, LS and LS.

2. Day 10 of storage

   The results of the approval criteria for alternative 1 are =LM, LV, LM, LM, LM, LM and LM.

   The results of the approval criteria for alternative 2 are =LM, LM, LM, LM, LM, LM and LM.

   The results of the approval criteria for alternative 3 are =LM, LM, LM, LM, LM, LM and LM.

   The results of the approval criteria for alternative 4 are =LM, LM, LS, LM, LS, LS and LM.

3. Day 20 of storage

   The results of the approval criteria for alternative 1 are =LM, LM, LS, LS, LS, LS and LS.

The results of the approval criteria for alternative 2 are =LS, LS, LS, LS, LM, LS and LS.
The results of the approval criteria for alternative 3 are =LM, LM, LM, LM, LS, LS and LS.
The results of the approval criteria for alternative 4 are =NT, NT, NT, NT, NT, DS and DS.

3.1.3. Determining the Tasters

Before calculating the approval process of a taster, we used Formula (3) to determine the value weights.

The value weights for Q1, Q2, Q3, Q4, Q5, Q6 and Q7 are DM, DS, NT, NT, LS, LM and LV, respectively.

We used Formula (4) to determine the process of approval of the tasters.

1. Day 1 of storage
   The result of the tasters' approval process for alternative 1 on day 1 is LS (Like slightly).
   The result of the tasters' approval process for alternative 2 on day 1 is LS (Like slightly).
   The result of the tasters' approval process for alternative 3 on day 1 is LS (Like slightly).
   The result of the tasters' approval process for alternative 4 on day 1 is LS (Like slightly).
2. Day 10 of storage
   The result of the tasters' approval process for alternative 1 on day 10 is LM (Like moderately).
   The result of the tasters' approval process for alternative 2 on day 10 is LM (Like moderately).
   The result of the tasters' approval process for alternative 3 on day 10 is LM (Like moderately).
   The result of the tasters' approval process for alternative 4 on day 10 is LS (Like slightly).
3. Day 20 of storage
   The result of the tasters' approval process for alternative 1 on day 20 is LS (Like slightly).
   The result of the tasters' approval process for alternative 2 on day 20 is LS (Like slightly).
   The result of the tasters' approval process for alternative 3 on day 20 is LS (Like slightly).
   The result of the tasters' approval process for alternative 4 on day 20 is NT (Neither like nor dislike).

The best results obtained from the sensory analysis of the kefir samples were obtained on day 10. On this day, the kefir samples enriched with volatile oils obtained the grade "Like moderately", and the control sample obtained the grade "Like slightly". On day 1, all sensory tests of the kefir samples obtained the grade "Like slightly". On day 20, the kefir samples enriched with volatile oils obtained the grade "Like slightly", and the control sample obtained the grade "Neither like nor dislike".

3.2. Texture Profile Analysis

Figure 1 shows the profile analysis of the kefir samples on the first day of storage. The control sample has the highest firmness of 152.53 g, and the kefir sample with volatile lavender oil has the lowest firmness of 125.56 g. The kefir sample with volatile mint oil has a firmness of 142.98 g, and the kefir sample with volatile fennel oil has a firmness of 131.56 g. A high firmness value can lead to increased syneresis over time. In terms of cohesiveness, the lowest value is recorded in the control sample of −25.67 g, and the highest value is recorded in the sample of kefir with volatile fennel oil of −17.76 g. The kefir sample with volatile lavender oil has a cohesiveness of −19.7 g, and the kefir sample with volatile mint oil has a cohesiveness of −22.53 g. Cohesion represents the negative values (which tend to infinity) of the retraction forces and represents the risk of syneresis. The control sample has the highest consistency of 3542.55 g·s, and the kefir sample with volatile fennel oil has the lowest consistency of 3034.3 g·s. The consistency of the kefir sample with volatile lavender oil is 3069.18 g·s, and the consistency of the kefir sample with volatile mint oil is 3178.04 g·s.

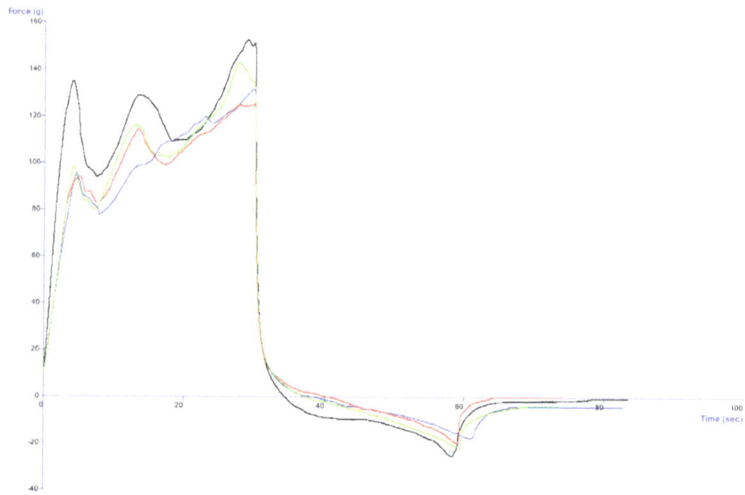

Figure 1. Analysis of the texture profile of kefir samples on the first day of storage.

where:

Red line—Cow's milk kefir enriched with encapsulated lavender volatile oil;
Green line—Cow's milk kefir enriched with encapsulated mint volatile oil;
Blue line—Cow's milk kefir enriched with encapsulated fennel volatile oil;
Black line—Control sample.

Figure 2 shows the analysis of the profile of the texture of kefir samples on the tenth day of storage. The sample of kefir with volatile lavender oil has the highest value of firmness of 160.34 g, and the sample of kefir with volatile fennel oil has the lowest value of firmness of 130.52 g. The kefir sample with volatile mint oil has a firmness of 141.21 g, and the control sample has a firmness of 149.84 g. The control sample has the lowest cohesiveness of −21.97 g, and the sample of kefir with volatile fennel oil has the highest cohesiveness of −16.41 g. The kefir sample with volatile lavender oil has a cohesiveness of −18.87 g, and the kefir sample with volatile mint oil has a cohesiveness of −20.82 g. The sample of kefir with volatile oil of lavender has the highest value of consistency of 4008.5 g·s, and the sample of kefir with volatile oil of fennel has the lowest value of consistency of 3263.12 g·s. The kefir sample with volatile mint oil has a consistency of 3530.25 g·s, and the control sample has a consistency of 3746.07 g·s.

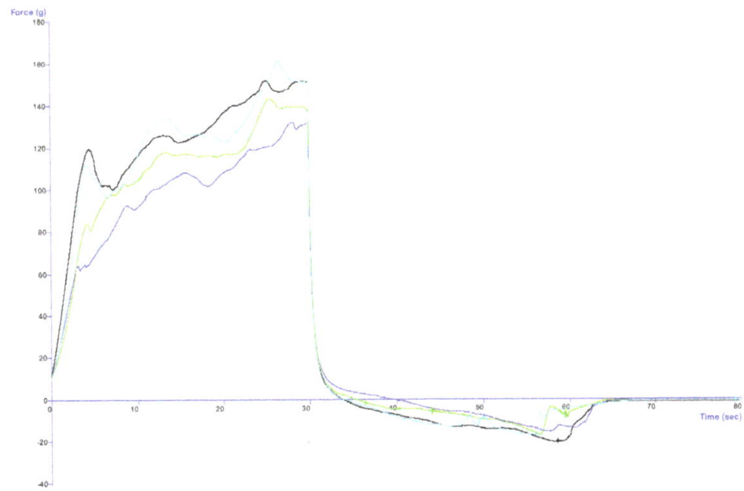

Figure 2. Analysis of the texture profile of kefir samples on the tenth day of storage.

where:

Light blue line—Cow's milk kefir enriched with encapsulated lavender volatile oil;
Green line—Cow's milk kefir enriched with encapsulated mint volatile oil;
Blue line—Cow's milk kefir enriched with encapsulated fennel volatile oil;
Black line—Control sample.

Figure 3 shows the profile analysis of the kefir samples on the twentieth day of storage. The kefir sample with volatile lavender oil has the highest value of firmness, 152.56 g, and the control sample has the lowest value of firmness, 109.58 g. The kefir sample with volatile mint oil has a firmness of 146.94 g, and the kefir sample with volatile fennel oil has a firmness of 147.43 g. The highest value of cohesiveness is recorded in the control sample of −12.95 g, and the lowest value is recorded in the kefir sample with volatile lavender oil of −22.89 g. The kefir sample with volatile mint oil has a cohesiveness of −20.82 g, and the kefir sample with volatile fennel oil has a cohesiveness of −18.86 g. The highest consistency is recorded in the kefir sample with volatile lavender oil, 3696.22 g·s, and the lowest consistency is recorded in the control sample of 2528.45 g·s. The kefir sample with volatile mint oil has a consistency of 3648.5 g·s, and the kefir sample with volatile fennel oil has a consistency of 3589.43 g·s.

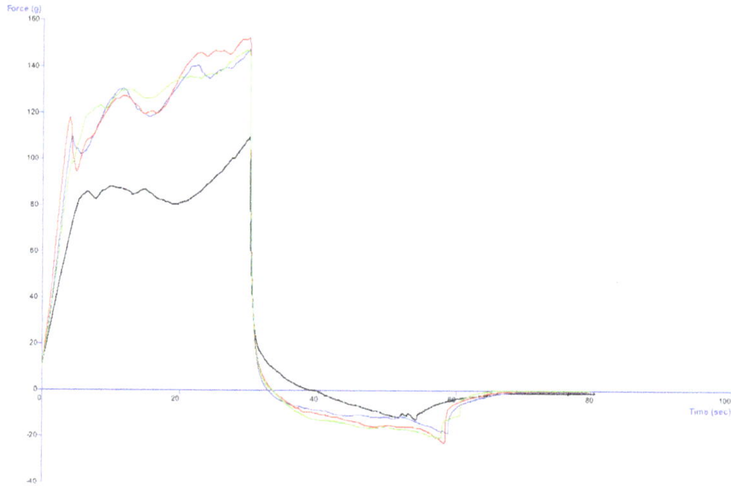

Figure 3. Analysis of the texture profile of kefir samples on the twentieth day of storage.

where:

Red line—Cow's milk kefir enriched with encapsulated lavender volatile oil;
Green line—Cow's milk kefir enriched with encapsulated mint volatile oil;
Blue line—Cow's milk kefir enriched with encapsulated fennel volatile oil;
Black line—Control sample.

4. Discussion

Four samples of cow's milk kefir were obtained, three with the addition of volatile oil of mint, lavender and fennel and one control sample. Studies have shown that volatile oils extracted from various herbs have antimicrobial and antioxidant properties. The volatile oils used were extracted with the help of the Neo-Clevenger device modified by Moritz, and the plants were harvested from crops in Sibiu County. In this study, the analysis of kefir samples from a sensory and textural point of view was considered.

For the sensory analysis, a team of seven amateur tasters was formed who regularly consume kefir, and the analysis periods were on the first day, on the 10th day and on the 20th day of storage. A non-numerical method based on several multi-personal approval criteria was used to interpret the results obtained. Following the sensory analysis, the results obtained show that for the kefir samples with volatile oils, the sensory characteristics are highlighted during storage; on day 10, the best grades are obtained, and by day 20, the characteristics begin to decrease, reaching those recorded on day 1 of the analysis. In the case of the control sample, the sensory characteristics are approximately stable in the first 10 days of storage, and then they begin to decrease, obtaining the lowest rating on day 20. The added volatile oils contributed to the kefir samples, as the sensory characteristics of these samples were much more appreciated compared to the control sample. The sensory characteristics of the kefir were influenced by the addition of volatile oils, improving them during its storage. Compared to the control sample whose characteristics deteriorated during storage, the kefir samples with volatile oils reached a peak of these characteristics on the tenth day of storage.

The analysis of the texture profile of the kefir samples was performed at room temperature (25 °C) using the texture analyzer TA.XT*puls*C. For this purpose, the firmness, consistency and cohesiveness of the four kefir samples were analyzed. The results obtained for the samples of kefir with encapsulated volatile oils are superior to those obtained in the case of the control sample, which indicates a higher elimination of whey during storage.

Thus, we can conclude that the added volatile oils bring not only a sensory aspect but also a physico-chemical aspect, because they influence the syneresis index.

The regular consumption of dairy products can prevent many cardiovascular diseases and digestive diseases. Kefir is a fermented dairy product highly appreciated by consumers, which is traditionally obtained with kefir grains consisting of a specific combination of yeast and bacteria.

Enrichment with bioactive components extracted from mint, fennel and lavender, in the form of encapsulated volatile oil, is a beneficial option for increasing the value of the product. The use of sodium alginate encapsulation ensures the stability of the volatile oils used. Due to the sensitivity of volatile oils to the action of environmental factors, it was decided to encapsulate them in sodium alginate and introduce them as spherical capsules to the dairy product. The amount of volatile oil extracted depends on the growing conditions and the soil and climatic conditions of the plants used. According to previous studies, it has been concluded that volatile oils have antimicrobial and antioxidant actions, representing an opportunity to use them in food products due to the health benefits brought to the consumer. Creating foods that contain natural antioxidants and antimicrobial compounds should be a priority in food management. Making foods that help boost the body's immunity or alleviate many chronic health problems is an effective and safe alternative to ensuring physical and mental health [35].

The use of volatile oils can eliminate the use of synthetic preservatives in dairy products because they have antimicrobial capacity. Food preservatives affect the health of consumers, often causing food poisoning [35].

In addition to the benefits of the products developed, they must be accepted by the consumer. The results obtained from the sensory and textural analysis certify an acceptance of kefir enriched with volatile oil of mint, fennel and lavender. All the results obtained in the case of these samples are superior to those obtained for the control sample in all three analysis periods.

5. Conclusions

Acidic dairy products are appreciated by consumers. They bring many benefits to the health of the consumer and can be consumed from an early age.

The kefir samples obtained were analyzed from a sensory point of view and a textural point of view. The sensory analysis was performed on the first day, on the 10th day and on the 20th day of storage of the kefir samples. A non-numerical method based on several multi-personal approval criteria was used to interpret the results obtained. The best results were obtained on the 10th day of storage, and the kefir samples enriched with volatile oils were much more appreciated compared to the control sample, to which volatile oils were not added.

The textural analysis analyzed the consistency, cohesiveness and firmness of the four kefir samples. The analysis period was identical to the one chosen for the sensory analysis. The results obtained for the samples of kefir with encapsulated volatile oils are superior to those obtained in the case of the control sample, which indicates a higher elimination of whey during storage.

We can conclude that the volatile oils added in both types of kefir positively influenced the sensory and textural characteristics of the finished product. The samples enriched with encapsulated volatile oils obtained superior results compared to the control sample in the case of both determinations performed.

All of these aspects show that the products analyzed, in addition to the beneficial action they bring to the consumer's health, are also accepted from a sensory point of view. The studied product is in accordance with the current trends due to the benefits for the health of the consumer by incorporating bioactive components.

Author Contributions: Conceptualization, M.A.T. and M.A.C.; methodology, M.A.T.; software, M.A.C.; validation, O.T., E.M., L.T. and V.B.; formal analysis, O.T.; investigation, M.A.T., M.A.C. and E.M.; resources, L.T. and V.B.; data curation, M.A.T. and O.T.; writing—original draft preparation,

M.A.C., M.A.T. and O.T.; writing—review and editing, E.M., L.T. and V.B.; visualization, M.A.T. and O.T.; supervision, M.A.T.; project administration, M.A.T.; funding acquisition, M.A.T. All authors have read and agreed to the published version of the manuscript.

Funding: This research was funded by the Lucian Blaga University of Sibiu and Hasso Plattner Foundation, grant number LBUS-IRG-2020-06/No. 2841—16 July 2020.

Institutional Review Board Statement: Not applicable.

Informed Consent Statement: Not applicable.

Data Availability Statement: Not applicable.

Acknowledgments: We would like to express our sincere gratitude to the Research Center in Biotechnology and Food Engineering (C.C.B.I.A.), the Lucian Blaga University of Sibiu and the Hasso Plattner Foundation for their support provided throughout the research period.

Conflicts of Interest: The authors declare no conflict of interest.

References

1. Aryana, K.J.; Olson, D.W. A 100-Year Review: Yogurt and other cultured dairy products. *Int. J. Dairy Sci.* **2017**, *100*, 9987–10013. [CrossRef] [PubMed]
2. Segal, R. *Caracteristicile Nutriționale Ale Produselor Lactate Fermentate. Produse Lactate Fermentate*; Academica Galați: Galați, Romania, 2005; pp. 303–327.
3. Jeyakumari, A.; Zynudheen, A.; Parvathy, U. Microencapsulation of Bioactive Food Ingredients and Controlled Release—A Review. *MOJ Food Process Technol.* **2016**, *2*, 214–224.
4. Ben Said, L.; Gaudreau, H.; Dallaire, L.; Tessier, M.; Fliss, I. Bioprotective Culture: A New Generation of Food Additives for the Preservation of Food Quality and Safety. *Ind. Biotechnol.* **2019**, *15*, 138–147. [CrossRef]
5. Bătușaru, C.M. Sustainability of the small business environment in Romania in the context of increasing economic competitiveness. *Manag. Sustain. Dev.* **2019**, *11*, 37–41. [CrossRef]
6. Cu Toții Resimțim Stresul Cauzat De Coronavirus, Dar Părinții Și Minoritățile Sunt Printre Cei Mai Puternic Afectați. Câteva Modele De Implicare Din România. Available online: https://www.libertatea.ro/stiri/stres-coronavirus-parinti-minoritati-3016858 (accessed on 2 July 2020).
7. Tanu, B.; Harpreet, K. Benefits of essential oil: A review. *J. Chem. Pharm.* **2016**, *8*, 143–149.
8. Whiley, H.; Gaskin, S.; Schroder, T.; Ross, K. Antifungal properties of essential oils for improvement of indoor air quality: A review. *Rev. Environ. Health* **2018**, *33*, 63–76. [CrossRef]
9. Tița, O.; Constantinescu, A.M.; Tița, A.M. Antioxidant and Antiseptic Properties of Volatile Oils from Different Medicinal Plants: A Review. *Int. J. Pharmacogn. Chin. Med.* **2019**, *3*, 1–13.
10. Sayed-Ahmad, B.; Straumite, E.; Šabovics, M.; Kruma, Z.; Merah, O.; Saad, Z.; Hijazi, A.; Talou, T. Effect of addition of fennel (*Foeniculum vulgare* L.) on the quality of protein bread. *Proc. Latv. Acad.* **2017**, *71*, 509–514. [CrossRef]
11. Zeng, H.; Chen, X.; Liang, J. In Vitro antifungal activity and mechanism of essential oil from fennel (*Foeniculum vulgare* L.) on dermatophyte species. *J. Med. Microbiol.* **2015**, *64*, 93–103. [CrossRef]
12. Al-Okbi, S.Y.; Hussein, A.M.S.; Elbakry, H.F.H.; Fouda, K.A.; Mahmoud, K.F.; Hassan, M.E. Health benefits of fennel, rosemary volatile oils and their nano-forms in dyslipidemic rat model. *Pak. J. Biol. Sci.* **2018**, *21*, 348–358. [CrossRef]
13. Thompson, A.; Meah, D.; Ahmed, N.; Conniff-Jenkins, R.; Chileshe, E.; Phillips, C.O.; Claypole, T.C.; Forman, D.W.; Row, P.E. Comparison of the antibacterial activity of essential oils and extracts of medicinal and culinary herbs to investigate potential new treatments for irritable bowel syndrome. *BMC Complement. Altern. Med.* **2013**, *13*, 1–19. [CrossRef]
14. Salma, S.; Ramakrishnan, L.; Vinothini, J. Screening of antibacterial activity of five different spices (ajwain, coriander, cumin, fennel, and fenugreek) against pathogenic bacterial strains. *Asian J. Pharm. Clin. Res.* **2018**, *11*, 11–13.
15. Pande, K.R.; Preetha, R. Essential oil of fennel seeds as natural preservative in butter and its shelf life assessment. *Chem. Asian J.* **2017**, *29*, 711–714. [CrossRef]
16. Caputo, L.; Reguilon, M.D.; Miňarro, J.; De Feo, V.; Rodriguez-Arias, M. Lavandula angustifolia essential oil and linalool counteract social aversion induced by social defeat. *Molecules* **2018**, *23*, 2694. [CrossRef]
17. Sokovicx, M.; Glamoćlija, J.; Marin, P.D.; Brkić, D.; Van Griensven, L.J.L.D. Antibacterial effects of the essential oils of commonly consumed medicinal herbs using an in vitro model. *Molecules* **2010**, *15*, 7532–7546. [CrossRef]
18. Woronuk, G.; Demissie, Z.; Rheault, M.; Mahmoud, S. Biosynthesis and therapeutic properties of lavandula essential oil constituents. *Planta Med.* **2011**, *77*, 7–15. [CrossRef]
19. Benny, A.; Thomas, J. Essential Oils as Treatment Strategy for Alzheimer's Disease: Current and Future Perspectives. *Planta Med.* **2019**, *85*, 239–248.
20. Zhao, Y.; Chen, R.; Wang, Y.; Qing, C.; Wang, W.; Yang, Y. In Vitro and in Vivo Efficacy Studies of *Lavender angustifolia* Essential Oil and Its Active Constituents on the Proliferation of Human Prostate Cancer. *Integr. Cancer Ther.* **2017**, *16*, 215–226. [CrossRef]

21. Kunicka-Styczyńska, A.; Śmigielski, K.; Prusinowska, R.; Rajkowska, K.; Kuśmider, B.; Sikora, M. Preservative activity of lavender hydrosols in moisturizing body gels. *Lett. Appl.* **2015**, *60*, 27–32. [CrossRef]
22. Andrys, D.; Kulpa, D.; Grzeszczuk, M.; Bihun, M.; Dobrowolska, A. Antioxidant and antimicrobial activities of Lavandula angustifolia Mill. field-grown and propagated in vitro. *Folia Hortic.* **2017**, *29*, 161–180. [CrossRef]
23. Oroian, C.; Covrig, I.; Odagiu, A.; Mălinaș, C.; Moldovan, C.; Fleșeriu, A. Effects of cultivation systems and environmental conditions on peppermint (*Mentha* × *piperita* L.) biomass yield and oil content. *Not. Bot. Horti Agrobot. Cluj Napoca* **2017**, *45*, 576–581. [CrossRef]
24. Kennedy, D.; Okello, E.; Chazot, P.; Howes, M.J.; Ohiomokhare, S.; Jackson, P.; Haskell-Ramsay, C.; Khan, J.; Forster, J.; Wightman, E. Volatile Terpenes and Brain Function: Investigation of the Cognitive and Mood Effects of *Mentha* × *Piperita* L. Essential Oil with In Vitro Properties Relevant to Central Nervous System Function. *Nutrients* **2018**, *10*, 1029. [CrossRef]
25. Ramos, R.S.; Rodrigues, A.B.L.; Farias, A.L.F.; Simões, R.C.; Pinheiro, M.T.; Ferreira, R.M.A.; Barbosa, L.M.C.; Souto, R.N.P.; Fernandes, J.B.; Santos, L.S.; et al. Chemical composition and in vitro antioxidant, cytotoxic, antimicrobial, and larvicidal activities of the essential oil of *Mentha piperita* L. (Lamiaceae). *Hindawi* **2017**, *2017*, 1–9.
26. Singh, R.; Shushni, M.A.M.; Belkheir, A. Antibacterial and antioxidant activities of *Mentha piperita* L. *Arab. J. Chem.* **2015**, *8*, 322–328. [CrossRef]
27. Bellikci-Koyu, E.; Sarer-Yurekli, B.P.; Akyon, Y.; Aydin-Kose, F.; Karagozlu, C.; Ozgen, A.G.; Brinkmann, A.; Nitsche, A.; Ergunay, K.; Yilmaz, E.; et al. Effects of Regular Kefir Consumption on Gut Microbiota in Patients with Metabolic Syndrome: A Parallel-Group, Randomized, Controlled Study. *Nutrients* **2019**, *11*, 2089. [CrossRef]
28. Dimidi, E.; Cox, S.R.; Rossi, M.; Whelan, K. Fermented Foods: Definitions and Characteristics, Impact on the Gut Microbiota and Effects on Gastrointestinal Health and Disease. *Nutrients* **2019**, *11*, 1806. [CrossRef]
29. Windayani, N.; Kurniati, T.; Listiawati, M. Psychochemical and Organoleptic Characteristics of Colostrum Kefir as Antibacterial. *J. Phys. Conf. Ser.* **2019**, *1175*, 1–6. [CrossRef]
30. *Farmacopeea Română*, 10th ed.; Medicală: București, Romania, 1993.
31. Nussinovitch, A. *Hydrocolloid Applications: Gum Technology in the Food and Other Industries*; Chapman & Hall: London, UK, 1997.
32. Fadhil, R.; Agustina, R.; Hayati, R. Sensory Assessment of Sauerkraut Using a Non-Numeric Approach Based on Multi-Criteria and Multi-Person Aggregation. *Bull. Transilv.* **2020**, *13*, 112–118. [CrossRef]
33. Ye, X.; Sui, Z. Physicochemical properties and starch digestibility of Chinese noodles in relation to optimal cooking time. *Int. J. Biol* **2016**, *84*, 428–433. [CrossRef]
34. Measuring the Texture of Dairy Products. Available online: https://www.azom.com/article.aspx?ArticleID=19386 (accessed on 17 October 2020).
35. Tița, O.; Constantinescu, M.A.; Tița, M.A.; Georgescu, C. Use of yoghurt enhanced with volatile plant oils encapsulated in sodium alginate to increase the human body's immunity in the present fight against stress. *Int. J. Environ.* **2020**, *17*, 7588. [CrossRef]

applied sciences

MDPI

Article

Valorisation of Spent Grain from Malt Whisky in the Spelt Pasta Formulation: Modelling and Optimization Study

Ancuța Chetrariu and Adriana Dabija *

Faculty of Food Engineering, Stefan cel Mare University of Suceava, 720229 Suceava, Romania; ancuta.chetrariu@fia.usv.ro
* Correspondence: adriana.dabija@fia.usv.ro; Tel.: +40-748-845-567

Abstract: Although durum wheat flour is conventionally used to produce pasta, in this study, emphasis was placed on the use of spelt flour in the formulation of the pasta recipe, with the replacement with spent grain obtained from distilleries for its content of fiber and protein. D-optimal design was used to optimize the influence of spent grain addition for the quality attributes of spelt pasta. In order to optimize the spelt pasta matrix, the spent grain content was varied (5%, 10%, 15%, and 20%) so that all responses were optimized (maximize cohesiveness, fracturability, proteins, total dietary fiber, total phenolic content, and antioxidant activity, minimize cooking loss, in-range firmness, and color paste). The optimal addition of spent grain in the spelt pasta recipe was 11.70%, yielding values with differences of less than 5% from the values predicted by the model and producing finished products with good nutritional properties without negative consequences on quality. Spent grain is a valuable byproduct that deserves to be used for fortification in order to obtain pro-health food. This study presents a formulation of spelt pasta with the addition of spent grain using mathematical modeling and statistical optimization.

Keywords: spelt pasta; optimization; spent grain; valorisation; pasta; pro-health food

Citation: Chetrariu, A.; Dabija, A. Valorisation of Spent Grain from Malt Whisky in the Spelt Pasta Formulation: Modelling and Optimization Study. *Appl. Sci.* **2022**, *12*, 1441. https://doi.org/10.3390/app12031441

Academic Editor: Suyong Lee

Received: 31 December 2021
Accepted: 27 January 2022
Published: 28 January 2022

1. Introduction

The agro-industry results in substantial quantities of by-products that have an increased content of organic compounds with significant environmental pollution impacts. On the other hand, the growing demand for food worldwide encourages the identification of alternatives at reasonable prices and nutritional qualities. Spent grain (SG) is the main by-product produced by the beer industry and the distillation industry after wort production, accounting for about 85% of the total by-products generated [1,2]. The whisky industry generates, on average, 8–15 L of effluent and 2.5–3.0 kg of spent grain for every liter of whisky produced, and brewing generates 0.2 kg of wet brewer's spent grain per liter of beer [2,3]. Spent grain is a valuable by-product rich in nutrients, such as dietary fibers (hemicellulose, cellulose, and lignin), digestible protein, monosaccharides (glucose, xylose, and arabinose), minerals, vitamins, and lipids, and the products enriched with it are called fortified foods [4–9]. Spent grain contains a substantial amount of bioactive compounds with high antioxidant capacities, such as hydroxycinnamic acids, especially ferulic and p-coumaric [10,11]. The composition of SG is affected by the differences in grain variety, harvest time and condition, the malting and mashing methods, and the type of adjunct grains, resulting in different variants of this by-product [12,13].

The capitalization of these by-products and the development of sustainable processes is an urgent necessity in the industry [12]. Spent grain is the insoluble fraction left after the malted barley is grounded and mixed with water to produce wort, i.e., the liquid fermentation medium [14]. Before using spent grain in foods, it needs to be dried and turned into flour, although its use has some limitations because of its brownish color and its flavor [3]. Dried spent grain has been studied and has been found to be applicable in various food products with health-promoting properties, such as bread, cookies, flakes, pasta, biscuits, breakfast cereals, and snacks [15–19].

Spelt (*Triticum aestivum* var. *spelta*) has been cultivated extensively in Europe since ancient times and has returned to the attention of researchers and consumers because of its nutritional qualities and its resistance and ability to adapt to climatic conditions [9]. Consumer feedback shows that spelt flour-based products are more digestible and have therapeutic effects compared to products based on common wheat [14]. Climate change is another reason for the interest in spelt wheat since it has shown more resilience to drought and other extreme weather conditions than common wheat [8].

In the food industry, spelt grain has great technological potential for specialty breads, organic food, and food products, with characteristics that differ from regular wheat products [20]. Spelt grain can be used in brewing as an unmalted adjunct during mashing [21]. Other examples of foods that can be obtained from spelt grain are pasta, baked goods, high-fiber breakfast cereals, snacks, crackers, and beverages [6,22,23].

Spelt flour is used as a raw material to improve the nutritional and functional properties of pasta [24,25]. Spelt flour contains a higher fiber quantity and phenolic compounds, with health benefits such as reduced risk of some types of diseases; lower concentrations of blood lipids; stable blood glucose and insulin levels; a positive impact on minimizing fatigue and energy loss and removing toxins and lowering cholesterol levels in the blood; reduced risk of cardiovascular disease and better control of diabetes; the absence of constipation; and better weight management [14].

Consumers need good quality and affordable ready-to-eat food products that have a good glycemic index and long shelf life [26]. Among the most consumed foods that meet these requirements is pasta [2,27]. Furthermore, the production of pasta is simple (it is obtained by the extrusion of a mixture of flour from durum wheat and water), and it has a healthy nutrient content and good sensory qualities. Researchers have improved the matrix of pasta by enriching the nutritional potential by mixing durum wheat flour with different flours (e.g., pseudo-cereal flours and legume flours) [4,6,28]. Today, the use of high-functionality ingredients for value addition to pasta is a developing field of research to achieve innovative product categories, among which is the use of ancient grain flours, such as spelt flour [7]. Pasta is a good source of energy with high amounts of carbohydrates, moderate amounts of protein, and low lipids [10]. Fortifying these foods is a challenge, not only in terms of increasing nutritional properties and health but also the effects on cooking, texture, and sensory properties. The current research does not provide much information about the use of spelt flour and spent grain flour for pasta production, nutritional value, and spelt grain and spent grain's effects on health [6,7].

- The response surface method is used to find the optimal response and changes its direction because of the design variables, which can be seen as a visual graph. An experimental design should take into account the design constraints [29]. Some advantages of optimization methods are the following: computational efficiency; better description of the factors' influence in the process, both alone or in combinations; the relationship between the responses and the factors; and the achievement of a sustainable processing industry [30].

This study aimed to optimize the effect of the fortification of spent grain flour on the chemical composition, nutritional values, and selected quality properties of novel spelt pasta formulation. The novelty of this study is its use of spelt flour with the addition of spent grain from the process of producing whisky in the formulation of pasta recipes. Most previous studies have focused on spent grain from the beer industry [9,31–34], with few papers focusing on spent grain from distilleries; this is why little information is available about the nutritional value and fortification.

2. Materials and Methods

2.1. Materials

Spent grain from whisky production was collected from a local factory, Alexandrion Group (Ploiesti, Romania); wet spent grain was stored at −18 °C, dried at 50 °C for 24 h, ground by a mill, and then sieved. Spent grain flour was obtained from a fraction of less than 200 μm and was stored in paper bags at room temperature until further use. Spelt flour was purchased from a local market and had a Romanian origin.

All chemicals used in this paper were of analytical grade and were purchased from Sigma Aldrich (St. Louis, MO, USA).

2.2. Pasta Processing

The pasta dough was mixed with a Kitchen Aid mixer (Whirlpool Corporation, Benton Harbor, MI, USA) by adding spelt flour and different percentages of spent grain (5–20%) to obtain 32% dough moisture. The pasta was modeled after a resting time of 45 min, during which time the dough temperature was maintained at 40 °C using a small macaroni mold. Pasta drying was performed in an air oven for 6 h at 40 °C [26].

2.3. Dough Texture

The dough texture profile analysis (TPA) was performed using a Perten TVT-6700 texturometer (Perten Instruments, Hägersten, Sweden). From the dough, 50 g balls were prepared, which were analyzed to determine the firmness and cohesiveness. The sample was subjected to double compression at 50% height with a 35 mm cylinder probe at a speed of 5.0 mm/s and a trigger force of 20 g [35]. All measurements were made in triplicate.

2.4. Dry Pasta Color

A Konica Minolta CR-400 colorimeter (Tokyo, Japan) was used to measure the dry pasta color using CIELab color space coordinates, where L* values describe black to white (0 to 100), a* is the degree of redness (positive) or greenness (negative), and b* is yellowness (positive) or blueness (negative) [36]. The color change equation (Equation (1)) is described below:

$$\Delta E = \sqrt{(\Delta L)^2 + (\Delta a)^2 + (\Delta b)^2} \quad (1)$$

where ΔL = L* sample − L* control, Δa = a* sample − a* control, and Δb = b* sample − b* control. All measurements were made in triplicate.

2.5. Dry Pasta Fracturability

A Perten TVT-6700 device (Perten Instruments, Hägersten, Sweden) equipped with an aluminum breake rig set adjusted to a 10 mm width was used to determine dry pasta fracturability using the maximum force F (g) needed to break a pasta piece. The test speed was 2 mm/s and the trigger force was 5 g, and measurements were made in duplicate [37].

2.6. Crude Proteins and Total Dietary Fiber

Protein determination was carried out using Kjeldahl nitrogen analysis with a 5.7 conversion factor (Velp Scientifica, Usmate, Italy) [37]. A Megazyme total dietary fiber assay kit (Megazyme, Ireland) was used to determine the total dietary fiber using the enzymatic method [37].

2.7. Total Phenolic Content

A sample of 2 g of raw pasta was ground and mixed with 20 mL of methanol 80% (*v*/*v*) and sonicated for 40 min in a sonication bath at 37 °C and 45 Hz; then, the mixture was centrifuged for 5 min at 4000 rpm [38]. Then, 0.2 mL of extract was mixed with 2 mL of Folin–Ciocalteau reagent, diluted to 1:10, and mixed with 1.8 mL of sodium carbonate 7.5% (*w*/*v*) in a tube. The mixture was left for 30 min at room temperature in the dark. The total polyphenolic content was determined at a 750 nm wavelength using a UV–VIS–

NIR Shimadzu 3600 (Shimadzu Corporation, Kyoto, Japan). The calibration curve of the polyphenols was performed by using gallic acid at concentrations of 10–200 mg/L with the regression coefficient $R^2 = 0.99872$ and the equation $y = 0.00949x + 0.02950$. The samples were analyzed in triplicate, and the results were expressed in μg gallic acid equivalents per gram (μg GAE/g).

2.8. Antioxidant Activity of Pasta

2,2-Diphenyl-1-picrylhydrazyl (DPPH) was used to assess the antioxidant activity. Following the methods of Iuga and Mironeasa [35], a sample of 2 g of raw pasta was ground and mixed with 20 mL of methanol 80% (v/v) in a sonication bath at 37 °C and 45 Hz for 40 min and then centrifuged for 5 min at 4000 rpm. Then, 2 mL of extract was mixed with 2 mL of DPPH solution 0.1 mM in methanol [30]. Absorbance was determined at 517 nm using a UV–VIS–NIR spectrophotometer (Shimadzu Corporation, Kyoto, Japan) against a blank sample. The antioxidant capacity was measured in triplicate using Equation (2).

$$\%\ \text{Inhibition of DPPH} = [(1 - \text{As}/\text{Ab})] \times 100 \quad (2)$$

where As = absorbance of the sample and Ab = absorbance of the blank sample.

2.9. Pasta Cooking Behavior

Cooking loss (CL) was determined gravimetrically by the evaporation of the water that resulted after boiling for an optimal cooking time of 10 g of pasta in 100 mL of water without salt addition. The residue was weighed and expressed as grams of matter loss per 100 g of pasta [34]. Cooking loss was determined in triplicate.

2.10. Optimization of Spent Grain Level and Model Validation

A trial version of Design-Expert software (Stat-Ease, Inc., Minneapolis, MN, USA) was used for experimental design development, data analysis, regression modeling, and optimization. To validate the model, pasta produced with the optimal level of spent grain was analyzed, and real values were compared to the control sample made with no spent grain addition. D-optimal design with one factor varied at four levels (5%, 10%, 15%, and 20%) was used for the assessment of the spent grain addition on spelt pasta quality. The model's fit was evaluated through a sequential Fisher test, coefficients of determination (R^2), and adjusted coefficient of determination (Adj.-R^2) at a 95% confidence level.

2.11. Sensory Analysis of Pasta

For organoleptic analysis, samples of pasta cooked at a ratio of 1:10 (pasta:water) and served without any accompaniment were given to a group of seven experts selected and trained in order to identify particular attributes and in terms of sensory vocabulary. In this respect, the preferential method with a nine-point evaluation scale was used. Among the parameters evaluated were appearance, color, smell, taste, firmness during chewing, and general acceptability. During sample evaluation, talking among panelists was not allowed.

2.12. Statistical Analysis

All measurements in the present study were made in triplicate. Results are presented as means ± standard deviation (SD). XLSTAT for Excel 2021 (Addinsoft, New York, NY, USA) was used for the statistical analysis of the data. To evaluate significant differences ($p < 0.05$) among samples, the t-test was used.

3. Results and Discussion

3.1. Model Fitting and Statistical Analysis

A D-optimal design was used, which has an independent variable with four levels and three replications at the center point. The D-optimal design was used for the optimization of the formulation of spelt pasta, and different parameters of the optimized formulation were evaluated. The linear and quadratic models that were fitted on each response were evaluated. The chemical composition of spelt flour is described in Table 1, which we obtained from a previous study [39].

Table 1. Chemical composition of spent grain flour and spelt flour [39].

Chemical Composition	Spent Grain Flour	Spelt Flour
lipids	7.11 ± 0.39	3 ± 0.01
fiber	22.67 ± 0.42	8 ± 0.05
protein	18.88 ± 0.37	14 ± 0.09
ash	3.47 ± 0.02	2.11 ± 0.04
moisture	5.04 ± 0.42	11.26 ± 0.08

Results are expressed as g reported at 100 g dry matter.

The mathematical models were significant and presented the responses accurately; in all cases, the F-value was significant ($p < 0.01$), and R^2 values were more than 0.87, according to the ANOVA results presented in Table 2.

Table 2. ANOVA results for spent grain pasta and dough.

Response	Model	F-Value	*p*-Value	R^2	Adj.-R^2
Cohesiveness	quadratic	14.6	<0.01	0.87	0.81
Firmness	linear	100.76	<0.01	0.95	0.94
Color	linear	15.18	<0.01	0.75	0.70
Fracturability	quadratic	105.81	<0.01	0.98	0.97
Crude proteins	quadratic	415.93	<0.01	0.99	0.99
Total dietary fiber	linear	247.60	<0.01	0.98	0.97
Total phenolic content	linear	364.43	<0.01	0.98	0.98
Antioxidant activity	quadratic	63.25	<0.01	0.96	0.95
Cooking loss	quadratic	253.08	<0.01	0.99	0.98

R^2 represents the coefficient of determination, defining the ratio of the variation of the answers that is explained by the model. The closer the R^2 is to 1, the better it fits the model. Adjusted R^2 is a correction of the value R^2 dependent on the number of degrees of freedom [40]. The color results, firmness, total dietary fiber, and total polyphenol content were fitted to the linear model with 75%, 95%, 98%, and98% of the data variation explained, respectively. The quadratic model explained 87%, 98%, 99%, and 99%, respectively, of the data variation for dough cohesivity, fracturability, crude proteins, and cooking loss.

3.2. Optimization of Parameters and Validation of the Models

Small particle size (200 µm in this study) did not give significant differences in the cohesiveness, with the linear term having the most significant influence in Equation (3) ($p < 0.01$). Multiple regression analysis was used to analyze the experimental data, and thus a quadratic polynomial equation was obtained as follows (*** $p < 0.001$; ** $p < 0.01$; * $p < 0.05$). A small decrease in cohesiveness with spent grain addition level increase was observed in Figure 1A. Cohesiveness is the ability of the material to withstand two successive compressions [41]. Petitot et al. (2010) showed in their study that the addition of flour to pasta with vegetables had no impact on cohesiveness [41].

$$\text{Cohesiveness} = 0.289 - 0.021\text{A}^{***} + 0.008\text{A}^2 \quad (3)$$

Firmness represents the cutting force required to penetrate pasta dough and was significantly increased by the linear term in Equation (4). The shaping of the pasta is influenced by the firmness of the dough, with a stronger dough desirable for shaping the short pasta to keep its shape [42,43]. A rise in firmness with an increase in spent grain addition level was observed (Figure 1B). Zhao et al. reported that a pasta enriched with legume flour (navy bean, pinto bean, lentil, and green pea) or protein concentrates caused an increase in pasta firmness [44]. Tazrart et al. showed that an increase in bean flour levels resulted in a slight increase in pasta dough firmness [45].

$$\text{Firmness} = 5861.87 + 533.081\text{A}^{***} \tag{4}$$

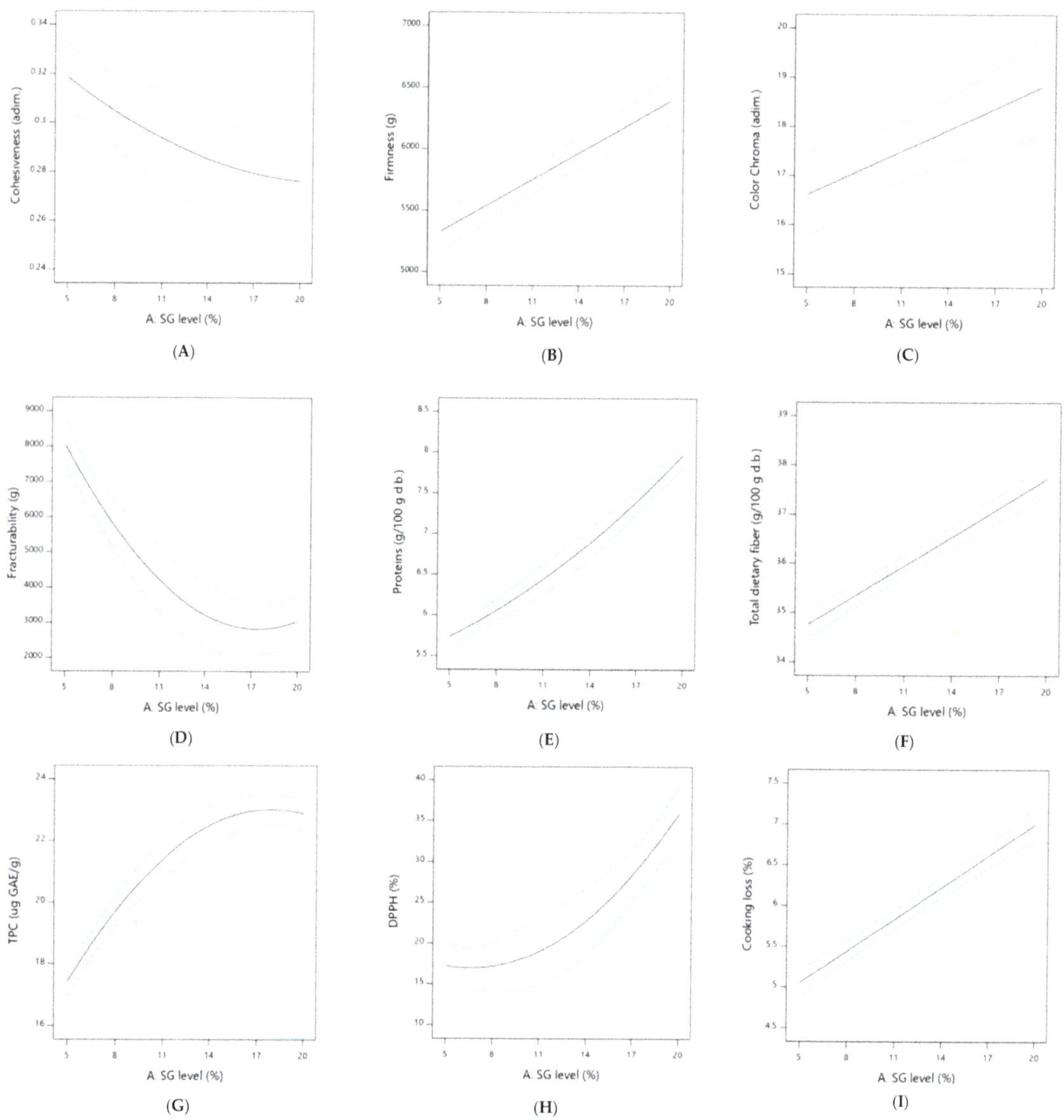

Figure 1. Effect of spent grain on (**A**) dough cohesiveness, (**B**) firmness of pasta dough, (**C**) pasta color, (**D**) pasta fracturability, (**E**) protein content, (**F**) total dietary fiber content, (**G**) total phenolic content, (**H**) antioxidant activity of pasta, and (**I**) pasta cooking loss.

The first factor governing food acceptability is the product appearance; in the case of spent grain incorporation, the color of the products is darker with increased incorporation. The same results were observed in other studies [34,46–48]. Products with high fiber contents have a positive feature of dark coloring, which consumers associate with products rich in fiber [34]. Pasta color was affected by spent grain incorporation; a significant increase in C* was observed with the additional level increase. The linear term had the greatest significant influence, as can be seen in Equation (5). Figure 1C shows the increase of color with spent grain addition.

$$\text{Color } (C^*) = 17.715 + 1.100A^{***} \quad (5)$$

Fracturability (g) is the maximum force needed to break a pasta piece [35,37]. The fracturability of dried pasta decreases with the increase of the addition of spent grain; the linear term had the greatest negative effect on the response, while the quadratic term had a significant positive effect on the dry pasta fracturability (Equation (6)). The decrease of fracturability with spent grain addition is shown in Figure 1D; the enriched fish pasta had the same decrease of fracturability, possibly because of the fiber content or fat composition of each assortment of the pasta [49]. Naibaho and Korzeniowska showed that the addition of spent grain in corn snacks decreased the fracturability of the products [15].

$$\text{Fracturability} = 3635.83 - 2505.19A^{***} + 1909.53A^{2***} \quad (6)$$

A protein content increase with spent grain addition can be observed in Figure 1E. Pasta proteins were significantly influenced by the linear and quadratic terms in Equation (7). This tendency for protein levels to increase with increased levels of added spent grain was also observed in other studies [45,50].

$$\text{Proteins} = 6.658 + 1.115A^{***} + 0.192A^{2**} \quad (7)$$

A total dietary fiber increase with spent grain addition can be observed in Figure 1F. The total dietary fiber of pasta was significantly influenced by the linear term in Equation (8). Nocente et al. obtained an increased total dietary fiber content with the increase of spent grain in pasta production [32]. The same results were obtained in other studies [14,31,45,46,50,51].

$$\text{Total dietary fiber} = 36.244 + 1.483A^{***} \quad (8)$$

The total phenolic content increased with spent grain addition (Figure 1G), and from Equation (9), it can be observed that the linear term had the highest positive effect and the quadratic term had a significant negative effect on the response. Spelt pasta can be used to produce precooked pasta with a high content of biologically active compounds compared to refined flour [14]. Pasta supplemented with native spent grain or fermented spent grain had an increase in phenolic content [34]. Spinelli et al. obtained fortified spent grain pasta with an increase in phenolic content [33].

$$\text{Total phenolic content} = 22.023 + 2.728A^{***} - 1.875A^{2***} \quad (9)$$

An antioxidant activity increase with spent grain addition can be observed in Figure 1H. The DPPH test was significantly influenced by the linear and quadratic terms in Equation (10). In their study, Reis and Abu-Ghannam obtained snacks with 10–40% spent grain addition, which increased the phenolic content and antioxidant activity [52]. Nocente et al. obtained spent grain pasta with a total antioxidant capacity of up to 19% compared with the control durum flour pasta [48].

$$\text{Antioxidant activity} = 20.654 + 9.327A^{***} + 5.901A^{2***} \quad (10)$$

A value below 12% of cooking loss indicates good quality products, and cooking loss is considered an indicator of the general cooking performance of pasta [53–55]. Cooking loss increased with the increase of spent grain addition (Figure 1I), but cooking losses were below the values reported for good quality durum wheat pasta (<7%) [56]. The linear term had the most significant positive effect, as can be seen in Equation (11). Cooking loss increases as cohesiveness decreases, according to Vital et al. [53]. Pasta rich in fiber tends to have a higher cooking loss, probably due to the weakening of the protein network by the presence of fiber content [32]. According to Rousta et al., as the amount of protein and fiber increased in the pasta recipe, the cooking loss increased [27].

$$\text{Cooking loss} = 6.026 + 0.972\text{A}^{***} \quad (11)$$

The predicted results for the responses were verified, and differences of $\leq$5% were obtained between the predicted and verified values.

3.3. *Optimization of Spent Grain Level and Model Validation*

Optimizing the spent grain level in the pasta recipe showed that spent grain can be added at a rate of 11.70% with desirability of 0.371 without affecting the quality characteristics, obtaining maximum nutritional benefits, applying the constraints (Figure 2).

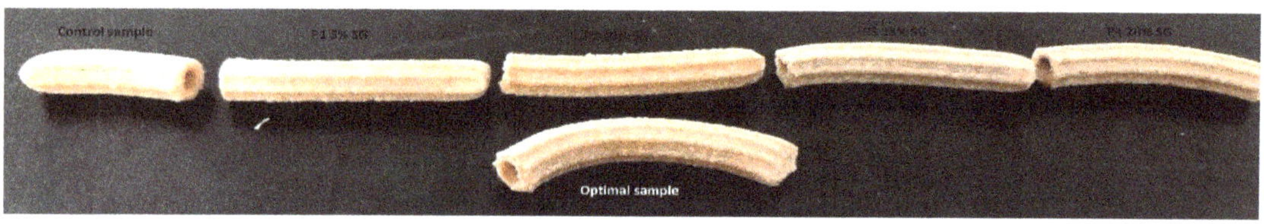

Figure 2. Pasta samples obtained in this research study.

The results obtained made it possible to locate the lower and upper limits according to the target, as shown in Table 3. The following objectives were chosen to take into account the measured quality characteristics: minimization of cooking losses; maximization of cohesiveness, fracturability, crude proteins, and total dietary fiber content; and total phenolic content and antioxidant activity, keeping in range the firmness and color of pasta.

Table 3. Optimization constraints for pasta recipe.

Name	Goal	Lower Limit	Upper Limit	Importance
A:SG level (%)	in range	5	20	3
Cohesiveness (adim.)	maximize	0.27	0.329	3
Firmness (g)	in range	5359	6437	3
Color chroma (adim.)	in range	15.53	18.88	3
Fracturability (g)	maximize	2935	8120	3
Crude proteins (g/100 g d.b.)	maximize	5.68	7.99	3
Total dietary fiber (g/100 g d.b.)	maximize	34.64	37.95	3
Total phenolic content (μg GAE/g)	maximize	17.21	23.03	3
Antioxidant activity (% inhibition)	maximize	15.6	35.55	3
Cooking loss (%)	minimize	5.01	6.992	3

A sample was made with the optimal value predicted to validate the model, and the answers were evaluated in triplicate; the predicted value and verified value of the optimal sample can be seen in Table 4.

A pasta sample was obtained with the optimal level of spent grain resulting after the optimization process. All responses were checked in triplicate, and the values of the experimental results were less than 5% different from the predicted ones. The addition of spent grain in the pasta recipe in the presented conditions allowed the production of pasta with a short preparation time and an attractive texture after hydration with warm water.

Table 4. Predicted and verified optimal sample.

Factor	Spent Grain Pasta		
	Predicted Value	Verified Value	Relative Deviation * (%)
SG level	11.70	11.70	
Cohesiveness (adim.)	0.291 [a]	0.302 [a]	3.64
Firmness (g)	5805.01 [a]	5786.06 [a]	−0.33
Color (adim.)	17.59 [a]	17.69 [a]	0.52
Fracturability (g)	3924.74 [a]	3976.25 [a]	1.30
Crude proteins (g/100 g d.b.)	6.54 [a]	6.67 [b]	1.96
Total dietary fiber (g/100 g d.b.)	36.08 [a]	36.42 [a]	0.92
Total phenolic content (µg GAE/g)	21.71 [a]	22.84 [a]	4.94
Antioxidant activity (% inhibition)	19.72 [a]	20.12 [a]	1.96
Cooking loss (%)	5.92 [a]	5.71 [a]	−3.71

* relative deviation = ((experimental value − predicted value)/experimental value) × 100, [a,b] indicates that differences among predicted and observed values are significantly different ($p < 0.05$).

3.4. Sensory Analysis of Pasta

The appearance of the pasta with spent grain addition was pleasant, obtaining a slightly higher score than the control sample obtained only from spelt flour. The panelists also appreciated the color and smell parameters positively compared to the reference sample. The sensory properties of cooked pasta are shown in Figure 3. The highest values were obtained for color, appearance, and smell parameters, while the taste parameter obtained the lowest score.

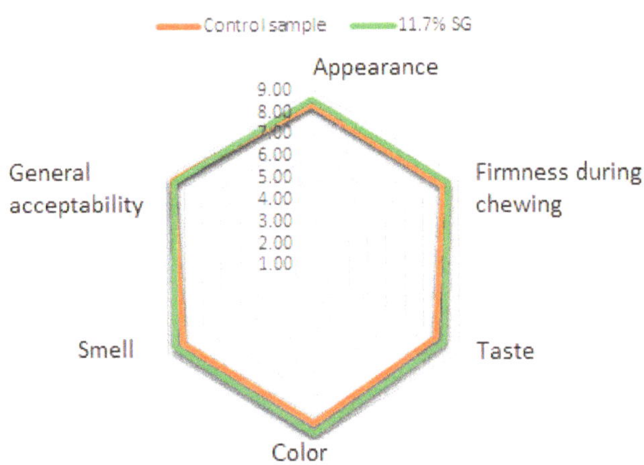

Figure 3. Sensory scores for pasta.

4. Conclusions

The purpose of this work was to establish the optimal value of spent grain in spelt pasta with the preservation in parameters of the desired responses. Spent grain is an important source of protein, fiber, and compounds with antioxidant properties. As the results show, the optimal value for spelt pasta formulation with a good balance between sensory and nutritional aspects was 11.70%. The optimal level of spent grain was used without compromising the acceptability of the product. Spelt pasta fortified with spent grain is included in products with a high fiber and protein content and with antioxidant activity and high polyphenol content. The color of the pasta obtained was acceptable, and the cooking losses were within the limit of 12%, which fits them into good quality products. These results show that spent grain can be used successfully in the recipe for fortified pasta, obtaining high-quality products. Spent grain flours can be used in food formulations because of their potential to improve the nutritional quality of the product and may have a lower glycemic index compared to pasta based on white flour from durum wheat. The valorization of spent grain can improve the sustainability of the brewing process and the whisky production process. With the help of experimental modeling, pasta recipes can be developed to meet consumer requirements or by directing recipes to certain categories of consumers, depending on needs.

Author Contributions: Conceptualization, A.C. and A.D.; Methodology, A.C.; Formal Analysis, A.C. and A.D.; Investigation, A.C.; Resources, A.D.; Writing—Original Draft Preparation, A.C.; Writing—Review and Editing, A.C. and A.D. All authors have read and agreed to the published version of the manuscript.

Funding: No funding was obtained for this work.

Institutional Review Board Statement: Not applicable.

Informed Consent Statement: Not applicable.

Data Availability Statement: The data presented in this study are available in this article.

Acknowledgments: We would like to thank Alexandrion Group Romania for providing the spent grain used in this work. The authors acknowledge financial support from Stefan cel Mare University of Suceava, Romania.

Conflicts of Interest: The authors declare no conflict of interest.

Sample Availability: Samples were available from Alexandrion Group Romania.

References

1. Naibaho, J.; Korzeniowska, M.; Wojdyło, A.; Figiel, A.; Yang, B.; Laaksonen, O.; Foste, M.; Vilu, R.; Viiard, E. Fiber modification of brewers' spent grain by autoclave treatment to improve its properties as a functional food ingredient. *LWT* **2021**, *149*, 111877. [CrossRef]
2. Parchami, M.; Ferreira, J.A.; Taherzadeh, M.J. Starch and protein recovery from brewer's spent grain using hydrothermal pretreatment and their conversion to edible filamentous fungi—A brewery biorefinery concept. *Bioresour. Technol.* **2021**, *337*, 125409. [CrossRef] [PubMed]
3. Jackson, S.A.; Kang, X.; O'Shea, R.; O'Leary, N.; Murphy, J.D.; Dobson, A.D.W. Anaerobic digestion performance and microbial community structures in biogas production from whiskey distillers organic by-products. *Bioresour. Technol. Rep.* **2020**, *12*, 100565. [CrossRef]
4. Czubaszek, A.; Wojciechowicz-Budzisz, A.; Spychaj, R.; Kawa-Rygielska, J. Baking properties of flour and nutritional value of rye bread with brewer's spent grain. *LWT* **2021**, *150*, 111955. [CrossRef]
5. Korcari, D.; Secchiero, R.; Laureati, M.; Marti, A.; Cardone, G.; Rabitti, N.S.; Ricci, G.; Fortina, M.G. Technological properties, shelf life and consumer preference of spelt-based sourdough bread using novel, selected starter cultures. *LWT* **2021**, *151*, 112097. [CrossRef]
6. Witczak, T.; Gałkowska, D. Sorption and thermal characteristics of ancient grain pasta of various compositions. *LWT* **2021**, *137*, 110433. [CrossRef]
7. Romano, A.; Ferranti, P.; Gallo, V.; Masi, P. New ingredients and alternatives to durum wheat semolina for a high quality dried pasta. *Curr. Opin. Food Sci.* **2021**, *41*, 249–259. [CrossRef]
8. Zamaratskaia, G.; Gerhardt, K.; Wendin, K. Biochemical characteristics and potential applications of ancient cereals—An underexploited opportunity for sustainable production and consumption. *Trends Food Sci. Technol.* **2021**, *107*, 114–123. [CrossRef]
9. Sahin, A.W.; Atzler, J.J.; Valdeperez, D.; Münch, S.; Cattaneo, G.; O'Riordan, P.; Arendt, E.K. Rejuvenated Brewer's Spent Grain: EverVita Ingredients as Game-Changers in Fibre-Enriched Bread. *Foods* **2021**, *10*, 1162. [CrossRef]
10. Patrignani, M.; Brantsen, J.F.; Awika, J.M.; Conforti, P.A. Application of a novel microwave energy treatment on brewers' spent grain (BSG): Effect on its functionality and chemical characteristics. *Food Chem.* **2021**, *346*, 128935. [CrossRef]
11. Mussatto, S.I.; Dragone, G.; Roberto, I.C. Brewers' spent grain: Generation, characteristics and potential applications. *J. Cereal Sci.* **2006**, *43*, 1–14. [CrossRef]
12. Wagner, E.; Pería, M.E.; Ortiz, G.E.; Rojas, N.L.; Ghiringhelli, P.D. Valorization of brewer's spent grain by different strategies of structural destabilization and enzymatic saccharification. *Ind. Crop. Prod.* **2021**, *163*, 113329. [CrossRef]
13. Kavalopoulos, M.; Stoumpou, V.; Christofi, A.; Mai, S.; Barampouti, E.M.; Moustakas, K.; Malamis, D.; Loizidou, M. Sustainable valorisation pathways mitigating environmental pollution from brewers' spent grains. *Environ. Pollut.* **2021**, *270*, 116069. [CrossRef] [PubMed]
14. Wójtowicz, A.; Oniszczuk, A.; Kasprzak, K.; Olech, M.; Mitrus, M.; Oniszczuk, T. Chemical composition and selected quality characteristics of new types of precooked wheat and spelt pasta products. *Food Chem.* **2020**, *309*, 125673. [CrossRef]

15. Naibaho, J.; Korzeniowska, M. Brewers' spent grain in food systems: Processing and final products quality as a function of fiber modification treatment. *J. Food Sci.* **2021**, *86*, 1532–1551. [CrossRef]
16. Gonçalves, S.D.; Quiroga, F.; Vilaça, A.C.; Lancetti, R.; Canallis, M.S.B.; de Andrade, M.H.C.; Ribotta, P.D. Physical–chemical evaluation of flours from brewery and macauba residues and their uses in the elaboration of cookies. *J. Food Process. Preserv.* **2021**, *45*, e15700. [CrossRef]
17. Grasso, S. Extruded snacks from industrial by-products: A review. *Trends Food Sci. Technol.* **2020**, *99*, 284–294. [CrossRef]
18. Vriesekoop, F.; Haynes, A.; van der Heijden, N.; Liang, H.; Paximada, P.; Zuidberg, A. Incorporation of Fermented Brewers Spent Grain in the Production of Sourdough Bread. *Fermentation* **2021**, *7*, 96. [CrossRef]
19. Nagy, V.; Diósi, G. Using brewer's spent grain as a byproduct of the brewing industry in the bakery industry. *Elelmvizsg. Kozl.* **2021**, *67*, 3339–3350. [CrossRef]
20. Escarnot, E.; Agneessens, R.; Wathelet, B.; Paquot, M. Quantitative and qualitative study of spelt and wheat fibres in varying milling fractions. *Food Chem.* **2010**, *122*, 857–863. [CrossRef]
21. Sterczyńska, M.; Stachnik, M.; Poreda, A.; Pużyńska, K.; Piepiórka-Stepuk, J.; Fiutak, G.; Jakubowski, M. Ionic composition of beer worts produced with selected unmalted grains. *LWT* **2021**, *137*, 110348. [CrossRef]
22. Świeca, M.; Dziki, D.; Gawlik-Dziki, U.; Różyło, R.; Andruszczak, S.; Kraska, P.; Kowalczyk, D.; Pałys, E.; Baraniak, B. Grinding and Nutritional Properties of Six Spelt (*Triticum aestivum* ssp. *Spelta* L.). *Cultivars. Cereal Chem.* **2014**, *91*, 247–254. [CrossRef]
23. Gałkowska, D.; Witczak, T.; Witczak, M. Ancient Wheat and Quinoa Flours as Ingredients for Pasta Dough—Evaluation of Thermal and Rheological Properties. *Molecules* **2021**, *26*, 7033. [CrossRef] [PubMed]
24. Filipović, J.; Pezo, L.; Filipović, V.; Brkljača, J.; Krulj, J. The effects of ω-3 fatty acids and inulin addition to spelt pasta quality. *LWT* **2015**, *63*, 43–51. [CrossRef]
25. Marconi, E.; Carcea, M.; Schiavone, M.; Cubadda, R. Spelt (*Triticum spelta* L.) Pasta Quality: Combined Effect of Flour Properties and Drying Conditions. *Cereal Chem.* **2002**, *79*, 634–639. [CrossRef]
26. Chetrariu, A.; Dabija, A. Quality Characteristics of Spelt Pasta Enriched with Spent Grain. *Agronomy* **2021**, *11*, 1824. [CrossRef]
27. Rousta, L.K.; Yazdi, A.P.G.; Khorasani, S.; Tavakoli, M.; Ahmadi, Z.; Amini, M. Optimization of novel multigrain pasta and evaluation of physicochemical properties: Using D-optimal mixture design. *Food Sci. Nutr.* **2021**, *9*, 5546–5556. [CrossRef] [PubMed]
28. Marconi, E.; Messia, M.C. *Pasta Made from Nontraditional Raw Materials: Technological and Nutritional Aspects*, 2nd ed.; AACC International, Inc.: Eagan, MN, USA, 2012; ISBN 9780128104323.
29. Bradley, N. *The Response Surface Methodology*; Department of Mathematical Sciences, Indiana University of South Bend: South Bend, IN, USA, 2007.
30. Mironeasa, S.; Mironeasa, C. Dough bread from refined wheat flour partially replaced by grape peels: Optimizing the rheological properties. *J. Food Process Eng.* **2019**, *42*, e13207. [CrossRef]
31. Sobukola, O.P.; Babajide, J.M.; Ogunsade, O. Effect of brewers spent grain addition and extrusion parameters on some properties of extruded yam starch-based pasta. *J. Food Process. Preserv.* **2013**, *37*, 734–743. [CrossRef]
32. Nocente, F.; Natale, C.; Galassi, E.; Taddei, F.; Gazza, L. Using Einkorn and Tritordeum Brewers' Spent Grain to Increase the Nutritional Potential of Durum Wheat Pasta. *Foods* **2021**, *10*, 502. [CrossRef]
33. Spinelli, S.; Padalino, L.; Costa, C.; Del Nobile, M.A.; Conte, A. Food by-products to fortified pasta: A new approach for optimization. *J. Clean. Prod.* **2019**, *215*, 985–991. [CrossRef]
34. Schettino, R.; Verni, M.; Acin-Albiac, M.; Vincentini, O.; Krona, A.; Knaapila, A.; Cagno, R.; Gobbetti, M.; Rizzello, C.; Coda, R. Bioprocessed Brewers' Spent Grain Improves Nutritional and Antioxidant Properties of Pasta. *Antioxidants* **2021**, *10*, 742. [CrossRef] [PubMed]
35. Iuga, M.; Mironeasa, S. Use of Grape Peels By-Product for Wheat Pasta Manufacturing. *Plants* **2021**, *10*, 926. [CrossRef] [PubMed]
36. Palavecino, P.M.; Bustos, M.C.; Alabí, M.B.H.; Nicolazzi, M.S.; Penci, M.C.; Ribotta, P.D. Effect of Ingredients on the Quality of Gluten-Free Sorghum Pasta. *J. Food Sci.* **2017**, *82*, 2085–2093. [CrossRef] [PubMed]
37. Iuga, M.; Mironeasa, S. Application of heat moisture treatment in wheat pasta production. *Food Control* **2021**, *128*, 108176. [CrossRef]
38. Oroian, M.; Ursachi, F.; Dranca, F. Ultrasound-Assisted Extraction of Polyphenols from Crude Pollen. *Antioxidants* **2020**, *9*, 322. [CrossRef] [PubMed]
39. Chetrariu, A.; Dabija, A. Spent Grain from Malt Whisky: Assessment of the Phenolic Compounds. *Molecules* **2021**, *26*, 3236. [CrossRef] [PubMed]
40. Nkeletela, L.B.; Ganongo-Po, F.B.; Kimbonguila, A.; Guenonie, S.I.; Matos, L.; Petit, J.; Scher, J.; Nzikou, J.M. Manufacturing of Composite Pasta by a Mixing Plan. *Food Nutr. Sci.* **2021**, *12*, 206–221. [CrossRef]
41. Petitot, M.; Boyer, L.; Minier, C.; Micard, V. Fortification of pasta with split pea and faba bean flours: Pasta processing and quality evaluation. *Food Res. Int.* **2010**, *43*, 634–641. [CrossRef]
42. Iuga, M.; Batariuc, A.; Mironeasa, S. Synergistic Effects of Heat-Moisture Treatment Regime and Grape Peels Addition on Wheat Dough and Pasta Features. *Appl. Sci.* **2021**, *11*, 5403. [CrossRef]
43. Palavecino, P.M.; Curti, M.I.; Bustos, M.C.; Penci, M.C.; Ribotta, P.D. Sorghum Pasta and Noodles: Technological and Nutritional Aspects. *Plant Foods Hum. Nutr.* **2020**, *75*, 326–336. [CrossRef] [PubMed]

44. Zhao, Y.H.; Manthey, F.A.; Chang, S.K.C.; Hou, H.-J.; Yuan, S.H. Quality Characteristics of Spaghetti as Affected by Green and Yellow Pea, Lentil, and Chickpea Flours. *J. Food Sci.* **2006**, *70*, s371–s376. [CrossRef]
45. Tazrart, K.; Zaidi, F.; Lamacchia, C.; Haros, M. Effect of durum wheat semolina substitution with broad bean flour (*Vicia faba*) on the Maccheronccini pasta quality. *Eur. Food Res. Technol.* **2016**, *242*, 477–485. [CrossRef]
46. Cappa, C.; Alamprese, C. Brewer's spent grain valorization in fiber-enriched fresh egg pasta production: Modelling and optimization study. *LWT* **2017**, *82*, 464–470. [CrossRef]
47. Heredia-Sandoval, N.G.; del Carmen Granados-Nevárez, M.; de la Barca, A.M.C.; Vásquez-Lara, F.; Malunga, L.N.; Apea-Bah, F.B.; Beta, T.; Islas-Rubio, A.R. Phenolic acids, antioxidant capacity, and estimated glycemic index of cookies added with brewer's spent grain. *Plant Foods Hum. Nutr.* **2020**, *75*, 41–47. [CrossRef]
48. Nocente, F.; Taddei, F.; Galassi, E.; Gazza, L. Upcycling of brewers' spent grain by production of dry pasta with higher nutritional potential. *LWT* **2019**, *114*, 108421. [CrossRef]
49. Ainsa, A.; Honrado, A.; Marquina, P.L.; Roncalés, P.; Beltrán, J.A.; Calanche M., J.B. Innovative Development of Pasta with the Addition of Fish By-Products from Two Species. *Foods* **2021**, *10*, 1889. [CrossRef]
50. Maga, J.; van Everen, K.E. Chemical and sensory properties of wholewheat pasta products supplemented with wheat-derived dried distillers grain (DDG). *J. Food Process. Preserv.* **1988**, *13*, 71–78. [CrossRef]
51. Bianchi, F.; Tolve, R.; Rainero, G.; Bordiga, M.; Brennan, C.S.; Simonato, B. Technological, nutritional and sensory properties of pasta fortified with agro-industrial by-products: A review. *Int. J. Food Sci. Technol.* **2021**, *56*, 4356–4366. [CrossRef]
52. Reis, S.F.; Abu-Ghannam, N. Antioxidant capacity, arabinoxylans content and in vitro glycaemic index of cereal-based snacks incorporated with brewer's spent grain. *LWT* **2014**, *55*, 269–277. [CrossRef]
53. Vital, A.C.P.; Itoda, C.; Crepaldi, Y.S.; Saraiva, B.R.; Matumoto-Pintro, P.T. Use of asparagus flour from non-commercial plants (residue) for functional pasta production: Asparagus flour for functional pasta production. *J. Food Sci. Technol.* **2020**, *57*, 2926–2933. [CrossRef] [PubMed]
54. Bertin, N.N.G.; Annie, T.N.; Cédric, K.V.; Ulrich, M.N.S.; Bilkissou, N.; Pascal, W.; Elie, F.; François-Xavier, D.E. Formulation of Plantain Flour-Based Pasta: Process Optimization and Sensory Evaluation. *Food Sci. Nutr. Res.* **2020**, *3*, 1–11. [CrossRef]
55. Sahin, A.W.; Hardiman, K.; Atzler, J.J.; Vogelsang-O'Dwyer, M.; Valdeperez, D.; Münch, S.; Cattaneo, G.; O'Riordan, P.; Arendt, E.K. Rejuvenated Brewer's Spent Grain: The impact of two BSG-derived ingredients on techno-functional and nutritional characteristics of fibre-enriched pasta. *Innov. Food Sci. Emerg. Technol.* **2021**, *68*, 102633. [CrossRef]
56. Manthey, F.A.; Schorno, A.L. Physical and Cooking Quality of Spaghetti Made from Whole Wheat Durum. *Cereal Chem.* **2002**, *79*, 504–510. [CrossRef]

applied sciences

MDPI

Article

Effects of Time and Temperature on Stability of Bioactive Molecules, Color and Volatile Compounds during Storage of Grape Pomace Flour

Carmela Gerardi, Miriana Durante, Maria Tufariello , Francesco Grieco and Giovanna Giovinazzo *

National Research Council (CNR)-Institute of Sciences of Food Production (ISPA), Via Lecce-Monteroni, 73100 Lecce, Italy; carmela.gerardi@ispa.cnr.it (C.G.); miriana.durante@ispa.cnr.it (M.D.); maria.tufariello@ispa.cnr.it (M.T.); francesco.grieco@ispa.cnr.it (F.G.)

* Correspondence: giovanna.giovinazzo@ispa.cnr.it

Abstract: Background: Grape pomace is highly attractive for the food industry as it contains numerous bioactive molecules relevant for human health. However, in order to exploit pomace flour as a functional food ingredient for food industry, it is important understand how long-term storage affects the stability of both bioactive molecules and volatile compounds, in addition to color. To this end, we analyzed whole pomace flour from red grape during a six-month storage period in the dark, either at 4 °C or 25 °C. Methods: The specific parameters monitored of grape pomace flour included: antioxidant activity (TEAC assay), total phenol content (Folin-Ciocalteu assay), phenol composition (high performance liquid chromatography), fatty acid composition (gas chromatography-mass spectrometry), volatile compound profiles (headspace-solid phase micro-extraction) and color. Results: Prolonged storage did not significantly affect total phenol content, antioxidant activity and characterized bioactive molecules (polyphenols, fatty acids). The only detected effect of storage was a slight whitening of the pomace flour and a small increase of volatile long chain esters and ketons after 6 months at 25 °C. Conclusions: The activity of several health-relevant bioactive compounds remained stable following storage of pomace flour for 6 months at 4 °C, supporting its possible use as a functional food ingredient.

Keywords: grape pomace flour; polyphenols; fatty acids; shelf life; functional food

Citation: Gerardi, C.; Durante, M.; Tufariello, M.; Grieco, F.; Giovinazzo, G. Effects of Time and Temperature on Stability of Bioactive Molecules, Color and Volatile Compounds during Storage of Grape Pomace Flour. *Appl. Sci.* **2022**, *12*, 3956. https://doi.org/10.3390/app12083956

Academic Editors: Georgiana Gabriela Codină, Adriana Dabija and Lăcrămioara Rusu

Received: 14 March 2022
Accepted: 12 April 2022
Published: 14 April 2022

1. Introduction

In conventional wine processing, large amounts of grape pomace (GP) are produced annually, accounting for 25% of the original fruit mass [1]. Although GP represents an excellent source of dietary fibers and polyphenols, this renewable natural resource is still underutilized in nutraceutical products [2,3]. However, deterioration of biologically active compounds caused by processing represents a major bottleneck for GP utilization in food production chains. Several studies have disclosed the possibility of creating value-added products by incorporating grape by-products in bakery products or pasta [4–6]. For example, the addition of GP to semolina has a positive impact on functional and physio-chemical properties of pasta such as adhesiveness, firmness and volatile profile. The inclusion of GP also leads to an increase in total polyphenol content and antioxidant activity and a decrease in the glycemic index as a result of an increase in resistant starch content [7]. As reported by Ungureann-Iuga and coworkers [8], grape peel powder also improves the technological properties of doughs used for the production of high quality gluten-free foods. The most economically viable approach of longterm storing GP is by reducing its highwater content (75–80%) by drying. That also leads to reduction in volume, thus lowering transportation costs. The drying method adopted is an important factor that affects both the content and activity of phenols, as well as functional properties of GP. In particular, high temperatures can compromise the bioactivity of thermally sensitive compounds. Thus, there is strong interest in the development of an economically and technically feasible technology for GP preservation at an industrial scale that results in a stable product with minimum presence of water and a high concentration of biomolecules. An optimal drying step could enhance the potential use of pomace as a powerful natural antioxidant ingredient in functional foods. To avoid the loss of bioactive compounds due to their thermal instability, freeze-drying is considered superior to oven drying [9]. However, this approach is time-consuming and too costly for industrial application [7]. However, as reported by Gerardi et al. [10] freeze or oven dried (50 °C) GP skin showed no significant difference in total phenol content. Moreover, it has been shown that heat treatment over long periods can favorize the extraction of low molecular weight molecules, thus increasing their bioavailability [11]. Oven drying at 50° is preferred because it is faster, more reproducible and it allows for the storage of grape pomace [10]. GP stabilization and grape pomace flour (GPF) storage conditions play a significant role in their subsequent utilization as functional ingredients [12]. In fact, phenols (such as catechins, soluble acids, flavonols, stilbenes and anthocyanins) and fatty acids present in GP exhibit antioxidative, cardioprotective, antidiabetic, gastroprotective and antilipemic effects [10]. The bioactive compounds that contribute to the positive health properties of grape-pomace-added foods are presumed to be poly and monounsaturated fats, fiber and antioxidants [13]. Moreover, previous studies [14,15] have demonstrated potential applications of polyphenols in food preservation and product shelf life. In previous investigations, GP obtained from winemaking of white or red grape cultivars was studied for antimicrobial and antioxidant properties [14,16]. The results suggested that the antioxidant and antimicrobial activities of different grape variety extracts were related to their phenolic profile. Given these results, flour pomace could be exploited for future applications in food, pharmaceutical and cosmetic industries. In this study, we prepared GPF as a functional food ingredient aimed at increasing the daily intake of dietary bioactive antioxidant compounds. To this end, we applied a simple dehydration technology to whole GP originating from individual varieties of grape and analyzed the content and stability of biomolecules present in GPF during a six month period at two different storage temperatures. We also assessed factors relating to shelf life such the stability of different classes of polyphenols and fatty acids, total phenols (T.P.) and antioxidant activity. Furthermore, to evaluate sensory changes over time, we monitored GP volatile profiles and color. To our knowledge, this is the first work that describes the effects of storage on the shelf life of GPF obtained from GP.

2. Materials and Methods

2.1. Reagents

Trans-resveratrol was obtained from by ICN Biomedicals (South Chillicothe Road, Aurora, OH, USA), whereas catechin, quercetin, quercetin-3-glucoside, epicatechin, rutin and oenin were purchased by Extrasynthese (Genay, France). All other compounds were provided by Sigma-Aldrich (St. Louis, MO, USA).

2.2. Raw Material and Sample Preparation

GP (skins and seeds) from the grape cultivar *Vitis vinifera* cultivar Negroamaro was obtained from a local winery. GP was dried in an oven at 50 °C, until constant weight (48 h in the dark) [10]. GPF was obtained by a laboratory sample mill (FOSS, Hillerød, Denmark). For the six-month storage study, GPF was stored in sealed polypropylene bags. GPF were prepared for sampling and stored either at 4 °C or 25 °C in the dark at ambient humidity. Samples were analyzed at time point (T0) and at monthly intervals for a period of six months (T1–T6).

2.3. GPF Polyphenol Extraction and HPLC Analysis

GPF samples (1 g) were extracted with methanol/ethanol/formic acid (75:20:5, *v/v/v*) in a ratio of 1:10 (GPF/solvent, *w/v*) [10]. Supernatants were collected in fresh tubes and stored at −20 °C until analysis. GPF extracts were quali-quantitatively examined by an

1100 Series HPLC system (Agilent) equipped with a Luna 5 µm C18(2) 100 Å column (250 × 4.6 mm) (Phenomenex, Torrance, CA, USA) as reported by Gerardi et al. [16]. The wavelengths used for quantification of phenol compounds were 280, 306, 320, 370 and 520 nm. The qualitative analysis of phenolic compounds was carried out on the basis of their retention times and spectroscopic spectrum. Weighted amounts of each standard compound were dissolved in 80% (*v/v*) methanol-water mixture to prepare the requested stock solutions. Each sample was independently injected into the HPLC column and then eluted using the above method in order to establish its chromatographic retention time and collecting UV spectrum. The working solutions were obtained by diluting the stock solutions with the methanol-water mixture (80% *v/v*), thus allowing to achieve for each compound a ten-point regression curve ($r^2 \geq 0.99$). Quantification of single compounds was achieved by employing a ten-point regression curve of the UV absorption data sampled at the wavelength of maximum absorbance of each analyte and expressed in µg/g.

2.4. High Performance Liquid Chromatography (HPLC) Characterization of Anthocyanins

To quantify and characterize the anthocyanin molecules in alcoholic extract from grape pomace flour we performed an HPLC analysis using an Agilent-1100 liquid chromatograph equipped with a DAD detector as described in Gerardi et al. [10]. The column was a C18 Luna (Phenomenex, 250 × 46 mm, 5 µm) in conjunction with a C18 guard cartridge column, both maintained at 30 °C temperature. The mobile phase was (A) H_2O/formic acid = 95/5 and (B) acetonitrile/formic acid = 95/5. The samples were eluted following a linear gradient: 1 min of isocratic elution with 6.7% B, 25 min of linear gradient from 6.7 to 16.7% B, 9 min of linear gradient from 16.7 to 55.6% B, 5 min of isocratic elution with 55.6% B, 3 min of linear gradient from 55.6 B to 80% B and 8 min of isocratic elution with 80% B. Flow rate: 0.7 mL/min. Chromatograms were acquired at 520 nm. Quantification of total anthocyanins was directly performed by HPLC/DAD using a five-point regression curve ($r^2 \geq 0.99$) generated through the use of malvidin 3-O-glucoside (oenin) as reference compounds and was expressed as oenin equivalents (OEs).

2.5. Total Polyphenols Content

A previously optimized Folin-Ciocalteu method was carried out to measure the total amount of polyphenols. The amount of total phenols in GPF extracts was assessed by determining the absorbance at 760 nm [16]. Results were expressed as milligram of Gallic Acid Equivalents per gram of GPF (mg GAEs/g of GPF).

2.6. TEAC Antioxidant Capacity Determination

The Trolox Equivalent Antioxidant Capacity (TEAC) assay is based on the scavenging capability of antioxidant molecules to reduce the radical cationic action of 2,2′-azinobis (3-ethylbenzoithiazolone 6-sulphonate) ($ABTS^{\bullet+}$). The analysis was performed as previously described [17]. To generate the $ABTS^{\bullet+}$ radical cation, ABTS (7 mM) was dissolved in water added with potassium persulfate solution (2.45 mM) and incubated at 25 °C in absence of light for 12–16 h. To produce the calibration curve, 20 µL of Trolox standard solution (from 0 to 25 µM) was added to the $ABTS^{\bullet+}$ solution diluted to an absorbance value (734 nm) of 0.40 ± 0.02. Absorbance was determined at 734 nm and obtained values were expressed as µmol Trolox Equivalents (TE)/g of GPF.

2.7. GPF Fatty Acids Extraction and GC Analysis

Total lipids were extracted from aliquots (0.1 g DW) with 4 mL of n-hexane and stirring (3000 rpm) overnight at 4 °C. Samples were centrifuged (6000× *g* 10 min) and a stream of nitrogen was used to vaporize the organic phase. Lipids were subjected to fatty acids derivatization, as previous reported [18]. Briefly, each sample was solubilized in a methanol solution (3 mL) of 0.5 M NaOH, incubated at 100 °C for 5 min and cooled to room temperature. Two mL of boron trifluoride in methanol (12% *w/v*) was added and boiled at 100 °C for 30 min. After cooling, 1 mL di n-hexane and 1 mL of NaCl (0.6% *w/v*) were added. The organic phase was diluted with n-hexane to 1:10 (*v/v*) and analyzed by GC-MS employing the Agilent 5977E Series GC/MS system (Agilent Technologies, Santa Clara, CA, USA) using a DB-Wax column (60 m, 0.25 mm i.d., 0.25 mm film thickness, Agilent) [18]. The GC parameters were as follows: the temperature of the column was 50 °C after injection for 1 min, then programmed at 25 °C/min to 200 °C, at 3 °C/min to 230 °C and maintained at constant temperature of 230 °C for 23 min. Split injection was conducted with a split ratio of 5:1, the flow-rate was 1.0 mL/min, carrier gas used was 99.999% pure helium, the injector temperature was 250 °C and the column head pressure was 40 psi for 0.4 min, constant pressure at 20 psi. The MS detection conditions were as follows: transfer line temperature 250 °C, mode Scan, source and quadrupole temperature 230 °C and 150 °C respectively, scanning method of acquisition, ranging from 46 to 500, for mass/charge (m/z) was optimized. Spectrum data were collected at 0.5 s intervals. Solvent cut time was set at 2 min and 40 min retention time sufficient for separating all the fatty acids. Compounds were identified by using online NIST-library spectra and published MS data. Furthermore, authentic standards were used to confirm MS data.

2.8. Analysis of Volatile Compounds

The Headspace-Solid phase micro-extraction (HS-SPME) technique was adopted to identify and quantify GPF volatile compounds. For each analysis, 0.5 g of sample was placed in a 20 mL glass vial, containing 0.5 g of sodium chloride and 2 mL of distilled water and then stirred for 20 min in a water bath (50 °C). The *n*-tridecane (100 mg/L) was used as internal standard A DVB/CAR/PDMS (Divinylbenzene/Carboxen/Polydimethylsiloxane) fiber was inserted, maintained for 40 min, and then it was removed and inserted in the GC injector to desorb the volatile molecules. Flour samples were analyzed with the 5977E Series GC/MS system (Agilent Technologies, Santa Clara, CA, USA) furnished with a DB-Wax column (60 m, 0.25 mm i.d., 0.25 mm film thickness). Oven temperature increased from 40 °C for 5 min to 50 °C at a rate of 3 °C/min, and finally to then 225 °C at a rate of 5 °C/min. The MS acquisition mode was full scan (40–450 m/z). The identity of molecules was determined by comparison of spectra with the NIST14 library and confirmed by injection of the pure reference standards when available. The semi-quantitative analysis was performed using the internal standard method [19].

2.9. Color and Moisture Evaluation

GPF color was measured during the six-month storage in the dark at 4 °C and 25 °C, using a Minolta CR-410 chromatometer (Konica Minolta Camera Co., Ltd., Osaka, Japan). The CIELAB color space was used to determine the parameters L^* from black (0) to white (100), a^* (red (+a) to green ($-a$) color) and b^* (yellow (+b) to blue ($-b$) color). Color measurement was replicated three times for each flour sample.

Grape pomace moisture was evaluated by utilizing a moisture measuring balance (AND MX50, A&D Company, Limited, Tokyo, Japan). The temperature for the measurement was 120 °C for 30 min. Moisture measurement was replicated three times for each flour sample.

2.10. Statistical Analysis

The statistical significance of the differences between the measured data was assessed by performing both a parametric method (two sample-*t*-test) and a non-parametric one (Mann-Whitney U Test) by means of the SigmaStat software Version 3.1 (Jandel Corp., Erkrath, Germany). The Principal Component Analysis and the heatmap were obtained using routines written in the MATLAB program design. Data are the mean ± standard deviation of three independent replicates (n = 3).

3. Results and Discussion

3.1. Stability of Polyphenols in Negroamaro Pomace Flour Extracts during Storage

GPF stored for six months in the dark at 4 and 25 °C, showed no significant differences in total anthocyanin and oenin content (Table 1).

In the chromatographic profile of the GPF extracts, we detected phenolic acids such as caffeic, caftaric, cutaric and gallic acids and these metabolites had a similar relative distribution during storage (Table 1).

Stilbenes content, and in particular resveratrol, shows significant differences among *V. vinifera* cultivars [20]. Table 1 shows the results related to the analysis of stilbene levels in Negroamaro GPF. It can be seen that trans-resveratrol levels increase during the six months storage period, both at 4 °C and 25 °C.

In grape, flavanols can be found as monomers (epicatechin 3-gallate, gallocatechin, catechin, epigallocatechin and epicatechin), oligomers and polymers (i.e., tannins and proanthocyanidins) that show antioxidant activity both in vitro and in vivo [21–23]. In this study we identified the flavanols catechin and epicatechin in GPF (Table 1). While storage at 4 °C did not affect their content significantly, we observed a slight but statistically significant decrease following storage for six months at 25 °C.

Flavonols are considered bioactive grape/wine compounds that are potentially relevant for human health and nutrition [24,25]. Moreover, in a process known as co-pigmentation anthocyanins can interact with flavonols, leading to an increase in red color. In our GPF we identified the flavonols quercetin and quercetin 3-glycoside (Table 1). In this study GPF prolonged storage do not affect flavonols content. Conversely, the concentration of other compounds such as stilbenes increased over time. This later is in agreement with other studies of food ingredients, that reported a storage-related increase in phenol content [26,27]. Storage can lead to changes in plant and cell tissue structure, as well as degradation of covalently bound phenolic molecules. In turn, that lead to an increase in solubility and consequently an improved extraction of such compounds. These results confirmed the hypothesis that GPF can be stored at both 4 °C and 25 °C without losing of several of its biological properties.

3.2. Assessment of Total Phenols and Antioxidant Activity of GPF during Storage at Different Temperatures

During six months, at 28-day intervals, we monitored variations in the concentration of phenolic compounds present in GPF using the Folin Ciocalteu assay. The obtained results indicated that the total phenol content in grape flour does not change significantly throughout the entire period either at 4 °C or 25 °C (Figure 1) remaining comparable to the control sample (T0).

Total antioxidant activity (AA) was determined using the TEAC/ABTS$^{\bullet+}$ assay. This assay, despite its limitations mainly due to the steric accessibility of ABTS$^{\bullet+}$, has been and still is used to monitor changes in antioxidant activity during food processing or storage, as the sample components with antioxidant activity are constant. We observed no significant changes in antioxidant activity across all timepoints analyzed during the six month storage period. The above findings highlighted a correlation with the stability of different polyphenol groups (in Table 1), and their synergic activity (Figure 2).

Table 1. Characterization of different classes of phenolic compounds in GPF samples stored for 6 months at 4 or 25 °C in the dark.

Time (Months)	Anthocyanins	Oenin	Caffeic Acid	Caftaric Acid	Cutaric Acid	Gallic Acid	Trans-Resveratrol	Catechin	Epicatechin	Quercetin	Querc. 3-Gluc.	Rutin
	(μgOE/g)	μg/g										
Temperature of storage 4 °C												
T0	790 ± 10	220 ± 20	0.77 ± 0.57	21.63 ± 4.58	2.95 ± 0.80	41.00 ± 0.72	8.81 ± 1.71	460 ± 30	800 ± 70	57.61 ± 0.50	28.12 ± 0.41	530 ± 7
T1	840 ± 30	240 ± 40	0.73 ± 0.084	22.45 ± 3.86	2.71 ± 0.54	38.40 ± 2.87	13.43 ± 0.23 *	460 ± 30	700 ± 20	59.72 ± 2.01	33.04 ± 5.18	580 ± 3 *
T2	970 ± 70 *	270 ± 50	1.35 ± 0.33	23.41 ± 2.34	3.08 ± 0.38	46.77 ± 1.55 *	12.85 ± 2.57 *	480 ± 80	680 ± 60	62.69 ± 0.22 *	42.18 ± 12.1	740 ± 10 *
T3	780 ± 1	220 ± 20	0.70 ± 0.35	21.59 ± 3.95	2.71 ± 0.82	41.60 ± 3.50	12.45 ± 1.12 *	470 ± 40	610 ± 60	58.67 ± 1.75	27.55 ± 0.30	540 ± 7
T4	850 ± 60	250 ± 50	1.38 ± 0.13	23.22 ± 2.28	3.29 ± 0.54	38.80 ± 0.66 *	8.95 ± 1.68	540 ± 70	720 ± 130	59.71 ± 3.16	36.98 ± 14.1	640 ± 5 *
T5	850 ± 10 *	250 ± 10	1.89 ± 0.78	22.29 ± 4.41	3.24 ± 0.90	49.35 ± 5.50	17.88 ± 4.01 *	520 ± 30	720 ± 120	66.04 ± 2.98 *	35.22 ± 4.98	680 ± 1 *
T6	820 ± 50	240 ± 30	1.46 ± 0.45	23.26 ± 4.11	3.41 ± 0.97	41.54 ± 6.07	17.53 ± 3.88 *	510 ± 40	660 ± 10	62.04 ± 1.19 *	34.45 ± 8.23	660 ± 6 *
Temperature of storage 25 °C												
T1	760 ± 10 *	220 ± 20	0. 94 ± 0.07	22.13 ± 4.49	2.97 ± 0.80	46.25 ± 3.41	9.80 ± 1.17	450 ± 10	710 ± 30	59.19 ± 0.92	27.89 ± 0.65	560 ± 6 *
T2	810 ± 40	240 ± 10	2.31 ± 0.89 *	23.77 ± 6.15	2.82 ± 0.95	51.16 ± 7.87	16.44 ± 4.45	490 ± 50	770 ± 110	64.78 ± 6.08	32.30 ± 4.56	640 ± 15 *
T3	740 ± 40	220 ± 30	2.39 ± 0.12 *	22.79 ± 3.37	2.76 ± 0.61	45.82 ± 1.25 *	8.94 ± 1.08	450 ± 20	740 ± 10	60.26 ± 2.52	29.79 ± 1.70	510 ± 4
T4	790 ± 40	210 ± 30	2.61 ± 0.16 *	22.21 ± 3.00	3.06 ± 0.66	44.10 ± 0.16 *	7.90 ± 2.09	410 ± 40	650 ± 10 *	60.44 ± 0.94 *	27.24 ± 3.57	560 ± 2 *
T5	850 ± 80	260 ± 50	2.09 ± 0.03 *	21.37 ± 3.23	2.99 ± 0.98	58.92 ± 12.9	6.51 ± 1.84	470 ± 40	710 ± 50	67.14 ± 10.9	36.50 ± 8.84	560 ± 6 *
T6	790 ± 40	230 ± 30	4.03 ± 0.09 *	23.89 ± 2.97	4.31 ± 0.16	55.04 ± 7.96 *	14.86 ± 0.32 *	430 ± 50	550 ± 80 *	64.90 ± 0.05 *	37.63 ± 11.1	690 ± 9 *

Data are the mean ± standard deviation of three independent replicates ($n = 3$). "*" indicates statistically significant differences ($p < 0.05$) between each treatment after T1, T2, T3, T4, T5 and T6 time (months) versus initial storage time (T0) as determined by the two sample-*t*-test and the Mann-Whitney U Test.

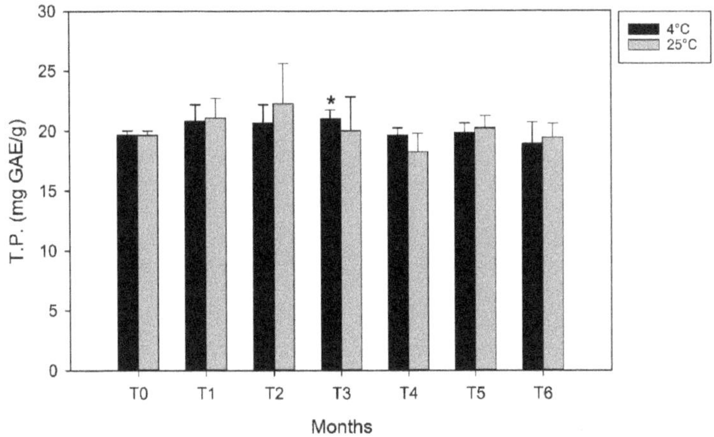

Figure 1. Comparison of Total Phenols content in Negroamaro GPF during six months storage at 4 °C and 25 °C in the dark. Data represent mean values ± standard deviation (n = 3). "*" indicates a timepoint that was significantly different ($p < 0.05$) relative to T0 as determined by the two sample-*t*-test and the Mann-Whitney U Test.

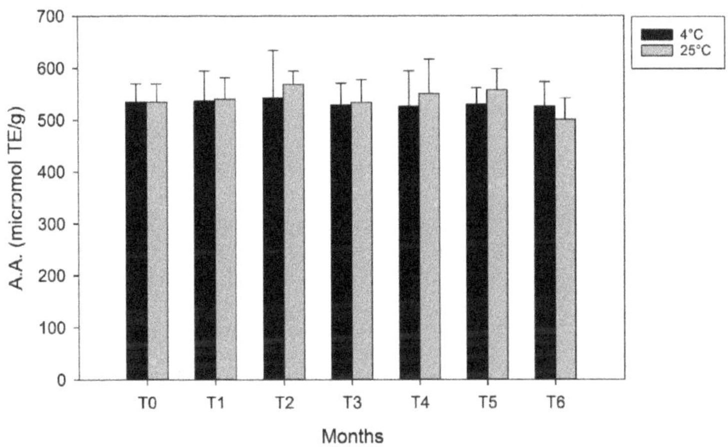

Figure 2. Comparison of TEAC values (AA) in Negroamaro GPF during the six month storage period at 4 °C and 25 °C in the dark. Data represent mean values ± standard deviation (n = 3).

3.3. Effect of Storage on Fatty Acids and Volatile Composition of Negroamaro GPF

Analysis of the FA profile shows that polyunsaturated fatty acids (PUFA) accounted for approximately 57% of the total percentage of fatty acids identified, followed by monounsaturated (MUFA, ~24%) and saturated (SFA, ~19%) acids (Table 2).

Linoleic acid (C18:2n6) represented approximately 54% of the total fatty acids, followed by oleic (C18:1 n9, ~24%), palmitic (C16:0, ~11%) and stearic (C18:0; ~7%) acid. During the storage period of six months, we observed a significant increase in SFA at 5 and 6 months ($p < 0.05$) in GPF stored at 4 °C or 25 °C, compared to T0 (Table 2). However, neither MUFA nor PUFA content changed significantly across the time period analyzed, indicating excellent stability with regard to fatty acid oxidation. The presence of antioxidants such as polyphenols in GPF could play an important role in protecting unsaturated fatty acids from oxidation. Likewise, olive pomace by-products, that are characterized by a high content of phenolic compounds, have also been shown to increase oxidative stability and improving the quality and shelf life of food products [28].

Table 2. Characterization and comparison of different fatty acids composition in GPF stored for six months at 4 and 25 °C in the dark.

	Time Storage (Months)													
	0	**1**	**2**	**3**	**4**	**5**	**6**	**0**	**1**	**2**	**3**	**4**	**5**	**6**
	% of Total Fatty Acids Identified													
Fatty Acids	**4 °C**							**25 °C**						
Palmitic acid (C16:0)	11.55	12.46	12.72	12.81	13.09	14.37	14.72 *	11.55	12.71	12.57	12.77	13.10	13.66	15.28
Palmitoleic acid (C16:1)	0.71	1.01	0.95	0.96	0.78	0.76	0.91 *	0.71	0.95 *	0.99 *	0.92 *	1.20 *	0.93 *	1.07 *
Stearic acid (C18:0)	6.81	7.43	7.04	7.22	7.19	9.49 *	9.52 *	6.81	7.23 *	7.03 *	7.94 *	7.63 *	9.12 *	9.96 *
Oleic acid (C18:1 n9)	23.57	23.64	23.09	23.50	23.76	21.50	21.09	23.57	23.17	23.31	23.82	23.67	22.02	20.97
Linoleic acid (C18:2 n6)	53.85	52.39	53.40	52.88	52.35	51.27	51.16	53.85	52.96	53.20	51.58	51.58	51.69	50.17
Linolenic acid (C18:3 n3)	2.97	2.47	2.29	2.14 *	2.12 *	1.99 *	2.08 *	2.97	2.44	2.40	2.26 *	2.20 *	2.06 *	2.00 *
Arachidic acid (C20:0)	0.54	0.60	0.51	0.49	0.71 *	0.62 *	0.52	0.54	0.54	0.50	0.71 *	0.62	0.52	0.55
SFA	**18.90**	**20.49**	**20.27**	**20.52**	**20.99**	**24.48 ***	**24.66 ***	**18.90**	**20.48**	**20.10**	**21.42 ***	**21.35 ***	**23.30 ***	**25.79 ***
MUFA	**24.28**	**24.65**	**24.04**	**24.46**	**24.54**	**22.36**	**22.10**	**24.28**	**24.12**	**24.30**	**24.74**	**24.87**	**22.95**	**22.04**
PUFA	**56.82**	**54.86**	**55.69**	**55.02**	**54.47**	**53.16**	**53.24**	**56.82**	**55.40**	**55.60**	**53.84**	**53.78**	**53.75**	**52.17**

Standard deviation is less than 10%. "*" indicates statistically significant differences ($p < 0.05$) between each treatment after T1, T2, T3, T4, T5 and T6 time (months) versus initial storage time (T0) as determined by the two sample-*t*-test and the Mann-Whitney U Test.

Although the total lipid content in GPF is not very high, if GPF is subjected to storage conditions at temperature higher than 4 °C, enzymatic and non-enzymatic processes could contribute to lipid oxidation leading to rancidity. Therefore, we investigated whether changes in SFA, MUFA and PUFA content significant contribution to the oxidation stability of grape pomace flour. The stability of oleaginous component in the stored whole grape pomace flour containing seeds was assessed. Fatty acids (FAs) have a variable number of carbons and double bonds [29]. Generally, in foods containing a fatty component then, lipid oxidation is one of the main factors linked to oxidative rancidity, the loss of essential fatty acids, and the development of unpleasant flavor and smells in food ingredients. Polyphenols perform their antioxidant action by: (i) inactivating free radical, (ii) scavenging and neutralizing reactive oxygen species (ROS), (iii) enhancing the electron transfer, (iv) preventing the oxidation propagation phase and avoiding the formation of peroxides [30,31].

We also characterized the volatile profile of Negroamaro GPF during six months of storage at different temperatures. In total, 30 volatile compounds were identified, quantified by triphasic fiber coupled GC-MS analysis and grouped according to the class of volatile (esters, alcohols, aldehydes, ketons, furans and hydrocarbons). As shown in Figure 3 the differences between samples collected at T0 and after six months at 4–25 °C (T6_4 °C and T6_25 °C, respectively) appear to be quantitative rather than qualitative [32].

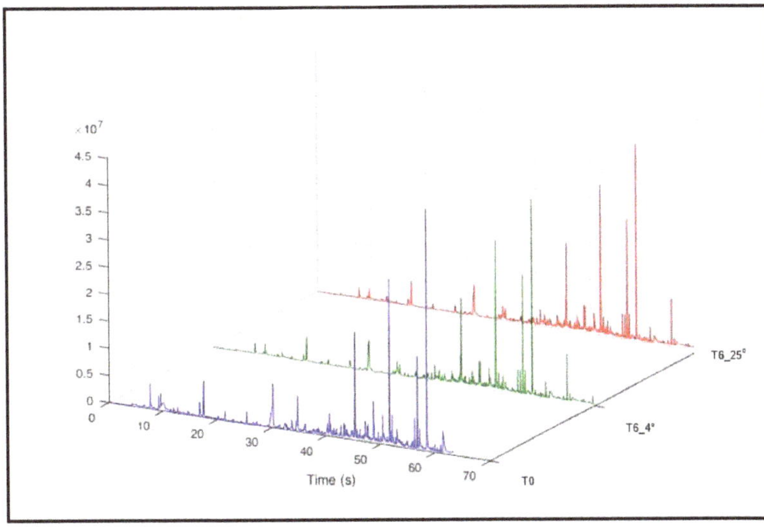

Figure 3. Total ion chromatogram of Negroamaro pomace flour at initial and final time point (T0 and T6, respectively) at 4 °C and 25 °C in the dark.

In all samples, esters, alcohols and aldehydes were the main groups of compounds identified in GP. Esters ranged from 51.65% of the total peak area in T0 sample to 63.38 and 77.40% in T6_4 °C and T6_25 °C, respectively. Alcohols ranged from 31.49% in T0 to 31.38 and 36.88% in T6_4 °C and T6_25 °C, respectively. Ethyl esters of fatty acids with a linear chain from 6 to 18 carbon atoms, are the molecules with the highest concentration detectable in all samples. The volatile fraction of Negroamaro grapes is also characterized by ethyl esters in free form whose content depends from ethanolysis of acetyl-CoA during fatty acids synthesis or degradation [33]. Among acetate esters, ethyl acetate showed a decrease during storage from 4.24 mg/kg at T0 to 1.34 and 1.011 at T6_4 °C and T_25 °C, respectively. Lower concentrations of ethyl acetate in processed samples may be due to glucose reduction via Maillard reactions, or inactivation by microorganisms produced during storage [34]. The detection of esters in control sample and their increase during the storage suggests that the fruity odor of the initial raw material is well preserved. While, the defined mechanism underlying the increase in esters is unknown [34], so it could be hypothesized that it results from oxidative breakdown of unsaturated fatty acids [35].

The aldehyde content varied from 11.13% at T0 to 15.94 and 20.93% in T6_4 °C and T6_25 °C samples, respectively. The large amount of linear aldehydes identified in GPF could result from oxidation degradation of unsaturated fatty acids, especially oleic, linoleic, linolenic [36], while the increase in aromatic aldehydes such as benzaldehyde and phenylacetaldehyde resulted from the degradation of aromatic amino acids during drying [37]. We found that 2-hexenal, a commonly used indicator of food rancidity, decreased during storage; i.e., from 0.41 mg/Kg in T0 to 0.06 and 0.08 mg/Kg in T6_4 °C and T6_25 °C samples, respectively. Furthermore, both values were lower than the concentration associated with rancid odors [38]. Moreover, we found no significant difference betweenT6_4 °C and T6_25 °C. It could be considered that constant antioxidant activity with a non-significant decrease in PUFA and MUFA contributes to reduce lipid oxidation.

While there is limited literature reporting changes in 2-hexenal content GP during storage, the decrease in 2-hexenal could be linked to oxidative breakdown processes. As C6 compounds are products of fatty acid oxidation, we also detected 1-hexanol in concentration varying from 0.26 mg/Kg at T0 to 0.14 and 0.15 mg/Kg in T6_4 °C and T6_25 °C, samples respectively. Volatile alcohols, in particular 1-hexanol and 1-octen-3-ol are produced at low temperature by oxidation of linoleic acid [39,40]. Therefore, it is probable that a decrease in C6 alcohols is linked to a decrease in the level of its precursor.

In the light of this, the low percentage of 2-hexenal and the absence of pentanal considered responsible for off flavor, the increase in esters and aromatic aldehydes during the six months of storage suggest a preservation of sensory quality. To identify main volatile compounds that discriminated GPF samples collected at two different times and temperatures, we performed a principal component analysis (PCA) on data obtained by SPME-GC/MS. As shown in Figure 4, two principal components (PC), PC1 and PC2 accounted for all variation; i.e., 60.23 and 39.77%, respectively.

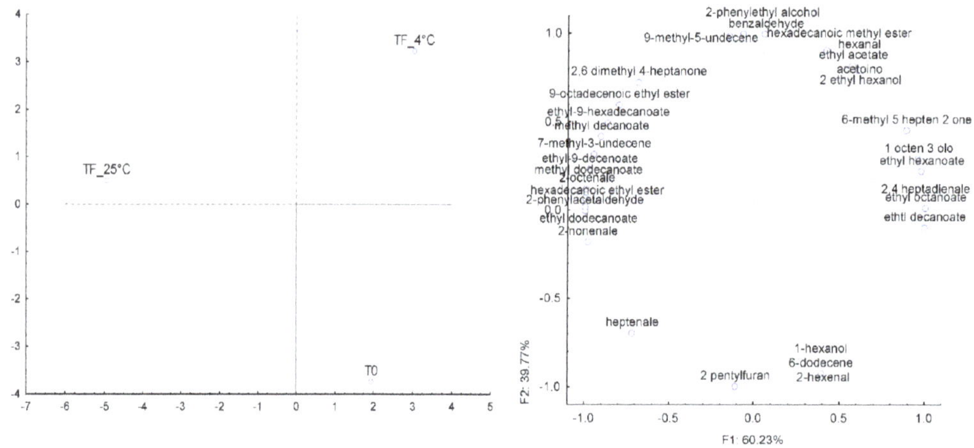

Figure 4. Principal Component Analysis of volatile compound concentrations detected in GPF stored for six months at 4 °C and 25 °C in the dark. PCA biplots on the PC1-PC2 plane combining score plots of major volatiles variables and the final sampling of GPF stored at 4° C (T6) and at 25 °C (T6).

PCA allowed us to differentiate the three samples collected at two different times and temperatures, based on the concentration of specific volatiles. PC1, which explained 60.23% of variance, clearly differentiated T0 and T6_4 °C samples from the T6_25 °C sample. On the other hand, PC2, that explained 39.77% of variation, separated the samples into three different groups. The T0 sample was located in the negative region of PC2 and positive PC1 for the higher values of C6 molecules. The T6_4 °C sample located in the positive region of PC1 and PC2, was characterized by an increase of specific esters (hexadecanoic methyl ester, ethyl acetate, ethyl hexanoate), alcohols (2-ethyl-1-hexanol and 1-octen-3-ol) and aldehydes (hexanal, 2,4-heptadienal). Finally, T6_25 °C sample is described by component negative of PC1 for the high content of long chain esters and ketones. In generally, alcohols, esters and aldehydes are derived from oxidation of unsaturated fatty acids [39,40]. Previous reports have indicated that higher storage temperature might also be responsible for loss of GPF phenol content [41,42]. However, the above results suggest the lack of any significant effect of temperature the quality of GPF.

The heatmap shown in Figure 5 illustrates the changes in the amounts of different volatiles at T0_4 °C or T6_25 °C relative to T. Alcohols and esters, the two main groups of molecules identified, showed significant changes from T0 to T6 at either temperature. Among aldehydes C6-C10, heptenale, 2-octenale, benzaldehyde, 2-nonenale and 2-phenylacetaldehyde increased during storage and with increasing temperature as previously reported [40]. The increase in the concentrations of alcohols and aldehydes could be associated with variation in the percentage of mono and polyunsaturated fatty acids detected. The formation of ketones, 2-phenylethylalcohol, 2-pentylfuran and some hydrocarbons seems to depend on both time and temperature. The two factors, time and temperature, led to a significant increase in alcohols and esters, the latter associated with sensory notes of fruitiness, while no significant variation was found of the remaining of molecules detected [43].

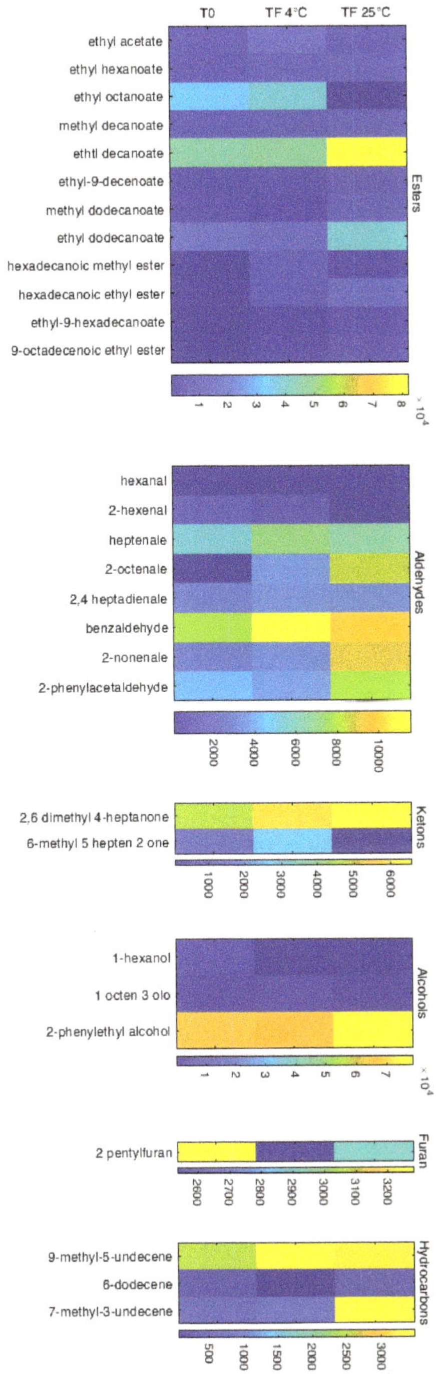

Figure 5. Heatmap of volatile compounds.

The evaluation of the volatile component, aldehydes were detected after six months of storage at the both temperatures, which could accompany the rancidity process of the GPF mostly at 25 °C. The concentrations of these compounds suggest a start of production of the volatile component, which is generally very subjective in individual perception.

3.4. Color Profile and Moisture Percentage of Negroamaro Flour Pomace after Six Months of Storage at Different Temperatures

GFP color is a quality parameter that affects both the appearance and consumer perception. L^*, a^* and b^* (see Materials and Methods section) values of GPF at T0 and T6 samples stored at 4 °C and 25 °C in the dark and air conditions, are shown in Table 3. GPF samples showed positive a^* and b^* values, indicating that a GPF is more red than green and more yellow than blue, respectively.

Table 3. Color and moisture of GPF stored for six months at 4 and 25 °C in the dark.

Time (Months)	Color Parameter			Moisture
	L^*	a^*	b^*	%
T0	44.23 ± 0.08	6.13 ± 0.06	4.89 ± 0.06	7.56 ± 0.06
		Temperature of storage 4 °C		
T6	44.48 ± 0.38	6.38 ± 0.20	5.07 ± 0.20	7.69 ± 0.08
		Temperature of storage 25 °C		
T6	45.61 ± 0.19 *	6.83 ± 0.13 *	5.59 ± 0.09 *	7.9 ± 0.09

Standard deviation is less than 10%. "*" indicates statistically significant differences ($p < 0.05$) between T6 time (months) versus initial storage time (T0) as determined by the two sample-*t*-test and the Mann-Whitney U Test.

Table 3 showed that storing pomace flour during six months lead to a significant increase of L^*, a^* and b^* values. On the other end storage of pomace flour for six months at 4 °C does not effect L^*, a^* and b^* values. Increase of L^*, a^* and b^* values suggests that storage at 25 °C of GPF leads to a mild whitening and an intensification of red and yellow color. This effect could be a direct consequence of the initial anthocyanins degradation in grape pomace during the storage at 25 °C [44,45]. With regards to the percentage of moisture, no significant differences were found between sampling time nor storage temperature (Table 3). Moisture content of food ingredients affect physical and chemical aspects related to freshness and stability during long term storage. Moreover, even slight deviations from a defined standard can adversely affect the physical properties of a food. Interestingly, in the case of wheat flour, fluctuation in moisture content during storage may be associated with changes in flavonoids and antioxidant activity [46].

4. Conclusions

This study is the first to assess the effect of two commonly employed storage temperatures on the chemical and technological properties of GPF. Our results show that the content of total phenols and other bioactive molecules such as polyphenols and fatty acids are largely preserved at both temperatures. However, minor changes in GPF color and volatile component content occur following a storage period of six months at 25 °C. In particular, GPF color becomes slight whitened and shows and small increase in the volatile long chain esters and ketones. Based on this data, we conclude that the optimal storage temperature for GPF storage is 4 °C. Furthermore, our results lend further support to include GPF as a sustainable natural ingredient in functional foods.

Author Contributions: Conceptualization, G.G., C.G. and F.G.; methodology, C.G., M.T. and M.D.; investigation, C.G., G.G., M.D. and M.T.; writing—original draft preparation, G.G.; writing—review and editing, C.G., G.G., M.D. and F.G.; project administration, C.G. and G.G.; funding acquisition, C.G. and G.G. All authors have read and agreed to the published version of the manuscript.

Funding: This research was funded by CNR project NUTR-AGE (FOE2019, DSB.AD004.271), and ValBioVit-SNSvS NP-2.75 project-"2.2 Circular Economy" MITE-Italy.

Institutional Review Board Statement: Not applicable.

Acknowledgments: The authors are grateful for technical support to L. D'Amico, to A.M. Pascali, and to Azienda Agricola Cantele (Guagnano, Lecce, Italy) for providing the grape pomaces. The authors are grateful to Lorenzo Palombi (CNR-IFAC) for the valuable assistance in the data statistical analyses and Gertrud Lund for English language revision.

Conflicts of Interest: Authors declare no conflict of interest.

References

1. Berbegal, C.; Spano, G.; Tristezza, M.; Grieco, F.; Capozzi, V. Microbial Resources and Innovation in the Wine Production Sector. *S. Afr. J. Enol. Vitic.* **2017**, *38*, 156–166. [CrossRef]
2. Ahmad, B.; Yadav, V.; Yadav, A.; Rahman, M.U.; Yuan, W.Z.; Li, Z.; Wang, X. Integrated Biorefinery Approach to Valorize Winery Waste: A Review from Waste to Energy Perspectives. *Sci. Total Environ.* **2020**, *719*, 137315. [CrossRef] [PubMed]
3. Beres, C.; Costa, G.N.S.; Cabezudo, I.; da Silva-James, N.K.; Teles, A.S.C.; Cruz, A.P.G.; Mellinger-Silva, C.; Tonon, R.V.; Cabral, L.M.C.; Freitas, S.P. Towards Integral Utilization of Grape Pomace from Winemaking Process: A Review. *Waste Manag.* **2017**, *68*, 581–594. [CrossRef] [PubMed]
4. Bender, A.B.B.; Speroni, C.S.; Salvador, P.R.; Loureiro, B.B.; Lovatto, N.M.; Goulart, F.R.; Lovatto, M.T.; Miranda, M.Z.; Silva, L.P.; Penna, N.G. Grape Pomace Skins and the Effects of Its Inclusion in the Technological Properties of Muffins. *J. Culin. Sci. Technol.* **2017**, *15*, 143–157. [CrossRef]
5. Gaita, C.; Alexa, E.; Moigradean, D.; Conforti, F.; Poiana, M.-A. Designing of High Value-Added Pasta Formulas by Incorporation of Grape Pomace Skins. *Rom. Biotechnol. Lett.* **2020**, *25*, 1607–1614. [CrossRef]
6. Iuga, M.; Mironeasa, S. Use of Grape Peels by-Product for Wheat Pasta Manufacturing. *Plants* **2021**, *10*, 926. [CrossRef]
7. Tolve, R.; Pasini, G.; Vignale, F.; Favati, F.; Simonato, B. Effect of Grape Pomace Addition on the Technological, Sensory, and Nutritional Properties of Durum Wheat Pasta. *Foods* **2020**, *9*, 354. [CrossRef]
8. Ungureanu-Iuga, M.; Dimian, M.; Mironeasa, S. Development and Quality Evaluation of Gluten-Free Pasta with Grape Peels and Whey Powders. *LWT* **2020**, *130*, 109714. [CrossRef]
9. Tseng, A.; Zhao, Y. Effect of Different Drying Methods and Storage Time on the Retention of Bioactive Compounds and Antibacterial Activity of Wine Grape Pomace (Pinot Noir and Merlot). *J. Food Sci.* **2012**, *77*, H192–H201. [CrossRef]
10. Gerardi, C.; D'amico, L.; Migoni, D., Santino, A.; Salomone, A.; Carluccio, M.A.; Giovinazzo, G. Strategies for Reuse of Skins Separated from Grape Pomace as Ingredient of Functional Beverages. *Front. Bioeng. Biotechnol.* **2020**, *8*, 645. [CrossRef]
11. Planinić, M. Influence of Temperature and Drying Time on Extraction Yield of Phenolic Compounds from Grape Pomace Variety "Portogizac". *Chem. Biochem. Eng. Q.* **2015**, *29*, 343–350. [CrossRef]
12. Teixeira, A.; Baenas, N.; Dominguez-Perles, R.; Barros, A.; Rosa, E.; Moreno, D.; Garcia-Viguera, C. Natural Bioactive Compounds from Winery by-Products as Health Promoters: A Review. *Int. J. Mol. Sci.* **2014**, *15*, 15638–15678. [CrossRef] [PubMed]
13. Giovinazzo, G.; Gerardi, C.; Uberti-Foppa, C.; Lopalco, L. Can Natural Polyphenols Help in Reducing Cytokine Storm in COVID-19 Patients? *Molecules* **2020**, *25*, 5888. [CrossRef] [PubMed]
14. Iuga, M.; Mironeasa, S. Potential of Grape Byproducts as Functional Ingredients in Baked Goods and Pasta. *Compr. Rev. Food Sci. Food Saf.* **2020**, *19*, 2473–2505. [CrossRef] [PubMed]
15. Oliveira, D.A.; Salvador, A.A.; Smânia, A.; Smânia, E.F.A.; Maraschin, M.; Ferreira, S.R.S. Antimicrobial Activity and Composition Profile of Grape (*Vitis vinifera*) Pomace Extracts Obtained by Supercritical Fluids. *J. Biotechnol.* **2013**, *164*, 423–432. [CrossRef]
16. Gerardi, C.; Pinto, L.; Baruzzi, F.; Giovinazzo, G. Comparison of Antibacterial and Antioxidant Properties of Red (Cv. Negramaro) and White (Cv. Fiano) Skin Pomace Extracts. *Molecules* **2021**, *26*, 5918. [CrossRef]
17. Gerardi, C.; Tommasi, N.; Albano, C.; Blando, F.; Rescio, L.; Pinthus, E.; Mita, G.; Prunus Mahaleb, L. Fruit Extracts: A Novel Source for Natural Food Pigments. *Eur. Food Res. Technol.* **2015**, *241*, 683–695. [CrossRef]
18. Durante, M.; Lenucci, M.S.; Marrese, P.P.; Rizzi, V.; De Caroli, M.; Piro, G.; Fini, P.; Russo, G.L.; Mita, G. α-Cyclodextrin Encapsulation of Supercritical CO_2 Extracted Oleoresins from Different Plant Matrices: A Stability Study. *Food Chem.* **2016**, *199*, 684–693. [CrossRef]
19. Tufariello, M.; Pati, S.; D'Amico, L.; Bleve, G.; Losito, I.; Grieco, F. Quantitative Issues Related to the Headspace-SPME-GC/MS Analysis of Volatile Compounds in Wines: The Case of Maresco Sparkling Wine. *LWT* **2019**, *108*, 268–276. [CrossRef]
20. Gatto, P.; Vrhovsek, U.; Muth, J.; Segala, C.; Romualdi, C.; Fontana, P.; Pruefer, D.; Stefanini, M.; Moser, C.; Mattivi, F.; et al. Ripening and Genotype Control Stilbene Accumulation in Healthy Grapes. *J. Agric. Food Chem.* **2008**, *56*, 11773–11785. [CrossRef]
21. Calabriso, N.; Massaro, M.; Scoditti, E.; Pellegrino, M.; Ingrosso, I.; Giovinazzo, G.; Carluccio, M. Red Grapes Skin Polyphenols Blunt Matrix Metalloproteinase-2 and -9 Activity and Expression in Cell Models of Vascular Inflammation: Protective Role in Degenerative and Inflammatory Diseases. *Molecules* **2016**, *21*, 1147. [CrossRef] [PubMed]

22. Calabriso, N.; Scoditti, E.; Massaro, M.; Pellegrino, M.; Storelli, C.; Ingrosso, I.; Giovinazzo, G.; Carluccio, M.A. Multiple Anti-Inflammatory and Anti-Atherosclerotic Properties of Red Wine Polyphenolic Extracts: Differential Role of Hydroxycinnamic Acids, Flavonols and Stilbenes on Endothelial Inflammatory Gene Expression. *Eur. J. Nutr.* **2016**, *55*, 477–489. [CrossRef] [PubMed]
23. Flamini, R.; Mattivi, F.; Rosso, M.; Arapitsas, P.; Bavaresco, L. Advanced Knowledge of Three Important Classes of Grape Phenolics: Anthocyanins, Stilbenes and Flavonols. *Int. J. Mol. Sci.* **2013**, *14*, 19651–19669. [CrossRef] [PubMed]
24. Giovinazzo, G.; Grieco, F. Functional Properties of Grape and Wine Polyphenols. *Plant Foods Hum. Nutr.* **2015**, *70*, 454–462. [CrossRef]
25. Lesjak, M.; Beara, I.; Simin, N.; Pintać, D.; Majkić, T.; Bekvalac, K.; Orčić, D.; Mimica-Dukić, N. Antioxidant and Anti-Inflammatory Activities of Quercetin and Its Derivatives. *J. Funct. Foods* **2018**, *40*, 68–75. [CrossRef]
26. Bolling, B.W.; Blumberg, J.B.; Oliver Chen, C.-Y. The Influence of Roasting, Pasteurisation, and Storage on the Polyphenol Content and Antioxidant Capacity of California Almond Skins. *Food Chem.* **2010**, *123*, 1040–1047. [CrossRef]
27. Kapcum, C.; Uriyapongson, J. Effects of Storage Conditions on Phytochemical and Stability of Purple Corn Cob Extract Powder. *Food Sci. Technol.* **2018**, *38*, 301–305. [CrossRef]
28. Difonzo, G.; Troilo, M.; Squeo, G.; Pasqualone, A.; Caponio, F. Functional Compounds from Olive Pomace to Obtain High-added Value Foods—A Review. *J. Sci. Food Agric.* **2021**, *101*, 15–26. [CrossRef]
29. Fahy, E.; Subramaniam, S.; Murphy, R.C.; Nishijima, M.; Raetz, C.R.H.; Shimizu, T.; Spener, F.; van Meer, G.; Wakelam, M.J.O.; Dennis, E.A. Update of the LIPID MAPS Comprehensive Classification System for Lipids. *J. Lipid Res.* **2009**, *50*, S9–S14. [CrossRef]
30. Radzikowska, U.; Rinaldi, A.O.; Çelebi Sözener, Z.; Karaguzel, D.; Wojcik, M.; Cypryk, K.; Akdis, M.; Akdis, C.A.; Sokolowska, M. The Influence of Dietary Fatty Acids on Lmmune Responses. *Nutrients* **2019**, *11*, 2990. [CrossRef]
31. Zhao, M.; Chiriboga, D.; Olendzki, B.; Xie, B.; Li, Y.; McGonigal, L.; Maldonado-Contreras, A.; Ma, Y. Substantial Increase in Compliance with Saturated Fatty Acid Intake Recommendations after One Year Following the American Heart Association Diet. *Nutrients* **2018**, *10*, 1486. [CrossRef] [PubMed]
32. Tufariello, M.; Fragasso, M.; Pico, J.; Panighel, A.; Castellarin, S.D.; Flamini, R.; Grieco, F. Influence of Non-Saccharomyces on Wine Chemistry: A Focus on Aroma-Related Compounds. *Molecules* **2021**, *26*, 644. [CrossRef] [PubMed]
33. Coletta, A.; Toci, A.T.; Pati, S.; Ferrara, G.; Grieco, F.; Tufariello, M.; Crupi, P. Effect of Soil Management and Training System on Negroamaro Wine Aroma. *Foods* **2021**, *10*, 454. [CrossRef] [PubMed]
34. Gasiński, A.; Kawa-Rygielska, J.; Mikulski, D.; Kłosowski, G.; Głowacki, A. Application of White Grape Pomace in the Brewing Technology and Its Impact on the Concentration of Esters and Alcohols, Physicochemical Parameteres and Antioxidative Properties of the Beer. *Food Chem.* **2022**, *367*, 130646. [CrossRef]
35. Wang, D.; Javed, H.U.; Shi, Y.; Naz, S.; Ali, S.; Duan, C.-Q. Impact of Drying Method on the Evaluation of Fatty Acids and Their Derived Volatile Compounds in 'Thompson Seedless' Raisins. *Molecules* **2020**, *25*, 608. [CrossRef]
36. Whitfield, F.B.; Mottram, D.S. Volatiles from Interactions of Maillard Reactions and Lipids. *Crit. Rev. Food Sci. Nutr.* **1992**, *31*, 1–58. [CrossRef] [PubMed]
37. Cirlini, M.; Dall'Asta, C.; Silvanini, A.; Beghè, D.; Fabbri, A.; Galaverna, G.; Ganino, T. Volatile Fingerprinting of Chestnut Flours from Traditional Emilia Romagna (Italy) Cultivars. *Food Chem.* **2012**, *134*, 662–668. [CrossRef]
38. Fritsch, C.W.; Gale, J.A. Hexanal as a Measure of Rancidity in Low Fat Foods. *J. Am. Oil Chem. Soc.* **1977**, *54*, 225–228. [CrossRef]
39. Chai, D.; Li, C.; Zhang, X.; Yang, J.; Liu, L.; Xu, X.; Du, M.; Wang, Y.; Chen, Y.; Dong, L. Analysis of Volatile Compounds from Wheat Flour in the Heating Process. *Int. J. Food Eng.* **2019**, *15*, 1–13. [CrossRef]
40. Cramer, A.-C.J.; Mattinson, D.S.; Fellman, J.K.; Baik, B.-K. Analysis of Volatile Compounds from Various Types of Barley Cultivars. *J. Agric. Food Chem.* **2005**, *53*, 7526–7531. [CrossRef]
41. Galmarini, M.V.; Maury, C.; Mehinagic, E.; Sanchez, V.; Baeza, R.I.; Mignot, S.; Zamora, M.C.; Chirife, J. Stability of Individual Phenolic Compounds and Antioxidant Activity during Storage of a Red Wine Powder. *Food Bioprocess Technol.* **2013**, *6*, 3585–3595. [CrossRef]
42. Tolun, A.; Artik, N.; Altintas, Z. Effect of Different Microencapsulating Materials and Relative Humidities on Storage Stability of Microencapsulated Grape Pomace Extract. *Food Chem.* **2020**, *302*, 125347. [CrossRef] [PubMed]
43. Tufariello, M.; Chiriatti, M.A.; Grieco, F.; Perrotta, C.; Capone, S.; Rampino, P.; Tristezza, M.; Mita, G.; Grieco, F. Influence of Autochthonous Saccharomyces Cerevisiae Strains on Volatile Profile of Negroamaro Wines. *LWT—Food Sci. Technol.* **2014**, *58*, 35–48. [CrossRef]
44. Augustine, S.; Kudachikar, V.B.; Vanajakshi, V.; Ravi, R. Effect of Combined Preservation Techniques on the Stability and Microbial Quality and Retention of Anthocyanins in Grape Pomace Stored at Low Temperature. *J. Food Sci. Technol.* **2013**, *50*, 332–338. [CrossRef]
45. Ayed, N.; Yu, H.-L.; Lacroix, M. Improvement of Anthocyanin Yield and Shelf-Life Extension of Grape Pomace by Gamma Irradiation. *Food Res. Int.* **1999**, *32*, 539–543. [CrossRef]
46. Zhang, Y.; Truzzi, F.; D'Amen, E.; Dinelli, G. Effect of Storage Conditions and Time on the Polyphenol Content of Wheat Flours. *Processes* **2021**, *9*, 248. [CrossRef]

applied sciences

MDPI

Article

Study of Antioxidant Activity of Garden Blackberries (*Rubus fruticosus* L.) Extracts Obtained with Different Extraction Solvents

Csilla Albert [1], Georgiana Gabriela Codină [2], Melinda Héjja [3], Csaba Dezső András [1], Ancuța Chetrariu [2] and Adriana Dabija [2,*]

[1] Department of Food Science, Faculty of Economics, Socio-Human Sciences and Engineering, Sapientia Hungarian University of Transylvania, 530104 Miercurea Ciuc, Romania; albertcsilla@uni.sapientia.ro (C.A.); andrascsaba@uni.sapientia.ro (C.D.A.)
[2] Faculty of Food Engineering, Stefan cel Mare University of Suceava, 720229 Suceava, Romania; codina@fia.usv.ro (G.G.C.); ancuta.chetrariu@fia.usv.ro (A.C.)
[3] Doctoral School of Nutrition and Food Sciences, University of Debrecen, 4032 Debrecen, Hungary; hejja.melinda@agr.unideb.hu
* Correspondence: adriana.dabija@fia.usv.ro; Tel.: +40-748-845-567

Abstract: Blackberries are rich in antioxidants due to their high polyphenol and vitamin content, making them an excellent source of health protection. It is well known that good quality juices and fruit extracts can be obtained only from high quality fruits. The aim of this research is to characterize the antioxidant properties of garden blackberries fruits grown in Turia (Romania). Extracts were made from blackberry fruits with four different solvents, and the antioxidant capacity was studied by applying DPPH and FRAP assay. The total phenolic content (with Folin–Ciocâlteu reagent), total anthocyanin content (with the pH differential method), and total flavonoid content (with aluminum chloride colorimetric method) were also measured. The quercetin and gallic acid content were also determined by HPLC-DAD. As the best results were obtained with 90% v/v acidified acetonitrile, the Hansen parameters analysis was performed for the acetonitrile-water solvent mixture as a solvent and cyanidin-3-*O*-glucoside as a solute.

Keywords: blackberry; antioxidant capacity; solvent selection; Hansen solubility parameters

Citation: Albert, C.; Codină, G.G.; Héjja, M.; András, C.D.; Chetrariu, A.; Dabija, A. Study of Antioxidant Activity of Garden Blackberries (*Rubus fruticosus* L.) Extracts Obtained with Different Extraction Solvents. *Appl. Sci.* **2022**, *12*, 4004. https://doi.org/10.3390/app12084004

Academic Editor: Anna Lante

Received: 15 March 2022
Accepted: 13 April 2022
Published: 15 April 2022

1. Introduction

Blackberries belong to the thorny shrubs (genus *Rubus*) group and are members of the rose family (*Rosaceae*). Wild species are found on five continents, and low-growing shrubs are common in temperate forests and tropical highlands. Some species prefer cool mountain habitats, others grow better in marshy tundra areas. Researchers have identified hundreds of species, but only the fruits of a few species are commercially available [1]. Blackberries (*Rubus fruticosus* L.) are widely cultivated edible fruits in Europe and North America. Their composition and nutritional value depend on several agro-geo-climatological factors, such as genotype, environmental conditions, agronomic practices, harvesting time, post-harvest storage, and processing techniques [2,3]. Among the most important bioactive compounds in blackberries—flavonoids, phenolic acids, tannins, anthocyanins, and vitamins—have been found [4,5]. The dietary consumption of blackberries has been associated with several health benefits, such as preventing and treating metabolic syndrome, supporting the digestive and immune systems, preventing inflammatory diseases and cardiovascular diseases, and providing protective effects against gastrointestinal cancers [3,4,6].

Blackberry presents multiple health benefits for food consumers due to its antioxidant activity and the polypharmacological effects of anthocyanins [7]. Blackberries are used as conventional raw materials in obtaining jam, compote, as well as unfermented and fermented beverages, but they may replace synthetic additives from foodstuffs such as colorants, stabilizers, etc. [8,9]. Additionally, the use of blackberries improves the nutritional value and shelf life of finished food. The anthocyanins from blackberries can be applied in food packaging systems and as food colorants, since natural dyes are rarely toxic as well as easy to prepare and pollution free [10–12]. The possible applications are shown in Figure 1.

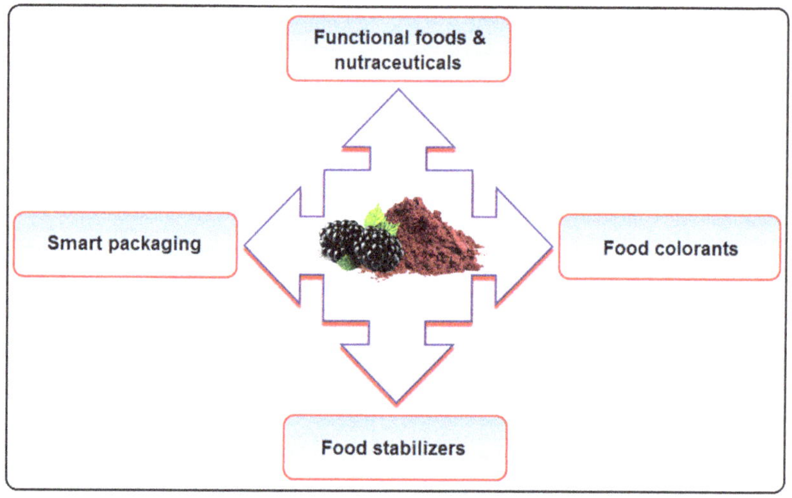

Figure 1. The main unconventional applications of blackberries in the food industry.

During fruit processing, a large by-product quantity is generated with an important bioactive compound content [3,5,13]. Through food by-products utilization, it is possible to simultaneously recover the valuable bioactive compounds and minimize the waste quantity. This way, the fruit processing industry has become more environmentally friendly by removing phenolic compounds from the waste, which has a bacteriostatic effect on degradation microorganisms [14]. There are also difficulties in recovering bioactive compounds from food industry by-products. Extractions involve the co-extraction of non-phenolic substances, such as sugars, organic acids, and proteins, which require subsequent purification processes. The temperature decreases solvent viscosity and surface tension, enhancing diffusion and extraction efficiency. However, high temperatures may accelerate the degradation of polyphenols, reduce antioxidant capacity, and produce solvent evaporation. Extraction of polyphenols is usually carried out at low pH, as, in acidic environments, these compounds take a neutral form, which is best suited for solubilization. However, excessive acidification may impair extraction, as the profile of native polyphenols may be distorted by the hydrolysis of simple glycosides. Extraction time is also crucial, as prolonged exposure to oxygen/light can degrade phenols. In addition, other polyphenols are susceptible to oxidation or volatility and therefore require short processes or conditions that protect the dissolved fraction from oxygen/light damage. Current extraction methods (ultrasound-assisted extraction, supercritical fluid extraction, hot-pressurized liquid extraction, microwave-assisted extraction, enzyme-assisted extraction) are generally efficient, sustainable, potentially cost-effective, and easier to scale up. Nevertheless, the optimal feasibility of extracting plant by-products is still far from being within reach. Efforts should be made to find the optimal compromise between maximum recovery and minimum disturbance of polyphenol structure [15].

In recent years, more and more studies have been frequently published looking at red berries' composition and antioxidant properties. Several methods have been adapted to determine antioxidants in plants, foods, dietary supplements, and food supplements. The total antioxidant capacity of food does not necessarily correlate with the value obtained by summing the levels of individual antioxidants from a composite food matrix. Antioxidant activity can be monitored by various assays with different mechanisms, including hydrogen atom transfer (HAT), single electron transfer (SET), reducing power, and metal chelation. Among the methods using the SET reaction mechanism, the FRAP (Ferric reducing antioxidant power), the Folin–Ciocâlteu method, and the DPPH (2,2-diphenylpicrylhydrazyl) radical scavenging method are widely used. Flavonoids and anthocyanins can also be quantified by UV-Vis spectrophotometry.

In parallel, extensive research is taking place worldwide to find the optimal extraction methods to obtain antioxidant-rich products for a range of berries. Although conventional solvent extraction is the most widespread technique for extracting antioxidant compounds from red berries, new non-conventional methods have emerged as environmentally friendly alternatives to the first method, such as ultrasound, microwave, and pressure-assisted extractions, applied alone or in conjunction with the use of solvents, to reduce energy and solvent requirements [16,17].

The traditional method for anthocyanins extraction is Soxhlet extraction with acidified ethanol or methanol. Acid addition is necessary because anthocyanins are reactive compounds and are sensitive to pH changes. Anthocyanins present instability and tend to change their color in response to changes in environmental parameters such as pH, heat, light, oxygen, and co-existing substances [18]. Anthocyanins can be found in different chemical forms that depend on the solution's pH (Figure 2).

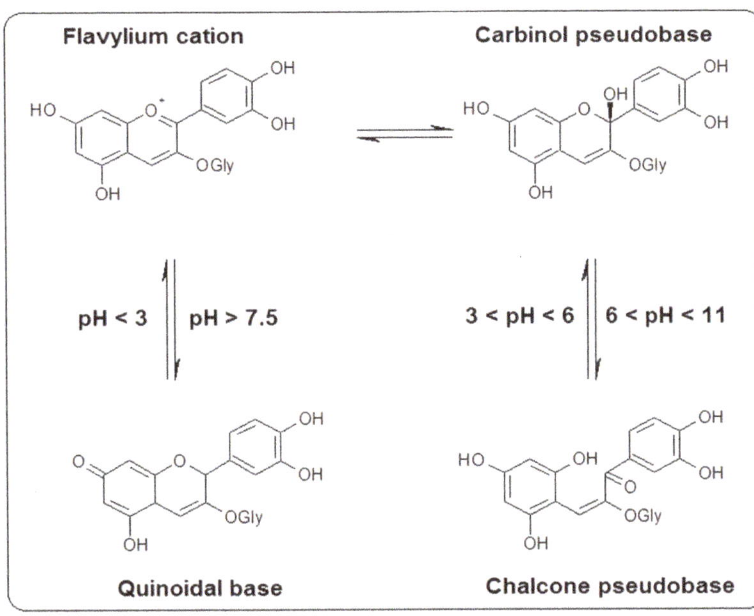

Figure 2. The structures of cyanidin-3-O-glucoside at different pHs.

The presence of ascorbic acid in an environment containing anthocyanins causes a loss of color, thus suggesting a direct interaction between the two molecules. At the same time, the presence of oxygen in the environment will favor faster degradation of anthocyanins by ascorbic acid, thus determining the formation of the polymeric pigment and the whitening of the anthocyanin pigment. The specific degradation process is unknown; however, adding ascorbic acid to anthocyanins increases the decomposition rate of both molecules. The presumed mechanisms by which this phenomenon would occur are the direct condensation of ascorbic acid with anthocyanins or the formation of hydrogen peroxide and oxidative cleavage [19].

The most commonly used extraction solvents are acidified methanol and ethanol, but in the food industry, ethanol is preferred because of the toxicity of methanol. Among ethanol and methanol extractions, many other extraction solvents have been mentioned in the literature, such as: ethyl acetate, chloroform, acetone/water 10–90% (v/v), n-hexane, isooctane [20]. Solvents with high polarity (ethanol and methanol) are favorable for use in the extraction of polar compounds such as phenolic compounds and flavonoids. Non-polar solvents, such as ether, or solvents with low polarity, such as chloroform and ester, are used in specific cases with a low frequency of use. In addition to the extraction method and solvent selection, one should pay great attention to the extraction temperature and pH. Different extraction temperatures may affect the types of polyphenols extracted. At high extraction temperatures, the formation of new compounds, known as products of the Maillard reaction, also involves the polyphenols extraction, and thermal degradation of the polyphenols may occur [21].

A solvent mixture can display more affinity with the chemical target than a pure solvent and can enhance the extraction yields. The solubility of the different compounds in variable solvents was predicted, initially, by comparing the Hildebrand solubility parameters of the solute and solvent. Still, this method had many discrepancies between the practice and theory. The Hildebrand solubility parameter is the square root of cohesive energy density, and this is mainly applicable for nonpolar (van der Waals) interactions.

Improvement was made by Hansen (2007) by introducing three partial solubility parameters (δ_D, δ_P, δ_H), characterizing the different physical interactions separately (dispersive, polar, hydrogen-bond) that can occur between the solvent and solute molecules [22]. Mathematically, the original Hildebrand parameter was treated, somehow, as a vector in three-dimensional space (having three components), with the value of the Hildebrand parameter representing the length of the vector, given by the square root of the sums of the square of its components, as in Equation (1):

$$\delta = \sqrt{\delta_D^2 + \delta_P^2 + \delta_H^2} \qquad (1)$$

Two components mix easily if their value of Hildebrand parameters differs only by 2 $MPa^{0.5}$. However, experience shows that it is also important that at least two of the Hansen parameters have close values to ensure good solubility. Therefore, comparison of the Hansen parameters would give more plausible predictions.

The aim of this research is to characterize the antioxidant properties of garden blackberries fruits grown in Turia, Romania. To obtain more accurate and comprehensive results, we used several solvents for extraction. Using Hansen's solubility parameters, we want to calculate the optimal concentration of a solvent mixture suitable for extracting anthocyanins.

2. Materials and Methods

In this study, frozen garden blackberries were used, collected from own cultivation in Turia, Covasna County, Romania (46°3′51.678″ N and 25°58′40.8828″ E) in September 2020. The village has a temperate continental climate (hot summers, cold winters), with average annual temperatures ranging from 2–7 °C. Annual rainfall in Turia varies between 500 and 1100 mm. The type of soil is eutricambosol soil. The blackberry fruits were hand-harvested at full maturity from randomly selected plants, the fruits without any damage were picked and stored at −20 °C until the analysis.

2.1. Chemicals

Ascorbic acid, Folin–Ciocâlteu's phenol reagent, 2,4,6-tris(2-pyridyl)-1,3,5-triazine (TPTZ), 2-diphenyl-1-picrylhydrazyl (DPPH), aluminum chloride, quercetin, and gallic acid were purchased from Sigma–Aldrich Chemie GmbH (Hamburg, Germany). The solvents (methanol, ethanol, acetonitrile, acetone, hydrochloric acid, acetic acid, and formic acid) were LiChrosolv HPLC grade, purchased from Merck KGaA (Darmstadt, Germany).

Sodium acetate, ferric chloride, sodium carbonate, and potassium chloride were of analytical grade and purchased from VWR International (Batavia, IL, USA).

2.2. Sample Preparation

The frozen garden blackberries were homogenized using a domestic blender, and 20 mL of the following solvents were added to 1 g of sample:

- Et: 80% (*v*/*v*) ethanol [23];
- A: 70% (*v*/*v*) acetone + 2% (*v*/*v*) acetic acid [24–26];
- Met: 60% (*v*/*v*) methanol + 3% (*v*/*v*) formic acid [27];
- ACN: 90% (*v*/*v*) acetonitrile + 10% (*v*/*v*) 6 molar HCl [28].

The samples were sonicated for 20 min for a better cell wall rupture, centrifuged at 3461 RCF (6000 RPM) for 10 min, and the supernatant was removed and used for further determinations. Each sample was independently extracted in triplicate, and analyses were performed on the same day. Among the utilized solvents, only ethanol and acetone (in some cases) are allowed for food industry purposes. However, our research aimed to map the antioxidant capacity of blackberries and not to produce an extract suitable for human consumption. Therefore, we used some other solvents that do not belong to the GRAS (Generally Recognized As Safe) category.

2.3. Determination of the Dry Matter Content of Blackberries

The dry matter content was determined using 2 g of the sample in an oven at 105 ± 2 °C (Memmert, Memmert Gmbh, Schwabach, Germany) until constant weight.

2.4. Antioxidant Activity Determination

2.4.1. 2,2-Diphenyl-1-Picrylhydrazyl (DPPH) Assay

For the measurement, first, a methanolic DPPH solution was prepared (9 mg DPPH dissolved in 100 mL methanol), and then, sample mixtures were prepared. To 1 mL DPPH solution, 20 μL blackberry extract and 980 μL distilled water were added. The mixture was incubated for 30 min at room temperature in the dark, and the absorbance was measured at λ = 517 nm on Varian Cary 50 UV–VIS spectrophotometer (Varian Co., Palo Alto, Santa Clara, CA, USA) [29]. Measurements were performed in triplicate from all samples. The percentage of the DPPH consumed was calculated with the following equation (Equation (2)):

$$E\% = 100 \times (A_0 - A_1)/A_0 \quad (2)$$

where: A_0—the initial absorbance, and A_1—the absorbance in the presence of the extract.

2.4.2. Ferric Reducing Antioxidant Power (FRAP) Assay

The antioxidant potential of blackberries extract was also determined using a FRAP assay measuring the change in absorbance at λ = 593 nm due to the formation of a blue-colored Fe^{2+}-tripyridyl-triazine compound from colorless oxidized Fe^{3+}-form by the action of electron-donating antioxidants [30]. The stock solutions required for the FRAP solution were prepared as follows: acetate buffer: 3.1 g sodium-acetate + 16 mL acetic acid + 1 L distilled water; TPTZ solution: 0.312 g TPTZ + 100 mL distilled water + 336 μL hydrochloric acid (37% *v*/*v*); Ferric chloride solution: 0.54 g $FeCl_3$ + 100 mL distilled water. The FRAP solution was prepared as follows: 25 mL acetate buffer + 2.5 mL TPTZ solution + 2.5 mL ferric chloride solution.

The absorbance was measured at λ = 593 nm (Varian Cary 50 UV–VIS spectrophotometer, Varian Co., Palo Alto, Santa Clara, CA, USA). Calibration was performed with ascorbic acid, and results were expressed as milligrams of ascorbic acid equivalents (AAE) per 100 g of the dry weight of berries. Triplicate measurements were performed from all samples.

2.5. Determination of Total Polyphenol Content (TPC)

The total polyphenol content was determined with the Folin–Ciocâlteu reagent, according to the method modified by Singleton [31], using gallic acid as a standard. There were 50 μL of the sample diluted with 200 μL of methanol: water (4:1) solution was mixed with 0.25 mL of Folin–Ciocâlteu reagent, and 1 min later, 1 mL of sodium carbonate (0.7 M) was added. The mixture was heated to 50 °C and kept at this temperature for 5 min. After cooling, the absorbance was measured at λ = 760 nm. Gallic acid was used as standard, and the results were expressed as milligrams of gallic acid equivalents (GAE) per 100 g of dry weight (DW) of blackberries. Measurements were performed in triplicate on each sample.

2.6. Determination of Total Flavonoid Content (TFC)

The aluminum chloride colorimetric method was used to determine the total flavonoid content of the sample. For calibration, the following concentrations of quercetin solution were used: 12.5; 25; 50; 100; 200 μg/mL quercetin.

For the sample mixtures, 1.2 mL of aluminum chloride solution (2% *m*/*v*) were added to 0.6 mL of blackberry extract and incubated for 60 min at room temperature. Then, absorbances were measured at λ = 420 nm wavelength on a Varian Cary 50 UV spectrophotometer (Varian Co., Palo Alto, Santa Clara, CA, USA) [32]. Results were expressed as mg quercetin per 100 g dry weight. Measurements were performed in triplicate on each sample.

2.7. Determination of Total Anthocyanin Content (TAC)

Total anthocyanin content was determined according to the pH differential method [31]. The sample mixtures were prepared as follows:

- 0.2 mL blackberry extract + 1.8 mL 0.025 M potassium chloride solution (pH = 1);
- 0.2 mL blackberry extract + 1.8 mL 0.4 M sodium-acetate solution (pH = 4.5).

The sample mixtures were measured serially at λ = 510, and λ = 700 nm wavelengths, and the results were calculated using Equations (3) and (4) [33]. Results are expressed as mg cyanidin-3-glucoside/100 g dry matter.

$$A = (A_{510\,nm} - A_{700\,nm})_{pH1.0} - (A_{510\,nm} - A_{700\,nm})_{pH4.5} \quad (3)$$

$$TAC\ (mg/L) = A \times MW \times DF \times 1000/\varepsilon \times l \quad (4)$$

where: A—absorbance; MW—molecular weight (449.2 g/mol); DF—dilution factor; ε—molar absorption coefficient (26,900 L/(mol·cm)); l = optical path length (cm).

2.8. Determination of Gallic Acid and Quercetin by HPLC-DAD

In the first step, 10 g of frozen blackberry sample was poured into 100 mL of methanol and was sonicated in an ultrasonic bath (ultrasound-assisted extraction, UAE) for 30 min. The supernatants were filtered through a 0.45 μm syringe filter. The analyses were operated on an Agilent 1260 HPLC system (Agilent, Santa Clara, CA, USA) with a photodiode array detector, with Betasil C_{18} (150 mm × 4.6 mm, 5 μm) analytical column (Thermo Scientific®, Waltham, MA, USA). The mobile phase was a binary solvent mixture of A (0.1% formic acid in water) and B (methanol). The solvents used were filtered through a 0.45 μm filter and degassed. The mobile phase flow rate was 0.6 mL·min^{-1}. The gradient program was: 0–2 min: 15% B; 2–7 min: 15–30% B; 11–15 min: 30–80% B; 15–20 min: 80–15% B (UV absorbance was monitored in the interval λ = 20–400 nm [34,35]. The calibration curve was constructed by plotting standard solutions' peak area against the concentration (5–100 μg/mL for gallic acid and 2.5–50 μg/mL). The determination coefficient (R^2) was 0.9918 for gallic acid and 0.9952 for quercetin. The LOD and LOQ were 0.85 and 2.17 μg/mL for gallic acid and 0.17 and 0.32 μg/mL for quercetin.

2.9. Calculation of a Solvent Mixture Optimal Concentration for Anthocyanin Extraction Using the Hansen Solubility Parameters

The theoretical optimal concentration of a solvent mixture (e.g., acetonitrile-water solution) could be calculated by using the δ_P-δ_H coordinate system. In this coordinate, the solvent-water solution is represented by a line, and the flavylium cation is shown by a point (P). The coordinate of the intersection of the perpendicular lines from the point P to the line will give the two Hansen parameter of the optimal solvent composition. The equation of the line for solvent-water mixture is given by:

$$\delta^s{}_P = m \cdot \delta^s{}_H + n \quad (5)$$

where: $\delta^s{}_P$—the polar Hansen parameter of the solvent mixture; $\delta^s{}_H$—the hydrogen-bond Hansen parameter of the solvent mixture; m-slope; n-intercept. The numerical values for the slope and intercept are determined by linear regression.

The point, representing the flavylium cation, is marked as $P(\delta^f{}_H, \delta^f{}_P)$.

The coordinates of the closest point of the line to the P point, representing the optimal solvent composition ($P^*(\delta^*{}_H, \delta^*{}_P)$), are given by:

$$\delta^*{}_H = (\delta^f{}_H + m(\delta^f{}_P - n))/(1 + m^2) \quad (6)$$

and

$$\delta^*{}_P = m \cdot \delta^*{}_H + n \quad (7)$$

As the dependency $V\%_S$-$\delta^s{}_H$ is also linear, given by $V\%_S = a \cdot \delta^s{}_H + b$ (where a-slope, b-intercept, both determined by linear regression), the optimal solvent composition could be expressed as:

$$V_s\% = a \cdot \delta^*{}_H + b \quad (8)$$

2.10. Statistical Analysis

Data are expressed as the means $\pm$ standard deviation for at least three independent measurements. Statistical analysis and graphical representation were made using Microsoft Excel 2016 and Statistica 8.0 (StatSoft Inc., Tulsa, OK, USA). The normal distribution of the experimental data was tested with Kolmogorov–Smirnov and Shapiro–Wilk tests. The normality test was performed on three variables from three independent measurements. The relationship between bioactive compounds and antioxidant capacity was tested using Pearson correlation in SPPS 19.0 (SPSS Inc., Chicago, IL, USA).

3. Results and Discussion

Ultrasound-assisted extraction allows low concentrations of solvents and lower temperatures to extract anthocyanins [36]. Therefore, we opted for ultrasound-assisted extraction, at room temperature, with different extraction solvents [17].

3.1. Dry Matter Content of Blackberries

The dry matter content of the blackberries was obtained as 13.55 $\pm$ 0.92%. Hereafter, the results are expressed in terms of dry matter content.

3.2. Total Anthocyanin, Flavonoid, and Polyphenol Concentrations

The total anthocyanin, flavonoid, and polyphenol concentrations in the blackberries tested are summarized in Table 1. The results show that the total anthocyanin content ranges from 642.96 $\pm$ 71.24 to 885 $\pm$ 59.81 mg cyanidin-3-glucoside/100 g DW (average 780.13 $\pm$ 98.54 mg cyanidin-3-glucoside/100 g DW), and the total flavonoid content ranges from 129.75 $\pm$ 9.5 to 240.93 $\pm$ 16.95 mg quercetin/100 g DW (average 183.38 $\pm$ 14.87 mg quercetin/100 g DW). The total polyphenol concentration varies from 2143 $\pm$ 321.21 to 3311.83 $\pm$ 54.38 mg gallic acid/100 g DW (average 2872.69 $\pm$ 231.65 mg gallic acid/100 g DW) depending on the solvent. The values obtained agree with several literature data points [23,26–28,37,38]; total phenolic compounds present in blackberries could vary based on variety, climatic conditions of the year, soil fertility, harvest time, or extraction method. The Folin–Ciocâlteu test for TPC has several advantages, including simplicity, reproducibility, and robustness. However, it has several disadvantages. Firstly, the test is sensitive to pH, temperature, and reaction time, and secondly, overestimation of TPC in the Folin–Ciocâlteu test is a serious concern, as reducing sugars and ascorbic acid can be present in high amounts in fruit extracts and can reduce the effect of the Folin–Ciocâlteu reagent, which distorts the TPC results.

Table 1. Total anthocyanin, flavonoid, and polyphenol content of blackberry extracts obtained with different solvents.

Extraction Solvent	TAC (mg C3G/100 g DW)	TAC (mg C3G/100 g FW)	TFC (mg Q/100 g DW)	TFC (mg Q/100 g FW)	TPC (mg GAE/100 g DW)	TPC (mg GAE/100 g FW)
ACN	885.08 ± 59.81	120.01 ± 8.11	240.93 ± 16.96	32.64 ± 2.29	3311.84 ± 54.84	449.08 ± 7.43
Met	773.30 ± 68.34	104.86 ± 9.26	129.75 ± 9.51	17.58 ± 1.28	2789.97 ± 211.79	378.32 ± 28.71
A	819.18 ± 194.8	97.67 ± 29.8	199.85 ± 19.45	27.08 ± 2.63	3245.97 ± 338.77	440.15 ± 45.93
Et	642.96 ± 71.24	87.18 ± 9.66	163.01 ± 13.57	22.09 ± 1.83	2143.00 ± 321.21	290.59 ± 43.55

C3G: cyanidin-3-glucoside; Q: quercetin; GAE: gallic acid equivalents; DW: dry weight; FW: fresh weight.

The use of acetonitrile is common in research and analytical measurements, but it is not allowed in industrial-scale extractions due to its toxicity and flammability. In our study, we also investigated this solvent because it is known to be an excellent HPLC solvent for the analysis of anthocyanins. Even though acetonitrile is rarely used to extract plant bioactive components, our results showed that these extracts had the highest antioxidant activity. Sellappan et al. (2002) also used acetonitrile extraction with acetic acid pickling. In their case, the total anthocyanin concentration of blackberries from the state of Georgia was 116.59 ± 8.58 mg C3G/100 g FW, while the total polyphenol concentration was 486.53 ± 97.13 mg GAE/100 g FW [28].

The least effective solvent was found to be 80 v/v% ethanol even though, in our previous study, 50–70 v/v% ethanol was the most effective solvent for blackcurrant by-products, and other literature data also showed 50 v/v% ethanol to be the most effective [39]. This result can be explained by the fact that the extraction was carried out without acidification. Celant et al. (2016) also used 80% ethanol as solvent, when the total anthocyanin concentration varied between 6.76 ± 0.03 and 9.42 ± 0.03 mg C3G/g FW, the total flavonoid concentration between 0.46 ± 0.02 and 1.14 ± 0.01 mg Q/g FW, while the total polyphenol concentration ranged from 8.23 ± 0.16 to 14.98 ± 0.54 mg GAE/g FW [23]. Cho et al. (2005) used methanolic extraction with formic acid acidification. In their case, the total polyphenol concentration ranged between 292.2 and 446.4 mg GAE/100 g FW, depending on the blackberry ssp., which is very similar to our results [27]. For extracts obtained with acetone extraction, our results and the literature data do not always agree; however, the results of Moyer et al. (2002) (average of blackberry cultivars in the state of Oregon) are pretty close to our values since, in their case, the total anthocyanin content was 164 ± 36 mg C3G/100 g FW, while the total polyphenol concentration was 577 ± 56.1 mg GAE/100 g FW [26].

Despite the anthocyanins belonging to the group of flavonoids, in this case, the detected flavonoid content was lower than that of anthocyanins, although this should be inverse. A similar trend was observed by Celant et al. (2016), indicating that the aluminum chloride colorimetric method cannot detect all flavonoids [23]. This theory is supported by Tabart et al. research. They found that the pH differential method is very specific for anthocyanins, but different anthocyanins respond very differently. In contrast, the aluminum chloride method seems appropriate only for flavanols. Some other flavonoids, such as flavanones, could also be detected, but with lower sensitivity. Anthocyanins were not detected [40].

The standard deviation indicates the reproducibility of the measurements, the average precision of the measured results, and the deviation from the mean. Studying the standard deviations (Figure 3), the highest value was observed for the acetonic extracts, especially for TPC, which could be mainly due to the high volatility of acetone and their poor water miscibility. This latter property may be important because of the high moisture content of the plant samples. Due to the limited immiscibility, the availability of the solute contained in the solvent is highly influenced by the mixing intensity of the suspension, through the variation of the diffusivity of the solvent through the moisture film, around the plant material particles.

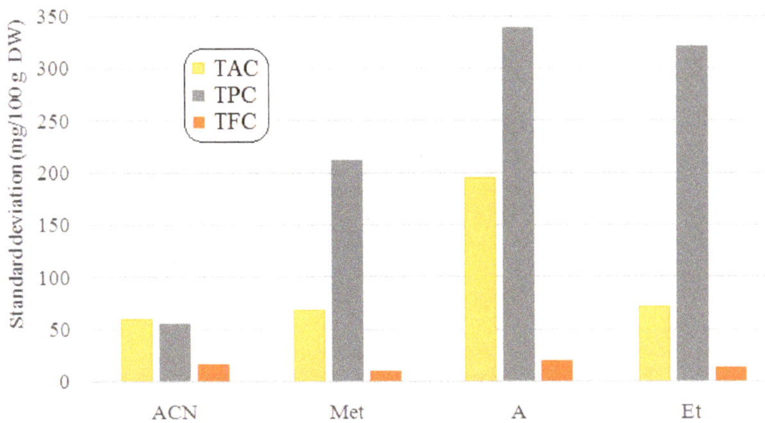

Figure 3. Comparison of standard deviations for different solvents and methods.

3.3. Antioxidant Capacity

In the determination of the DPPH radical binding activity, it was observed that the antioxidant capacity is also affected by the solvents, with the extract prepared with acetonitrile showing the highest antioxidant capacity by both FRAP and DPPH methods. The results are shown in Table 2. This could explain that components extracted with solvents of higher polarity are less effective at binding the DPPH radical than their lower polarity counterparts. However, changes in the polarity of solvents alter their ability to leach certain antioxidants, thus affecting the measurements [23]. This tendency is also evident in the present case, as acetonitrile has a relatively low polarity (0.460) and was found to be the best solvent. In contrast, ethanol has a higher polarity (0.654) and, therefore, inefficiency [41].

Table 2. The antioxidant capacity of blackberry extracts, obtained with different solvents, measured by three different methods.

Extraction Solvent	E% (DPPH Assay)	FRAP (mg AAE/100 g DW)	FRAP (mg AAE/100 g FW)
ACN	76.4078 ± 1.43	3547.68 ± 154.76	481.06 ± 20.98
Met	63.1294 ± 3.45	2455 ± 141.31	332.92 ± 19.16
A	71.9624 ± 3.04	2853.13 ± 413.98	386.88 ± 56.13
Et	47.2017 ± 2.47	1682.22 ± 357.37	228.11 ± 48.46

The obtained data suggest that the garden blackberries have excellent antioxidant potential due to their high polyphenol and anthocyanin content. The total antioxidant concentration values agree with Koczka et al.'s results, where the total antioxidant concentration of blackberries averaged 150 μmol ascorbic acid/100 g dry matter (2.64 g ascorbic acid/100 g dry matter) [42].

The standard deviations also follow the trend in the present case. Figure 4 shows that the highest standard deviation is for acetone, which is due to its higher volatility in an aqueous mixture, in comparison to the other used solvents.

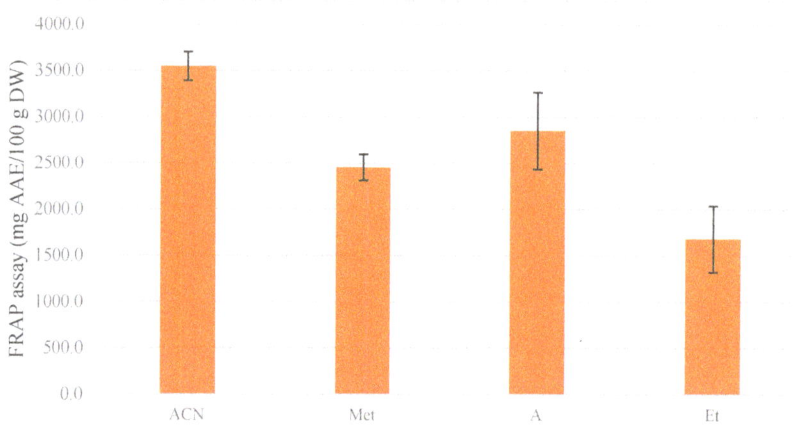

Figure 4. FRAP assay of blackberries extracts in mg AAE/g DW.

3.4. Statistical Correlations

Blackberries contain many phytochemicals, and to elucidate which components are responsible for the main antioxidant effect, we examined whether the data were normally distributed, and then, we applied the Pearson correlation between the antioxidant capacity and the total anthocyanins, flavonoids, and polyphenols. Table 3 shows that, in all cases, the distribution is normal, the values obtained for the Z-test fall within the 95% confidence interval (−1.96; +1.96), and both the Kolmogorov–Smirnov and Shapiro–Wilk tests also show that the distribution is normal at $p = 0.05$ significance level. Since the distribution is normal, the Pearson correlation could be applied.

Table 3. The normality testing results of the measurement error distribution.

N = 12		TAC	TPC	TFC	DPPH Assay	FRAP Assay
Skewness		−0.3	−0.682	0.211	−0.669	−0.158
Standard error		0.637	0.637	0.637	0.637	0.637
Z test $z = \frac{\text{Skewness}}{\text{Standard error}}$		−1.459	−1.07	0.331	−1.05	−0.248
Kurtosis		−1.049	−0.294	−1.176	−1.050	−0.769
Standard error		1.232	1.232	1.232	1.232	1.232
Z test $z = \frac{\text{Kurtosis}}{\text{Standard error}}$		−1.292	−0.238	−0.954	−0.852	−0.624
Kolmogorov-Smirnov	Statistics	0.105	0.162	0.123	0.177	0.172
	Significance	0.200	0.200	0.200	0.200	0.200
Shapiro-Wilk	Statistics	0.956	0.947	0.959	0.883	0.951
	Significance	0.850	0.592	0.769	0.096	0.648

In Table 4, the Pearson correlation analysis results are summarized. The result shows that there is a significant positive correlation between DPPH radical scavenging activity and total polyphenol content (r = 0.95; $p < 0.01$), a similar correlation has been reported in the literature [23,28,37,43,44]. Furthermore, it was shown that DPPH stable radicals are mainly able to be reduced by more reactive components than polyphenolic compounds. They are the main components responsible for the high antioxidant capacity of the extracts. This trend is also characteristic of the FRAP method (r = 0.836; $p < 0.01$), as pointed out by other authors such as Koczka et al., by the polyphenol's Fe^{3+} ion reducing ability [42].

Table 4. The results of Pearson linear correlation analysis.

		TAC	TPC	TFC	DPPH Assay	FRAP-Assay
DPPH radical scavenging activity	Pearson correlation	0.614 **	0.950 **	0.630 *	1	0.903 **
	Significance level (2-tailed)	0.004	0.000	0.028	-	0.000
	N	12	12	12	12	12
FRAP-assay	Pearson correlation	0.491	0.836 **	0.732 **	0.903 **	1
	Significance level (2-tailed)	0.065	0.001	0.007	0.000	-
	N	12	12	12	12	12

* The correlation is significant at 0.05 significance level (2-tailed). ** The correlation is significant at 0.01 significance level (2-tailed).

The correlation between total anthocyanins-total flavonoids, as well as between the DPPH–FRAP results are also significant. Still, the correlations are looser in comparison to the radical scavenging activity-total polyphenol content, indicating that anthocyanins and flavonoids are only moderately responsible for the antioxidant effect: a phenomenon that has been observed by other researchers [28,42].

Using linear regression in Microsoft Excel, the Figure 5 graphically illustrates the positive linear correlation between total polyphenol content and the DPPH scavenging activity of the extracts ($R^2 = 0.9026$).

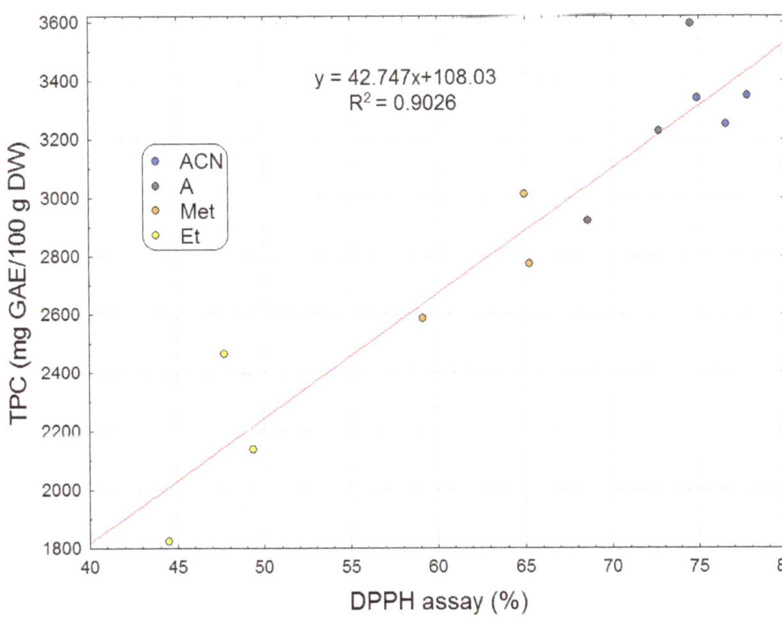

Figure 5. Relationship between radical-binding activity and total polyphenol content of the extracts obtained with different solvents.

3.5. Determination of Quercetin and Gallic Acid

The quercetin content was 0.465 ± 0.058 mg/100 g fresh weight, while the gallic acid content was 88.25 ± 0.478 mg/100 g fresh weight. The values for quercetin content from the literature are quite variable. Still, Ungureanu et al. (2020) obtained similar values [38], as in their case, the concentration of quercetin is 2.96 ± 0.05 mg/100 g dry weight, while Cho et al. (2005) obtained 0.63 ± 0.01 mg/100 g fresh weight [27]. This can be explained by the result obtained by Ungureanu et al. that higher precipitation and lower temperature led to an increase in the quercetin and gallic acid content of blackberries [38]. For gallic acid, our results were relatively close to the values obtained by Jacques et al. (2010): 113.13 mg/100 g fresh weight [45].

3.6. Theoretical Study of the Effect of Acetonitrile-Water Concentration on Anthocyanin Extraction Based on Hansen Solubility Parameters

Theoretical tools and models are often used to predict the solubility of bioactive molecules. These tools are phase solubility studies, quantitative structure-property relationships, calculation of Hansen solubility parameters by the group contribution method, the Flory–Huggins interaction parameter, the real solvent conductive screening model (COSMO-RS), and the COSMO segmental activity coefficient model. These models can be applied to the extraction process of anthocyanins to predict which solvent is most effective against which molecule. However, these models have varying degrees of complexity, and there is no literature available to evaluate which model would be preferable for this purpose. The Hansen solubility parameters are the most suitable for finding the most optimal organic solvent-water mixture ratio. Finding the most optimal ratio of organic solvent to water is important because it gives a higher dissolution rate. Using this mixture will shorten the analysis time, and the amount of anthocyanins extracted will be as high as possible, giving the closest value to the true antioxidant content. Due to the good solubility, the amount of solvent used can also be significantly reduced.

The difference in solvent efficiency could be explained by the terms of Hansen and Hildebrand solubility parameters of the solvents and solutes, as summarized in Table 5. Cyanidin and cyanidin-3-*O*-glucoside were chosen for solutes because they are the dominant anthocyanins in blackberry [46].

Table 5. Hansen and Hildebrand parameters of the main bioactive components of blackberry [47,48] and the used solvents [21].

Solvents and Solutes	**Hansen Parameters**			**Hildebrand Parameter**
	ä$_D$ (Dispersive) MPa$^{0.5}$	**ä$_P$ (Polar) MPa$^{0.5}$**	**ä$_H$ (H-Bond) MPa$^{0.5}$**	**ä MPa$^{0.5}$**
Acetone	15.50	10.40	7.00	19.93
Acetonitrile	15.30	18.00	6.10	24.39
Ethanol	15.80	8.80	19.40	26.52
Methanol	15.10	12.30	22.30	29.60
Water	15.60	16.00	42.30	47.80
Cyanidin	27.15	18.01	27.06	42.35
Cyanidin-3-*O*-glucoside	25.70	19.48	28.08	42.76

Both acetone and acetonitrile are polar solvents (Table 5), and their intermolecular hydrogen bond forming ability is low. Acetonitrile has the highest polar Hansen parameter among the used solvents and, therefore, is better able to extract more polar active compounds, some of which may have strong antioxidant activity, although, more precise correlations could be made only when the individual components in the extracts are known. In the case of aliphatic alcohols, strong hydrogen bonds are formed between the solvent molecules (polar protic solvents), making them suitable solvents, but in the case of alcohols, solubility decreases with increasing the carbon chain length, which explains why methanol is a better solvent than ethanol [49].

As the best results were obtained with 90% v/v acidified acetonitrile, the Hansen parameters analysis was performed for the acidified acetonitrile-water solvent mixture as a solvent and cyanidin-3-O-glucoside as a solute. The addition of acid shifts the equilibrium towards the more stable flavylium form and conforms to Figure 2, changing the δ_H value of cyanidin-3-O-glucoside.

Similar to malvidin, the difference between the Hansen solubility parameters of the flavylium and carbinol pseudobase forms of cyanidin is expected to be equal, as the structural changes between the tautomers don't affect the side structures that differentiate the two anthocyanidins [48]. Accordingly, the decrease in the HSPs during the carbinol-flavylium transformation, caused by acidification, could be approximately:

$\Delta\delta_P \approx 0$ MPa$^{0.5}$, $\Delta\delta_D = 1.85$ MPa$^{0.5}$, and $\Delta\delta_H = 6$ MPa$^{0.5}$, respectively.

The Hansen solubility parameters (HSP) of solvent mixtures differ from pure solvents. Still, they could be the volumetric fraction-weighted average of the components' HSPs, even for solvents in a supercritical state, as seen in Equation (9) [50].

$$\delta_{j,mix} = \Sigma^{n}{}_{i=1}\Phi_i\delta_{j,i} \quad (9)$$

where: $\delta_{j,mix}$ the HSPs of the solvent mixture; Φ_i the volumetric fraction of i-th component; n is the number of components of the mixture, and the j index denoting the different solubility parameters as follows: j = D, P, H$^-$ dispersion, polar and hydrogen bond partial HSPs.

Knowing the HSPs of water and acetonitrile, the solubility parameters of the mixture were calculated for the whole concentration range, as in Equation (10):

$$\delta_{j,mix} = \Phi_W\delta_W + \Phi_{AcN}\delta_{AcN} \quad (10)$$

In water-acetonitrile mixtures, the dispersion Hansen parameter varies minimally ($D\delta_{D,mix} \approx 0$ MPa$^{0.5}$), as changing the acetonitrile concentration would change the solubility of cyanidin-3-O-glucoside only through the other two members. The δ_P values of water and acetonitrile are also quite close ($\Delta\delta_{P,mix} < 2$ MPa$^{0.5}$), but the variation of δ_H had a wide variation range ($\Delta\delta_{H,mix} \leq 36.2$ MPa$^{0.5}$).

The value of the polar component of the flavylium cation ($\delta_P = 19.48$ MPa$^{0.5}$) doesn't change in comparison to the natural cyanidin-3-O-glucoside before the acidification, but this value is near to the values of the acetonitrile-water mixtures ($\delta_P \approx 16$–18 MPa$^{0.5}$). The value of the dispersive component of the flavylium cation ($\delta_D = 23.85$ MPa$^{0.5}$) is closer than of the original form ($\delta_D = 25.7$ MPa$^{0.5}$) to the values of the acetonitrile-water mixtures ($\delta_D \approx 15.4$ MPa$^{0.5}$), but the difference remains considerable. In this situation, the solvent optimization task is, mainly, to find the acetonitrile-water ratio with the δ_H value closest to the δ_H value of the flavylium ion ($\delta_H = 22.08$ MPa$^{0.5}$). As for the acetonitrile, the hydrogen-bond component had a low value ($\delta_H = 6.1$ MPa$^{0.5}$), and for water, it had a very high value ($\delta_H = 42.3$ MPa$^{0.5}$). Obviously, with the increase in the water amount, the optimum could be reached. In Figure 6, it could be observed that acidification decreases the cyanidin-3-O-glucoside point distance, substantially, to the acetonitrile-water curve. The hydrogen-bond component of the solvent mixture increases significantly with water content increase, but the polar component decreases slightly and has an opposite effect on distance. Therefore, the optimum concentration could differ from $c \approx 56\%$ v/v acetonitrile ($\delta_H \approx 22$ MPa$^{0.5}$) and should be calculated by an exact mathematical method. From Figure 6, we obtained data that, for the extraction of cyanidin-3-O-glucoside, predicted the optimal theoretical concentration of acetonitrile-water mixture will be $c_{opt} = 56.2\%$ v/v, which is in good agreement with the previous estimation.

For extractions at a higher temperature, such as Soxhlet and, especially, for pressurized liquid extraction (PLE), the Hansen parameters should be calculated, taking into account the temperature dependence of these parameters [50,51].

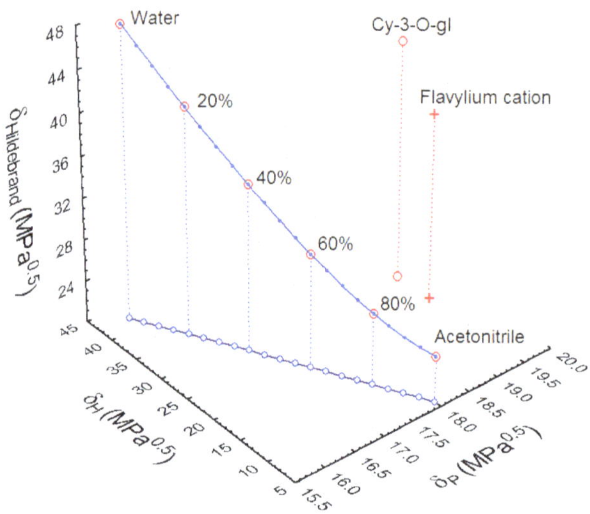

Figure 6. The Hansen space representation of solubility parameters of the acetonitrile-water mixture (solvent) and cyanidin-3-*O*-glucoside in natural and protonated form (solute).

4. Conclusions

The results obtained in our study show that the garden blackberry from Turia, Romania presented high antioxidant capacity. With our results, we would like to emphasize the nutritional role of blackberries. The obtained data suggest that the blackberries growing and harvested in our region are suitable for the production of premium functional foods. With modern and efficient extraction methods, pure blackberry anthocyanin extracts can be produced. Agricultural and food processing wastes from the blackberry industry are potential sources of anthocyanins. Blackberry has been used as a beneficial food and food ingredient, and it can contribute even more through modern extraction methods.

Our measurements showed that the best solvent is the acidified acetonitrile-water mixture. Still, it is known that it is only suitable for analytical tests and is not approved for use in the food industry. Based on theoretical calculations, 56.2% v/v acetonitrile would be the optimal solvent for the main anthocyanin. Acetonitrile is a widely used solvent in chromatography, so data on the solubility of different bioactive substances in this solvent can be very useful. The resulting extract can be used for quantitative chromatographic analysis without any other sample preparation

The Hansen parameters for different anthocyanidins show quite similar values, so our method can be further developed and extended to other solvent mixtures and other anthocyanins.

Author Contributions: Conceptualization, C.A., M.H. and C.D.A.; methodology, C.A.; software, A.C.; validation, C.A., G.G.C. and A.D.; formal analysis, C.A. and M.H.; investigation, C.A., M.H. and C.D.A.; resources, C.A., M.H. and C.D.A.; data curation, A.C.; writing—original draft preparation, C.A. and A.C.; writing—review and editing, C.A. and A.C.; visualization, C.A., G.G.C. and A.D.; supervision, G.G.C. and A.D. All authors have read and agreed to the published version of the manuscript.

Funding: This research received no external funding.

Institutional Review Board Statement: Not applicable.

Informed Consent Statement: Not applicable.

Data Availability Statement: The data presented in this study are available in this article.

Conflicts of Interest: The authors declare no conflict of interest.

References

1. Bushway, L.; Pritts, M.; Handley, D. Raspberry and Blackberry Production Guide for the Northeast, Midwest, and Eastern Canada (NRAES-35). Natural Resource, Agriculture, and Engineering Service (NRAES). 2008. Available online: https://hdl.handle.net/1813/66930 (accessed on 15 February 2022).
2. Yang, W.; Guo, Y.; Liu, M.; Chen, X.; Xiao, X.; Wang, S.; Gong, P.; Ma, Y.; Chen, F. Structure and function of blueberry anthocyanins: A review of recent advances. *J. Funct. Foods* **2022**, *88*, 104864. [CrossRef]
3. Zia-Ul-Haq, M.; Riaz, M.; de Feo, V.; Jaafar, H.Z.E.; Moga, M. *Rubus fruticosus* L.: Constituents, biological activities and health related uses. *Molecules* **2014**, *19*, 10998–11029. [CrossRef] [PubMed]
4. Nile, S.H.; Park, S.W. Edible berries: Bioactive components and their effect on human health. *Nutrition* **2014**, *30*, 134–144. [CrossRef] [PubMed]
5. Skrovankova, S.; Sumczynski, D.; Mlcek, J.; Jurikova, T.; Sochor, J. Bioactive compounds and antioxidant activity in different types of berries. *Int. J. Mol. Sci.* **2015**, *16*, 24673–24706. [CrossRef]
6. Cassidy, A.; Bertoia, M.; Chiuve, S.; Flint, A.; Forman, J.; Rimm, E.B. Habitual intake of anthocyanins and flavanones and risk of cardiovascular disease in men. *Am. J. Clin. Nutr.* **2016**, *104*, 587–594. [CrossRef]
7. Shahbazi, R.; Sharifzad, F.; Bagheri, R.; Alsadi, N.; Yasavoli-Sharahi, H.; Matar, C. Anti-Inflammatory and Immunomodulatory Properties of Fermented Plant Foods. *Nutrients* **2021**, *13*, 1516. [CrossRef]
8. Lorenzo, J.M.; Pateiro, M.; Domínguez, R.; Barba, F.J.; Putnik, P.; Kovačević, D.B.; Franco, D. Berries extracts as natural antioxidants in meat products: A review. *Food Res. Int.* **2018**, *106*, 1095–1104. [CrossRef]
9. Golovinskaia, O.; Wang, C.K. Review of functional and pharmacological activities of berries. *Molecules* **2021**, *26*, 3904. [CrossRef]
10. Rai, D.K.; Tzima, K. A Review on Chromatography–Mass Spectrometry Applications on Anthocyanin and Ellagitannin Metabolites of Blackberries and Raspberries. *Foods* **2021**, *10*, 2150. [CrossRef]
11. Ravichandran, K.S.; Krishnaswamy, K. Sustainable food processing of selected North American native berries to support agroforestry. *Crit. Rev. Food Sci. Nutr.* **2021**. [CrossRef]
12. Santhi, V.P.; Sriramavaratharajan, V.; Murugan, R.; Masilamani, P.; Gurav, S.S.; Sarasu, V.P.; Ayyanar, M. Edible fruit extracts and fruit juices as potential source of antiviral agents: A review. *J. Food Meas. Charact.* **2021**, *15*, 5181–5190. [CrossRef]
13. Takó, M.; Kerekes, E.B.; Zambrano, C.; Kotogán, A.; Papp, T.; Krisch, J.; Vágvölgyi, C. Plant phenolics and phenolic-enriched extracts as antimicrobial agents against food-contaminating microorganisms. *Antioxidants* **2020**, *9*, 165. [CrossRef] [PubMed]
14. Dormán, G.; Flachner, B.; Hajdú, I.; András, D.C. Target identification and polypharmacology of nutraceuticals. In *Nutraceuticals: Efficacy, Safety and Toxicity*; Gupta, R.C., Lall, R., Srivastava, A., Eds.; Elsevier: Amsterdam, The Netherlands, 2021; pp. 315–343.
15. Gil-Martín, E.; Forbes-Hernández, T.; Romero, A.; Cianciosi, D.; Giampieri, F.; Battino, M. Influence of the extraction method on the recovery of bioactive phenolic compounds from food industry by-products. *Food Chem.* **2022**, *378*, 131918. [CrossRef] [PubMed]
16. Hidalgo, G.I.; Almajano, M.P. Red fruits: Extraction of antioxidants, phenolic content, and radical scavenging determination: A review. *Antioxidants* **2017**, *6*, 7. [CrossRef]
17. Lefebvre, T.; Destandau, E.; Lesellier, E. Selective extraction of bioactive compounds from plants using recent extraction techniques: A review. *J. Chromatogr. A* **2021**, *1635*, 461770. [CrossRef]
18. Enaru, B.; Drețcanu, G.; Pop, T.D.; Stănilă, A.; Diaconeasa, Z. Anthocyanins: Factors affecting their stability and degradation. *Antioxidants* **2021**, *10*, 1967. [CrossRef]
19. Huang, Y.; Zhou, S.; Zhao, G.; Ye, F. Destabilisation and stabilisation of anthocyanins in purple-fleshed sweet potatoes: A review. *Trends Food Sci. Technol.* **2021**, *116*, 1141–1154. [CrossRef]
20. Ignat, I.; Volf, I.; Popa, V.I. A critical review of methods for characterisation of polyphenolic compounds in fruits and vegetables. *Food Chem.* **2011**, *126*, 1821–1845. [CrossRef]
21. Antony, A.; Farid, M. Effect of temperatures on polyphenols during extraction. *Appl. Sci.* **2022**, *12*, 2107. [CrossRef]
22. Hansen, C.M. *Hansen Solubility Parameters: A User's Handbook*, 2nd ed.; CRC Press: Boca Raton, FL, USA, 2007.
23. Celant, V.M.; Braga, G.C.; Vorpagel, J.A.; Salibe, A.B. Composição fenólica e atividade antioxidante dos extratos aquoso e etanólico de amora-preta. *Rev. Bras. De Frutic.* **2016**, *38*. [CrossRef]
24. Boeing, J.S.; Barizão, É.O.; e Silva, B.C.; Montanher, P.F.; de Cinque Almeida, V.; Visentainer, J.V. Evaluation of solvent effect on the extraction of phenolic compounds and antioxidant capacities from the berries: Application of principal component analysis. *Chem. Cent. J.* **2014**, *8*, 48. [CrossRef] [PubMed]
25. Beattie, J.; Crozier, A.; Duthie, G.G. Potential health benefits of berries. *Curr. Nutr. Food Sci.* **2005**, *1*, 71–86. [CrossRef]
26. Moyer, R.A.; Hummer, K.E.; Finn, C.E.; Frei, B.; Wrolstad, R.E. Anthocyanins, phenolics, and antioxidant capacity in diverse small fruits: *Vaccinium*, *Rubus*, and *Ribes*. *J. Agric. Food Chem.* **2002**, *50*, 519–525. [CrossRef] [PubMed]

27. Cho, M.J.; Howard, L.R.; Prior, R.L.; Clark, J.R. Flavonol glycosides and antioxidant capacity of various blackberry and blueberry genotypes determined by high-performance liquid chromatography/mass spectrometry. *J. Sci. Food Agric.* **2005**, *85*, 2149–2158. [CrossRef]
28. Sellappan, S.; Akoh, C.C.; Krewer, G. Phenolic compounds and antioxidant capacity of Georgia-grown blueberries and blackberries. *J. Agric. Food Chem.* **2002**, *50*, 2432–2438. [CrossRef]
29. Nour, V.; Trandafir, I.; Cosmulescu, S. Antioxidant capacity, phenolic compounds and minerals content of blackcurrant (*Ribes nigrum* L.) leaves as influenced by harvesting date and extraction method. *Ind. Crop. Prod.* **2014**, *53*, 133–139. [CrossRef]
30. Alam, M.N.; Bristi, N.J.; Rafiquzzaman, M. Review on in vivo and in vitro methods evaluation of antioxidant activity. *Saudi Pharm. J.* **2013**, *21*, 143–152. [CrossRef]
31. Singleton, V.L.; Orthofer, R.; Lamuela-Raventós, R.M. Analysis of total phenols and other oxidation substrates and antioxidants by means of Folin-Ciocalteu reagent. In *Methods in Enzymology*; Academic Press: Cambridge, MA, USA, 1999; p. 299.
32. Chandra, S.; Khan, S.; Avula, B.; Lata, H.; Yang, M.H.; Elsohly, M.A.; Khan, I.A. Assessment of total phenolic and flavonoid content, antioxidant properties, and yield of aeroponically and conventionally grown leafy vegetables and fruit crops: A comparative study. *Evid.-Based Complement. Altern. Med.* **2014**, *2014*, 253875. [CrossRef]
33. Lee, J.; Durst, R.W.; Wrolstad, R.E. Determination of total monomeric anthocyanin pigment content of fruit juices, beverages, natural colorants, and wines by the pH differential method: Collaborative study. *J. AOAC Int.* **2005**, *88*, 1269–1278. [CrossRef]
34. Aye, T.; Onoomar, T.; Malai, S. Simultaneous determination of gallic acid and quercetin in leaf extract of *Madhuca longifolia* from Myanmar by HPLC-DAD: Method development and validation. *Key Eng. Mater.* **2020**, *859*, 51–56.
35. Shirazi, O.U.; Khattak, M.M.A.K.; Shukri, N.A.M. Chromatographic evaluation of gallic acid, catechin and quercetin in methanolic extracts of selected formulations of spices and herbs. *Prog. Nutr.* **2019**, *21*, 236–251.
36. Wang, W.; Jung, J.; Tomasino, E.; Zhao, Y. Optimization of solvent and ultrasound-assisted extraction for different anthocyanin rich fruit and their effects on anthocyanin compositions. *LWT-Food Sci. Technol.* **2016**, *72*, 229–238. [CrossRef]
37. Croge, C.P.; Cuquel, F.L.; Pintro, P.T.M.; Biasi, L.A.; de Bona, C.M. Antioxidant capacity and polyphenolic compounds of blackberries produced in different climates. *HortScience* **2019**, *54*, 2209–2213. [CrossRef]
38. Ungureanu, C.R.M.; Lupitu, A.I.; Moisa, C.; Rivis, A.; Copolovici, L.O.; Poiana, M.A. Investigation on high-value bioactive compounds and antioxidant properties of blackberries and their fractions obtained by home-scale juice processing. *Sustainability* **2020**, *12*, 5681. [CrossRef]
39. Bae, I.Y.; An, J.S.; Oh, I.K.; Lee, H.G. Optimized preparation of anthocyanin-rich extract from black rice and its effects on in vitro digestibility. *Food Sci. Biotechnol.* **2017**, *26*, 1415–1422. [CrossRef] [PubMed]
40. Tabart, J.; Kevers, C.; Pincemail, J.; Defraigne, J.-O.; Dommes, J. Evaluation of spectrophotometric methods for antioxidant compound measurement in relation to total antioxidant capacity in beverages. *Food Chem.* **2010**, *120*, 607–614. [CrossRef]
41. Reichardt, C.; Welton, T. *Solvents and Solvent Effects in Organic Chemistry*, 4th ed.; Wiley-VCH: Weincheim, Germany, 2005.
42. Koczka, N.; Stefanovits-Bányai, É.; Prokaj, E. Element composition, total phenolics and antioxidant activity of wild and cultivated blackberry (*Rubus fruticosus* L.) fruits and leaves during the harvest time. *Not. Bot. Horti Agrobot. Cluj-Napoca* **2018**, *46*, 538. [CrossRef]
43. Koca, I.; Karadeniz, B. Antioxidant properties of blackberry and blueberry fruits grown in the Black Sea Region of Turkey. *Sci. Hortic.* **2009**, *121*, 447–450. [CrossRef]
44. Wang, S.Y.; Lin, H.S. Antioxidant activity in fruits and leaves of blackberry, raspberry, and strawberry varies with cultivar and developmental stage. *J. Agric. Food Chem.* **2000**, *48*, 140–146. [CrossRef]
45. Jacques, A.C.; Pertuzatti, P.B.; Barcia, M.T.; Zambiazi, R.C.; Chim, J.F. Estabilidade de compostos bioativos em polpa congelada de amora-preta (Rubus fruticosus) cv. Tupy. *Química Nova* **2010**, *33*, 1720–1725. [CrossRef]
46. Ponder, A.; Hallmann, E.; Kwolek, M.; Średnicka-Tober, D.; Kazimierczak, R. Genetic differentiation in anthocyanin content among berry fruits. *Curr. Issues Mol. Biol.* **2021**, *43*, 36–51. [CrossRef] [PubMed]
47. Phan, K.; Van Den Broeck, E.; Raes, K.; De Clerck, K.; Van Speybroeck, V.; De Meester, S. A comparative theoretical study on the solvent dependency of anthocyanin extraction profiles. *J. Mol. Liq.* **2022**, *351*, 118606. [CrossRef]
48. Srinivas, K.; King, J.W.; Monrad, J.K.; Howard, L.R.; Hansen, C.M. Optimization of subcritical fluid extraction of bioactive compounds using Hansen solubility parameters. *J. Food Sci.* **2009**, *74*, E342–E354. [CrossRef] [PubMed]
49. Kumoro, A.C.; Retnowati, D.S.; Budiyati, C.S. Solubility of delphinidin in water and various organic solvents between (298.15 and 343.15) K. *J. Chem. Eng. Data* **2010**, *55*, 2603–2606. [CrossRef]
50. András, C.D.; Mátyás, L.; Ráduly, B.; Salamon, R.V. Increasing the prediction efficiency of Hansen solubility parameters in supercritical fluids. *Period. Polytech. Chem. Eng.* **2019**, *63*, 286–293. [CrossRef]
51. Mihalovits, M.; Körösi, M.; Székely, E. New formula for the hydrogen-bonding Hansen component of methanol, ethanol, and n-propanol for non-ambient conditions—Application in gas antisolvent fractionation-based optical resolution. *ACS Omega* **2021**, *6*, 18964–18974. [CrossRef]

applied sciences

MDPI

Article

The Effect of Sea Salt with Low Sodium Content on Dough Rheological Properties and Bread Quality

Andreea Voinea, Silviu-Gabriel Stroe, Sorina Ropciuc, Adriana Dabija and Georgiana Gabriela Codină *

Faculty of Food Engineering, Stefan cel Mare University of Suceava, 720229 Suceava, Romania; andreea.musu@fia.usv.ro (A.V.); silvius@fia.usv.ro (S.-G.S.); sorina.ropciuc@fia.usv.ro (S.R.); adriana.dabija@fia.usv.ro (A.D.)

* Correspondence: codina@fia.usv.ro; Tel.: +40-745-460-727

Abstract: The aim of this study was to analyze the effects of the addition of sea salt with low sodium content (SS) in a refined wheat flour at the levels of 0.3%, 0.6%, 0.9% 1.2% and 1.5% on the rheological properties of the dough during mixing, extension, pasting and fermentation and the bread quality in terms of bread physical properties, crumb and crust color, texture and sensory characteristics. According to the data we obtained, the SS presented a strengthening effect on the dough network by increasing its stability, dough development time, energy and resistance. Moreover, the SS addition resulted in an increase in dough extensibility, to a delay of the gelatinization process and an increase of the falling number value. The bakery products obtained with the SS were of a higher quality compared to the control sample, presenting better physical and textural characteristics, a darker color and being more appreciated by consumers with the increased level of SS addition in the wheat flour. According to the sodium content from the bread recipe, the bread samples obtained may be classified as products with a very low sodium content of up to a 0.6% SS addition in the wheat flour or with a low sodium content if at least 0.9% SS is contained in the bread recipe.

Keywords: sea salt; rheological behavior; bread making; bread quality; sodium content

Citation: Voinea, A.; Stroe, S.-G.; Ropciuc, S.; Dabija, A.; Codină, G.G. The Effect of Sea Salt with Low Sodium Content on Dough Rheological Properties and Bread Quality. *Appl. Sci.* **2022**, *12*, 4344. https://doi.org/10.3390/app12094344

Academic Editor: Andrea Salvo

Received: 27 March 2022
Accepted: 23 April 2022
Published: 25 April 2022

1. Introduction

The World Health Organization (WHO) promotes and encourages the food industry to reduce the salt (sodium chloride, NaCl) content in foods to achieve a maximum salt intake of 5 g per day for adults [1]. Worldwide, one-third of all deaths are due to cardiovascular diseases such as heart attacks and strokes. High blood pressure, a major risk factor for these diseases, is directly related to high daily sodium intake. Many consumers are unaware that more than 70% of salt consumption comes from processed foods, which are a major component of many diets, which may explain why many consumers are not concerned about excessive salt (sodium) in their diets [2–4]. Cardiovascular diseases are the leading cause of death in the world, with Europe alone estimated to have more than four million deaths each year, accounting for almost half of all deaths [5]. However, it seems that no European country meets the WHO recommended level for salt consumption ($\leq$5 g/day); therefore, a salt reduction strategy of 16% is currently being applied over 4 years to reach a sodium chloride intake of less than 5 g per day by 2025 [6,7]. Although consumers are now becoming aware of the negative health effects of excess salt, they do not have much information about the link between salt and sodium, sodium intake in processed foods and recommended consumption. In developed countries, consumer awareness of proper nutrition and nutritionally healthy behavior increases with increasing education. Unfortunately, in middle- and low-income countries, this correlation does not apply, so regardless of the level of education they have access to, awareness of the negative effect of excessive sodium consumption on health is low [8]. Since salt (sodium chloride) is one of the four main ingredients that are used in bread making, and bread is one of the most consumed foods in the world nowadays, people are trying to find solutions to reduce its salt content [9,10]. Studies have shown that the addition of sodium chloride has an important effect on the development of the structure of gluten, on the fermentation process of the dough and the level of water activity in bakery products [11]. The addition of salt increases dough strength, presumably by affecting the distribution of charges on the protein. Salt anions in a dough system reduce the repulsion between the protein chains, favoring their interaction due to the fact that they can bind with the positive charges on the gluten proteins [12]. It affects dough rheological properties because it generates an additional conductivity due to the ion migration. According to Fanari et al. [13], salt addition has increased compliance values of the dough samples by having a negative effect on dough deformability and, as a consequence, it had an increased effect on dough strength. However, it seems that this effect is only when salt addition is in a concentration higher than 0.5%. Despite this, when salt addition in wheat flour was higher than 1.5%, it reduced dough elasticity and machinability stiffening of it in a significant way [12]. Still, it seems that salt's effect on dough mixing depends on the flour variety. For example, some studies reported a decrease of the storage (G′) and loss modulus (G″) [14], while others reported an increase of their values when salt was added in wheat flour [15]. Salt also has a controlling effect on fermentation, affecting the color of the crumbs and crust. Bread made without salt has a bland and unacceptable taste. However, it is very important not to use too much salt, as it masks the flavors created by fermentation and the addition of other ingredients. In addition to its contribution to taste, sodium plays an important role in bread acceptability by consumers [16]. In the absence of the addition of sodium chloride in a bread recipe, the activity of yeast is stimulated. This leads to a significant increase in the height of the dough, which cannot hold the gas that is formed very well, leading to products with low volume and porosity [17]. Moreover, the shelf life of the bread decreased, the crust became light-colored, the flavor was reduced and the taste became more of yeast. Although there are different methods reported for sodium chloride replacement or reduction in bread recipes, there are some difficulties, especially due to the salty taste produced by it [3,5,18]. The most used methods to reduce sodium chloride in bread making are by its substitution with other salts such as potassium chloride, magnesium salts and calcium salts, which are effective due to the fact that these salts may have a similar effect as sodium chloride on dough rheology and therefore on bread quality [3,6,19]. The aim of this study was to replace the sodium chloride with sea salt with low sodium content (SS), which is a combination between different chloride salts, namely sodium, potassium and magnesium, and to evaluate its impact on dough rheology and bread quality. This type of sodium chloride substitution may be useful in order to reduce the sodium content from bread in order to comply with World Health Organization recommendations and consumer demands. A few different studies related to the effect of sea salt on dough rheology and bread making have been previously made. However, to our knowledge, none of the previously reported studies were similar to our study. This is due to a multitude of factors. First of all, a wide variety of sea salt may exist with several chloride salt contents and sometimes different compounds that may contribute to the bread flavor [20]. Moreover, different combinations between sea salt and different ingredients that may improve bread sensory characteristics have also been studied [21,22], but their combined effects on dough rheology and bread quality were very different than those obtained with a single type of ingredient addition to wheat flour. Moreover, the studies previously made had also reported the effect of sea salt on dough behavior or bread quality in a limited way. For example, Arena et al. [23] reported the effect of low-sodium sea salt containing approximately 65% NaCl, 30% KCl, 1% $MgSO_4$ and 0.5% $CaSO_4$ only on durum wheat bread quality. Miller and Jeong [2] reported the effect of different commercial sea salts with high (48% sodium), medium (21.5% sodium) and low sodium contents (17.8% sodium) on dough rheology and bread quality (only consumer acceptance test). Simsek and Martinez [9] analyzed the effect of sea salt, of which the composition was not mentioned, on dough mixing, extension and bread quality (with no sensory analysis). Our study completes the previous studies made in a more complex manner, analyzing the effect of a sea salt with a low sodium content as sodium chloride (max. 7%) and other types of chloride salts ($MgCl_2$ and KCl) on dough rheological behavior during all bread-making processes (mixing, extension, pasting and fermentation) and bread quality (physical, textural, color parameters and sensory characteristics). It can offer a solution for bread producers to obtain bakery products of a good quality with a reduced sodium content that may have an important effect on daily sodium intake.

2. Materials and Methods

2.1. Materials

Refined wheat flour (harvest 2019) of 650 type provided by S.C. Mopan S.A. (Suceava, Romania) was used. The sea salt with low sodium content (SS) obtained from the Dead Sea was used. The sea salt presented the following composition: sodium as sodium chloride (max. 7%), magnesium chloride (31 ÷ 35%), potassium chloride (21 ÷ 27%) and water insoluble max. 0.1%. The wheat flour used in this study was analyzed for the following characteristics according to the Romanian and international standard methods: ash content according to ICC 104/1, moisture content according to ICC 110/1, protein content according to ICC 105/2, falling number according to ICC 107/1, wet gluten according to ICC106/1 and gluten deformation index according to SR 90:2007 [24,25]. The wheat flour data obtained were: 0.65 g/100 g ash content, 14.0 g/100 g moisture content, 12.67 g/100 g protein content, 442 s falling number value, 30 g/100 g wet gluten content and 6 mm gluten deformation value.

2.2. Dough Rheological Properties during Mixing and Extension

In order to analyze dough rheological properties during mixing and extension, a farinograph, 300 g capacity (Brabender, Duigsburg, Germany), and an extensograph (Brabender, Duigsburg, Germany) were used according to ICC methods 115/1 and 114/1, respectively [24]. The mixing values analyzed were water absorption (WA), dough stability (ST), dough development time (DDT) and degree of softening after 10 min (DS). The extension values analyzed were resistance to extension (R50), extensibility (Ext), maximum resistance to extension (Rmax), extensograph ratio number (R/E) at a proving time of 135 min and energy.

2.3. Dough Viscometric Rheological Properties

Dough viscometric rheological properties were analyzed using the amylograph (Brabender OGH, Duigsburg, Germany) and the Falling Number (Perten Instruments AB, Hägersten, Sweden) devices according to ICC methods 126/1 and 107/1, respectively [24]. The following parameters were determined: gelatinization temperature (T_g), temperature at peak viscosity (T_{max}), peak viscosity (PV_{max}) and falling number value.

2.4. Dough Rheological Properties during Fermentation

Dough rheological properties during fermentation were determined by using the rheofermentometer (Chopin Rheo, type F3, Villeneuve-La-Garenne Cedex, France) device according to the American Association of Cereal Chemists (AACC) method 89-01.01 [26]. The following parameters were determined: maximum level of gas production (H'm), volume of gas retained in the dough at the end of the test (VR), total volume of CO_2 production (VT) and retention coefficient (CR).

2.5. Bread Making

Bread samples were made by using as raw materials refined wheat flour of 650 type, sodium sea salt with low sodium content, deionized water according to the water absorption value and compressed yeast of Saccharomyces cerevisiae type in a level of 3% reported to the wheat flour used. The sea salt with low sodium content (SS) was added to the bread recipe in the following levels: 0.0% (control sample), 0.3% (SS_0.3), 0.6% (SS_0.6), 0.9% (SS_0.9), 1.2% (SS_1.2) and 1.5% (SS_1.5). All ingredients were mixed in a laboratory mixer (Kitchen Aid, ARTISAN 4.8 L Tilt-Head Stand Mixer 5KSM175PS Whirlpool Corporation,

Tulsa, OK, USA) at a low speed for 15 min until the dough presented elastic properties, began to detach from the wall of the mixing vat, the moisture from its surface disappeared and the surface of the dough became smooth and dry. After the dough was mixed, it was then divided, shaped and fermented in a leavening chamber (PL2008, Piron, Cadoneghe, Padova, Italy) for 40 min at 30 °C and 85% relative humidity. The dough was baked in an oven (Caboto PF8004D, Cadoneghe, Padova, Italy) for 50 min at 180 °C.

2.6. Bread Samples Analysis

2.6.1. Physical, Color, Texture Profile Analysis, Sensory Evaluation of Bread Samples

The bread's physical characteristics such as specific volume, elasticity and porosity were determined according to SR 91: 2007 methods [27]. Color characteristics L^*, a^* and b^* of crumb and crust of the bread were determined using a CR-700 colorimeter (Konica Minolta, Tokyo, Japan). Textural parameters such as gumminess, firmness, chewiness, elasticity and cohesiveness were determined using the TVT-6700 texture analyzer (Perten Instruments, Hägersten, Sweden). Sensory evaluation of the bread samples was performed using a nine-point hedonic scale by a panel of thirty semi-trained judges. The acceptance test was carried out with students from Stefan cel Mare University, who were screening based on their sensory acuity according to ISO 3972, ISO8586-1 and ISO8586-2 of the sensory laboratory of the Faculty of Food Engineering in an individual panel [28–30]. Bread samples were served at a temperature of approximately 20 °C (room temperature) in black dishes with dimensions of 10 cm × 3 cm. Every sample was labelled with random 3-digit codes. The bread appearance was evaluated for bread samples cut in half. The bread's sensory characteristics that were evaluated were appearance, color, flavor, texture, taste, smell and overall acceptability using a 9-point hedonic scale with the following meaning: 1 represents dislike extremely, 2 represents dislike very much, 3 represents dislike, 4 represents dislike slightly, 5 represent neither like nor dislike, 6 represents like slightly, 7 represents like, 8 represents like very much and 9 represents like extremely. Evaluators cleaned their mouths with water between every bread sample analysis, and no information was given to the evaluators about bread samples.

2.6.2. Sodium Analysis of Bread Samples

Chemical composition analysis of bread samples was carried out by atomic absorption spectroscopy (AAS) method, according to SR EN 14082:2003: 10 g ± 10 mg of sample was weighed in the crucible, the samples were calcined starting with the temperature of 100 °C, then the temperature was raised by 50 °C/hour up to 450 °C. The samples were kept overnight at 450 °C. Ash digestion was performed with concentrated HCl (6 mol/L). Evaporation of the acid was performed on the electric hob, and the residue was dissolved with 10 mL of HNO_3 (0.1 mol/L). The samples were diluted with deionized water. The absorbance of samples was measured by atomic absorption spectroscopy (AAS)-AA-6300-Shimadzu (Shimadzu Corporation, Kyoto, Japan) in order to determine the sodium concentrations. In order to achieve the calibration curve, solutions with 0, 1, 2, 3, 4, 5 and 6 mg Na/L stock solution were used. The atomization of the elements present in the samples was performed by a C_2H_2 flame; subsequently, the atoms interacting with the specific light of the element were provided by a hollow cathode lamp (Na, 589 nm). The analysis method of samples was configured to measure each sample five times, then to average three of them.

2.7. Statistical Analysis

Statistical software XLSTAT (free trial version, Addinsoft, NY, USA) was used for data processing. Data were expressed in triplicate as mean ± standard deviation. In order to evaluate the differences between means of the data, the one-way ANOVA was applied using Tukey's test at a 5% significance level. At $p < 0.05$, p values were considered statistically significant differences.

3. Results

3.1. Dough Rheological Properties during Mixing and Extension

The dough rheological mixing properties were shown in Table 1. As can be seen, the addition of the SS gradually decreased the water absorption and the degree of softening after 10 min by up to 6.9% and 26.3% for the sample with the highest level of SS addition in wheat flour compared to the control one. It can be seen that the development time and the dough stability increased with the increased level of SS addition in the wheat flour. For the samples with 1.5% SS addition, the ST values have been doubled, whereas for the DT this value has significantly increased ($p < 0.05$) up to 42.1%.

Table 1. Farinograph parameters of the dough samples with different levels of sea salt with low sodium content (SS) additions.

Dough Samples	WA (%)	DT (min)	ST (min)	DS (UB)
Control	60.5 ± 0.2 ^{a}	1.9 ± 0.1 ^{bc}	2.00 ± 0.1 ^{e}	76 ± 3.2 ^{a}
SS_0.3	59.5 ± 0.1 ^{b}	2.0 ± 0.05 ^{c}	2.9 ± 0.1 ^{e}	68 ± 0.5 ^{b}
SS_0.6	59.0 ± 0.1 ^{bc}	2.2 ± 0.05 ^{abc}	3.2 ± 0.1 ^{cd}	66 ± 1.5 ^{b}
SS_0.9	58.4 ± 0.2 ^{c}	2.4 ± 0.1 ^{ab}	3.5 ± 0.05 ^{bc}	62 ± 2.5 ^{bc}
SS_1.2	57.2 ± 0.1 ^{d}	2.5 ± 0.2 ^{a}	3.7 ± 0.05 ^{b}	60 ± 0.5 ^{cd}
SS_1.5	56.3 ± 0.2 ^{e}	2.7 ± 0.05 ^{a}	4.2 ± 0.1 ^{a}	56 ± 1.5 ^{d}

WA: water absorption; DT: development time; ST: dough stability; DS: degree of softening after 10 min. The results are the mean ± standard deviation (n = 3). Dough samples contain sea salt with low sodium content (SS). $^{a–e}$: mean values in the same column followed by different letters are significantly different ($p < 0.05$).

All dough rheological properties during extension increased with the increased level of the SS addition in the wheat flour, as can be seen from Table 2. Therefore, the highest values for dough rheological properties during extension were recorded for the sample with a 1.5% SS addition in the wheat flour of which all rheological values significantly increased ($p < 0.05$) up to 24.77% for R_{50}, up to 28.69% for Ext, up to 27.87% for Rmax and up to 66.66% for E compared to the control sample.

Table 2. Extensograph parameters at the proving time of 135 min of the dough samples with different levels of sea salt with low sodium content (SS) additions.

Dough Samples	R_{50} (BU)	Ext (mm)	R_{max} (BU)	E (cm^2)
Control	327 ± 2.3 ^{f}	115 ± 1.3 ^{e}	226 ± 2.1 ^{e}	57 ± 1.0 ^{f}
SS_0.3	346 ± 1.5 ^{e}	122 ± 2.5 ^{d}	231 ± 1.5 ^{e}	66 ± 1.0 ^{e}
SS_0.6	352 ± 2.1 ^{d}	128 ± 1.0 ^{cd}	245 ± 1.5 ^{d}	75 ± 1.5 ^{d}
SS_0.9	374 ± 1.5 ^{c}	132 ± 1.4 ^{bc}	254 ± 2.1 ^{c}	81 ± 1.0 ^{c}
SS_1.2	397 ± 1.5 ^{b}	137 ± 1.0 ^{b}	277 ± 1.0 ^{b}	88 ± 1.0 ^{b}
SS_1.5	408 ± 0.7 ^{a}	148 ± 1.0 ^{a}	289 ± 1.0 ^{a}	95 ± 0.7 ^{a}

R_{50}: resistance to extension up to 50 mm; Ext: extensibility; R_{max}: maximum resistance; E: energy. The results are the mean ± standard deviation (n = 3). Dough samples contain sea salt with low sodium content (SS). $^{a–f}$: mean values in the same column followed by different letters are significantly different ($p < 0.05$).

3.2. Dough Viscometric Rheological Properties

The effects of the addition of SS in the wheat flour on the rheological properties of viscosity are presented in Table 3. A significant increase ($p < 0.05$) of the gelatinization temperature (T_g) up to 2.4% and of the maximum viscosity (PV_{max}) up to 12% was observed with the increased level of the SS addition in the wheat flour. Moreover, a significant increase ($p < 0.05$) of the temperature at maximum viscosity (T_{max}) up to 1.4% was observed for the sample with an addition level of 1.5% SS in the wheat flour compared to the control one. The falling number value also increased by 10.24% for the sample with the highest level of SS addition in the wheat flour compared with the control one.

Table 3. Viscometric rheological properties of the dough samples with different levels of sea salt with low sodium content (SS) additions.

Dough Samples	Tg (°C)	PV_{max} (BU)	T_{max} (°C)	FN (s)
Control	62.3 ± 0.12 ^{b}	1081 ± 4.7 ^{f}	88.4 ± 0.18 ^{c}	322 ± 2.0 ^{a}
SS_0.3	63.3 ± 0.15 ^{a}	1092 ± 2.5 ^{e}	88.7 ± 0.05 ^{bc}	328 ± 1.5 ^{b}
SS_0.6	63.4 ± 0.05 ^{a}	1107 ± 3.5 ^{d}	89.0 ± 0.15 ^{abc}	342 ± 1.0 ^{c}
SS_0.9	63.4 ± 0.15 ^{a}	1153 ± 1.5 ^{c}	89.2 ± 0.15 ^{abc}	349 ± 1.0 ^{d}
SS_1.2	63.6 ± 0.12 ^{a}	1198 ± 2.0 ^{a}	89.4 ± 0.20 ^{ab}	352 ± 1.5 ^{de}
SS_1.5	63.8 ± 0.15 ^{a}	1211 ± 2.0 ^{b}	89.7 ± 0.05 ^{a}	355 ± 1.0 ^{e}

T_g: gelatinization temperature; PV_{max}: peak viscosity; T_{max}: temperature at peak viscosity; FN: falling number value. The results are the mean ± standard deviation (n = 3). Dough samples contain sea salt with low sodium content (SS). $^{a–f}$: mean values in the same column followed by different letters are significantly different ($p < 0.05$).

3.3. Dough Rheological Properties during Fermentation

Dough rheological properties during fermentation are shown in Table 4. As can be seen, the values obtained on the rheofermentometer device showed that the maximum height of gas production (H'm), the total volume of CO_2 produced during fermentation (VT) and the volume of gas retained in the dough at the end of the test (VR) increased significantly ($p < 0.05$) with the increased level of SS addition in the wheat flour up to 0.6%, whereas at high levels they decreased. The retention coefficient decreased with the increased level of SS addition in the wheat flour.

Table 4. Rheological properties during fermentation of the dough samples with different levels of sea salt with low sodium content (SS) additions.

Dough Samples	H'm (mm)	VT (mL)	VR (mL)	CR (%)
Control	67.3 ± 1.8 ^{e}	1287 ± 8.3 ^{e}	1066 ± 7.3 ^{e}	82.8 ± 1.3 ^{a}
SS_0.3	82.7 ± 1.1 ^{c}	1551 ± 6.8 ^{c}	1272 ± 6.1 ^{a}	82.0 ± 1.4 ^{a}
SS_0.6	89.3 ± 1.2 ^{a}	1694 ± 4.1 ^{a}	1290 ± 4.2 ^{b}	76.1 ± 1.2 ^{c}
SS_0.9	85.1 ± 1.2 ^{b}	1631 ± 5.1 ^{b}	1238 ± 3.4 ^{c}	75.9 ± 1.1 ^{c}
SS_1.2	78.6 ± 1.5 ^{d}	1557 ± 5.5 ^{c}	1169 ± 3.5 ^{d}	75.0 ± 1.1 ^{b}
SS_1.5	78.0 ± 1.5 ^{d}	1506 ± 4.5 ^{d}	1043 ± 3.4 ^{e}	69.2 ± 1.4 ^{c}

H'm: maximum height of gaseous production; VT: total volume of CO_2 produced during fermentation; VR: volume of the gas retained in the dough at the end of the test; CR: retention coefficient. The results are the mean ± standard deviation (n = 3). Dough samples contain sea salt with low sodium content (SS). $^{a–e}$: mean values in the same column followed by different letters are significantly different ($p < 0.05$).

3.4. Physical Characteristics of the Bread Samples

As can be seen from Table 5, significant differences ($p < 0.05$) were obtained between the control sample and bread samples with different levels of SS addition in the wheat flour. Moreover, the physical characteristics for the bread sample with a 1.5% sodium chloride addition (CS_1.5) had been determined. The highest physical characteristics were obtained for bread samples with a 0.6% SS addition in the wheat flour. However, when high levels of SS were added in the bread recipe, the bread physical characteristics began to decrease. This decrease is in agreement with the results reported by Arena et al. [23], who concluded that salts inhibit the activity of yeast, which leads to a decrease in CO_2 production and therefore to the bread's physical characteristics. The bread sample with a 1.5% sodium chloride addition in the wheat flour presented a significant ($p < 0.05$) lower specific volume compared to the bread samples with a sea salt addition in the bread recipe. However, this physical characteristic was higher than those obtained for the control sample with no salt addition in the wheat flour. The elasticity values were very close to those of the control sample, whereas the porosity did not present significant ($p < 0.05$) differences to the values obtained for the samples with a 0.9 and 1.2% sea salt addition in the wheat flour.

Table 5. Physical characteristics of the bread samples with different levels of sea salt with low sodium content (SS) additions.

Bread Samples	Specific Volume (cm^3/100 g)	Porosity (%)	Elasticity (%)
Control	325.25 ± 0.42 ^{a}	63.53 ± 0.31 ^{a}	95.08 ± 0.32 ^{ab}
SS_0.3	347.54 ± 0.86 ^{d}	69.30 ± 0.27 ^{d}	96.33 ± 0.23 ^{d}
SS_0.6	349.12 ± 0.12 ^{e}	72.67 ± 0.31 ^{e}	97.66 ± 0.21 ^{e}
SS_0.9	340.43 ± 0.64 ^{c}	68.09 ± 0.42 ^{c}	96.18 ± 0.32 ^{d}
SS_1.2	339.52 ± 1.03 ^{c}	67.99 ± 0.22 ^{c}	95.78 ± 0.42 ^{c}
SS_1.5	347.54 ± 0.86 ^{d}	66.97 ± 0.18 ^{b}	95.46 ± 0.23 ^{bc}
CS_1.5	332.97 ± 1.20 ^{b}	67.88 ± 0.12 ^{c}	95.00 ± 0.15 ^{a}

The results are the mean ± standard deviation (n = 3). Bread samples contain sea salt with low sodium content, SS and 1.5% sodium chloride (CS_1.5). $^{a–e}$: mean values in the same column followed by different letters are significantly different ($p < 0.05$).

From the point of view of the color of the crumbs, it can be seen from the Table 6 that the samples with SS addition in the wheat flour have lower brightness (L^* value), yellow (b^* value) and a higher tint of red (a^* value) compared to the control sample. From the point of view of the color of the bread crust, it can be seen that the samples with the addition of SS have the same color parameters tendency as in the case of the breadcrumbs with the increased level of SS addition in the wheat flour. According to the color data obtained, it can be concluded that the bread samples become darker by adding SS in the bread recipe compared to the control sample.

Table 6. Color parameters of the bread samples with different levels of sea salt with low sodium content (SS) additions.

Bread Samples	Crust Color			Crumb Color		
	L^*	a^*	b^*	L^*	a^*	b^*
Control	69.48 ± 0.32 ^{a}	−4.40 ± 0.22 ^{ab}	19.58 ± 0.27 ^{a}	74.94 ± 0.41 ^{a}	2.97 ± 0.11 ^{c}	36.81 ± 0.27 ^{b}
SS_0.3	68.34 ± 0.51 ^{d}	−4.32 ± 0.14 ^{b}	19.32 ± 0.32 ^{b}	73.65 ± 0.71 ^{b}	3.34 ± 0.21 ^{b}	25.55 ± 0.42 ^{a}
SS_0.6	66.05 ± 0.24 ^{c}	−4.21 ± 0.22 ^{b}	18.94 ± 0.41 ^{c}	72.78 ± 0.37 ^{c}	3.52 ± 0.14 ^{f}	24.83 ± 0.31 ^{b}
SS_0.9	65.65 ± 0.72 ^{b}	−3.99 ± 0.12 ^{ab}	18.36 ± 0.27 ^{d}	71.09 ± 0.18 ^{d}	3.78 ± 0.31 ^{d}	22.99 ± 0.34 ^{c}
SS_1.2	64.34 ± 0.58 ^{e}	−3.97 ± 0.31 ^{a}	18.32 ± 0.42 ^{d}	69.49 ± 0.71 ^{e}	3.97 ± 0.22 ^{a}	22.47 ± 0.22 ^{d}
SS_1.5	63.55 ± 0.92 ^{f}	−3.92 ± 0.42 ^{a}	17.67 ± 0.11 ^{e}	68.03 ± 0.31 ^{f}	4.22 ± 0.11 ^{e}	22.11 ± 0.22 ^{d}

The results are the mean ± standard deviation (n = 10). Bread samples contain sea salt with low sodium content (SS). $^{a–f}$: mean values in the same column followed by different letters are significantly different ($p < 0.05$).

The firmness and gumminess textural parameters of the bread samples decreased up to a level of 0.9% SS addition in the wheat flour, after which these values increased as it may be seen from the Table 7. The cohesiveness and resilience parameters decreased with the increased level of SS addition in the wheat flour up to 35.93% and 35.07%, respectively, for the bread sample with the highest level of SS addition in the wheat flour compared to the control one.

As can be seen from Figure 1, from a sensory point of view, in general the sample with the addition of 1.5% SS in the wheat flour was the most appreciated. It received the maximum score for overall acceptability, flavor, taste and smell. However, for color and texture characteristics, the sample with 1.2% SS addition in the wheat flour was the most appreciated, whereas for appearance the sample with 0.6% SS in the bread recipe received the maximum score value.

Table 7. Texture parameters of the bread samples with different levels of sea salt with low sodium content (SS) additions.

Bread Samples	Firmness (N)	Gumminess (N)	Cohesiveness (Adimensional)	Resilience (Adimensional)
Control	23.83 ± 0.16 [a]	15.29 ± 0.11 [a]	0.64 ± 0.01 [ab]	1.34 ± 0.02 [b]
SS_0.3	18.33 ± 0.32 [b]	11.08 ± 0.21 [b]	0.65 ± 0.01 [ab]	1.32 ± 0.01 [b]
SS_0.6	18.24 ± 0.22 [b]	10.29 ± 0.32 [c]	0.56 ± 0.01 [b]	1.14 ± 0.02 [c]
SS_0.9	14.31 ± 0.56 [c]	9.21 ± 0.37 [d]	0.52 ± 0.01 [a]	1.07 ± 0.03 [a]
SS_1.2	21.08 ± 0.22 [a]	14.31 ± 0.16 [c]	0.49 ± 0.02 [c]	0.99 ± 0.02 [d]
SS_1.5	21.47 ± 0.11 [a]	14.90 ± 0.21 [a]	0.41 ± 0.01 [d]	0.87 ± 0.02 [e]

The results are the mean ± standard deviation ($n = 3$). Bread samples contain sea salt with low sodium content (SS). [a–e] mean values in the same column followed by different letters are significantly different ($p < 0.05$).

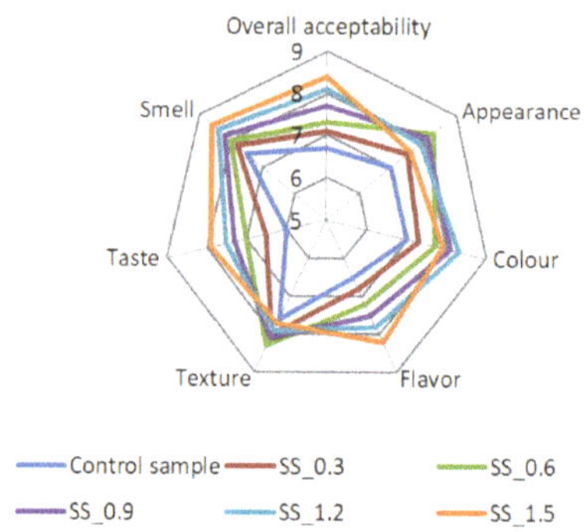

Figure 1. Sensory characteristics of the bread samples.

3.5. Sodium Content of the Bread Samples

The sodium content of the bread samples with sea salt addition in the wheat flour is shown in Table 8. As can be seen, the sodium value increased with the increased level of SS addition in the wheat flour. This increase is a significant one ($p < 0.05$) of ten times more for the sample with a 1.5% addition in the wheat flour compared to the control one.

Table 8. Sodium content of the bread samples with different levels of sea salt with low sodium content (SS) additions.

Bread Samples	Na (mg/100 g)
Control	8.14 ± 0.2 [f]
SS_0.3	23.22 ± 0.1 [e]
SS_0.6	38.35 ± 0.1 [d]
SS_0.9	53.75 ± 0.07 [c]
SS_1.2	67.25 ± 0.1 [b]
SS_1.5	82.5 ± 0.1 [a]

The results are the mean ± standard deviation ($n = 3$). Bread samples contain sea salt with low sodium content (SS). [a–f] mean values in the same column followed by different letters are significantly different ($p < 0.05$).

4. Discussion

Dough rheological properties with different levels of SS addition in the wheat flour provide information about dough behavior during mixing, extension, pasting and fermentation underlining the technological impact of SS on bread making. The SS addition decreased the water absorption (WA) value in agreement with the results reported by others [21,22]. This decrease is due to sea salt's ionic nature, which favors a higher association between gluten proteins as a consequence of their hydrophobic interactions, which increased by SS addition in the wheat flour. This leads to a decrease of the water-uptake ability of the wheat flour and therefore to the farinograph water absorption value. The SS addition in the wheat flour has a strength effect on wheat flour dough by increasing its stability (ST) and dough development time (DT) and by decreasing the degree of softening after 10 min (DS). This effect is due to a more compact gluten in the SS presence that may be attributed to the higher aggregation between gluten proteins, which presents almost 40% hydrophobic amino acids in their structure [31]. Our results are in agreement with other studies that have been reported by different research [31–33]. All the extensograph values increased with the increased level of SS addition in the wheat flour. These data also confirm the strengthening effect that SS has on wheat dough. SS increases dough resistance and energy, meaning that higher forces are necessary to stretch it. However, SS addition also increases dough extensibility. This may be due to the fact that SS ions affect the protein hydration from the dough system, competing for water with them. The fact that SS cannot hold water during all dough preparation periods may lead to an increase of free water amount from the dough system, causing an increase of its extensibility during proofing time. Similar data have also been reported by previous studies when different types of salts were added in the dough recipe [19,21,22,34,35]. During pasting, SS increases all the amylograph values and the falling number index. According to Nogueira et al. [36], the increase of the dough peak viscosity may be due to the suppression of the enzymatic activity in the SS presence. Another explanation is that SS may affect starch granule behavior, which remains intact before being broken down for a longer period of time. Moreover, the gelatinization temperature and temperature at peak viscosity increase with the increased level of SS addition in the wheat flour, indicating a delay of starch gelatinization and of dough maximum viscosity. The increase of dough viscosity is directly connected with falling number values, which are proportionally increasing with the increase of dough viscosity [37]. According to our data, SS addition in the wheat flour leads to an increase of dough viscosity and consequently to an increase of falling number values, these results being in agreement with those reported in other studies [21,22,38]. During fermentation, the maximum height of gaseous production (H'm), total carbon dioxide production (VT) and volume of the gas retained at the end of the test (VR) increased up to a level of 0.6% SS addition in the wheat flour, after which these values decreased. These behaviors are due to the yeast activity, which ferments the carbohydrates present in the dough system and also affects the dough's capability to retain the gas formed during the fermentation process. The yeast activity is expressed through the carbon dioxide released, which increases the H'm, VT and VR values. A higher value of these parameters indicates a more intense fermentation process by yeast of the fermentable sugars present in the wheat flour dough [39]. However, when high levels of SS are present in the wheat dough, the yeast activity is inhibited, probably due to the plasmolysis action of the yeast cells. Higher levels of SS addition in the wheat flour will induce an osmotic pressure on the yeast cells, which will lead to a decrease of the yeast activity and therefore of the H'm, VT and VR values [22]. The retention coefficient is related to VT and VR values being the ratio between VR and VT. A slight decrease of this parameter with the increased level of SS addition in the wheat flour, even when VT and VR are increased up to a 0.6% SS addition in wheat flour, indicates that the VR increase is lower than the VT one in the dough system. The data obtained by us were similar to those reported by others, who also concluded that higher levels of chloride salt addition in a wheat flour may lead to a decrease of the rheofermentometer values [19,21,22,40,41].

Bread's physical characteristics were improved up to a 0.6% SS addition in the wheat flour, after which these values decreased. However, these values were higher than those recorded for the control sample, which indicates the fact that for all bread samples that SS addition in the wheat flour improved bread quality. This improvement was a predictable one since SS addition increased the gas formed and dough height during the fermentation process, as the rheofermentometer data indicated. These data were in agreement with other studies that also reported an improvement of bread physical characteristics by chloride salts addition in wheat flour [17,23,41]. Moreover, compared to the bread sample with the 1.5% sodium chloride addition in the wheat flour, the samples with SS addition in the bread recipe presented higher values for bread physical characteristics, meaning that these samples were of a better quality. The sea salt is a mix of different chloride salts such as magnesium, potassium and sodium ones. According to Tuhumury et al. [34], the effects of KCl are very similar to those of NaCl on dough behavior due to its closeness in the Hofmeister series. However, the magnesium chloride, which represent 31 ÷ 35% from the sea salt that was in the highest level from the chloride salts, may cause a soluble effect on gluten proteins because they are situated in the destabilizing zone of the Hofmeister series. It can destabilize folded proteins, weakening the dough system [42]. On dough behavior, magnesium divalent cations had a similar effect as those with any salt addition [34]. Along with sodium and potassium, a chloride salt mix improves dough behavior to a higher level than in the case when only sodium chloride was added in the dough recipe, such as the data previously published by Voinea et al. [21]. This will affect bread quality, which will generally present lower values for the bread's physical characteristics than those obtained when sea salt was added in the bread recipe. The bread color characteristics indicate a darker color for samples with the increased level of SS addition in the bread recipe. This effect is due to the Maillard reaction, which takes place between amino acids and reducing sugars. The presence of SS will lead to a less intense fermentation of yeast in the initial stages of bread making, a fact that will allow a higher number of sugars and amino acids in the dough in the baking phase. SS will slow down the yeast activity, a fact that will increase the number of free sugars from the dough system, which will be involved in the Maillard reaction in the last stage of bread making [3,43]. From a textural point of view, the SS addition decreases the cohesiveness and resilience values and at high levels increases the firmness and gumminess ones. The increase in firmness and gumminess at high levels of SS addition in the wheat flour may be explained by gluten protein behavior, which becomes more compact by SS addition in the wheat flour. More of these variations of textural values may be correlated with a specific volume of bread sample values [44], which increased at high levels of SS addition and decreased at low ones. In general, the bread sensory characteristics were improved by SS addition in the wheat flour. This sensory improvement may have multiple explanations. First of all, SS addition has an important contribution to the Maillard reaction. Besides the fact that during this reaction the bread darkens in color, flavor compounds are also formed which contribute to the sensory perception of the bread [43,45]. Moreover, it is well known that SS decreases the amount of free water in the bread, a fact that will decrease the flavor compound's volatility [3]. Moreover, SS addition improves the bread's physical characteristics, which presented higher values than the control sample, a fact that will influence sensory perception of the bread by consumers. The most important effect that SS has on the bread's sensory characteristics is its salty taste, an effect which the Na^+ ions from its content have on consumers [46]. Therefore, the increased amount of SS addition in the bread recipe may improve the taste perception of the bread by consumers. SS is a salt with low sodium content. However, increasing its amount in the bread recipe will increase the sodium content from the bread samples. According to our data, bakery products with levels up to a 0.6% SS addition in the wheat flour can be classified according to the European Commission (EC) Regulation no. 1924/2006 as products with a very low sodium content because their value is less than 0.04 g/100 g of sodium. Bread products with the addition levels of 0.9, 1.2 and 1.5% SS incorporated into the bread recipe may be classified as products with low sodium content because they have values lower than 0.12 g/100 g of sodium [3,7]. Generally, the sodium chloride from bread varied between 1.00–1.50 g/100 g [22]. Our formulation leads to backery products with 10–15 times lower sodium than those obtained in regular bread products, which may have a major impact on daily sodium intake.

5. Conclusions

Bread and other bakery products are some of the major sources of sodium in the daily diet of the population. Therefore, most initiatives to reduce sodium consumption focus on reducing the sodium chloride content in these products. In many European countries, bread, together with a wide range of bakery products, accounts for about 30% of the daily dietary sodium intake. The reduction of sodium from bread by using sea salt with low sodium content may be an alternative for bakery producers. The effect of SS from a rheological point of view was similar to those obtained by using sodium chloride salt in bread making. Its addition resulted in a strengthening effect on wheat flour dough by increasing its stability, dough development time, energy, dough resistance and decreasing the degree of softening after 10 min. During heating, SS addition increased peak viscosity, gelatinization temperature, temperature at maximum peak viscosity and falling number values. In terms of fermentation behavior, the SS addition improved the rheofermentometer values, which presented maximum values to a 0.6% SS addition in the wheat flour. The bread quality was better appreciated from the sensory point of view with the increased level of SS addition in the wheat flour. The bread with SS incorporated into the bread recipe presented higher physical values and lower textural characteristics compared to the control sample. Depending on the dose of SS added in the wheat flour, the bread samples may be marketed as bread samples with a very low sodium content or with low sodium content if SS exceeds a 0.9% addition in the bread recipe.

Author Contributions: Conceptualization, A.V. and G.G.C.; methodology, S.R.; software, S.-G.S.; validation, S.-G.S., A.V. and G.G.C.; formal analysis, G.G.C.; investigation, S.R.; resources, A.V.; data curation, A.V.; writing—original draft preparation, A.V.; writing—review and editing, G.G.C.; visualization, A.D.; supervision, G.G.C.; project administration, A.V.; funding acquisition, G.G.C. All authors have read and agreed to the published version of the manuscript.

Funding: This work was funded by the Ministry of Research, Innovation and Digitalization within Program 1—Development of national research and development system, Subprogram 1.2—Institutional Performance—RDI excellence funding projects, under contract no. 10PFE/2021.

Institutional Review Board Statement: Not applicable.

Informed Consent Statement: Not applicable.

Data Availability Statement: Not applicable.

Conflicts of Interest: The authors declare no conflict of interest.

References

1. Zandstra, E.H.; Lion, R.; Newson, R.S. Salt reduction: Moving from consumer awareness to action. *Food Qual. Prefer.* **2016**, *48*, 376–381. [CrossRef]
2. Miller, R.A.; Jeong, J. Sodium reduction in bread using low-sodium sea salt. *Cereal Chem.* **2014**, *91*, 41–44. [CrossRef]
3. Codină, G.G.; Voinea, A.; Dabija, A. Strategies for Reducing Sodium Intake in Bakery Products, a Review. *Appl. Sci.* **2021**, *11*, 3093. [CrossRef]
4. Barbosa, S.S.; Sousa, L.C.M.; de Oliveira Silva, D.F.; Pimentel, J.B.; Evangelista, K.C.M.d.S.; Lyra, C.d.O.; Lopes, M.M.G.D.; Lima, S.C.V.C. A Systematic Review on Processed/Ultra-Processed Foods and Arterial Hypertension in Adults and Older People. *Nutrients* **2022**, *14*, 1215. [CrossRef] [PubMed]
5. He, F.J.; Campbell, N.R.C.; MacGregor, G.A. Reducing salt intake to prevent hypertension and cardiovascular disease. *Rev. Panam. Salud Publica* **2012**, *32*, 293–300. [CrossRef] [PubMed]
6. Belc, N.; Smeu, I.; Macri, A.; Vallauri, D.; Flynn, K. Reformulating foods to meet current scientific knowledge about salt, sugar and fats. *Trends Food Sci. Tehnol.* **2019**, *84*, 25–28. [CrossRef]
7. Silow, C.; Axel, C.; Zannini, E.; Arendt, E.K. Current status of salt reduction in bread and bakery products—A review. *J. Cereal Sci.* **2016**, *72*, 135–145. [CrossRef]
8. Kovac, B.; Blaznik, U. Systematic Reduction of Excessive Salt Intake. In *Salt in the Earth*; Cengiz Çinku, M., Karabulut, C., Eds.; IntechOpen: London, UK, 2019; pp. 1–16. ISBN 978-1-78984-635-5.
9. Simsek, S.; Martinez, M.O. Quality of dough and bread prepared with sea salt or sodium chloride. *J. Food Process Eng.* **2016**, *39*, 44–52. [CrossRef]
10. Miocinovic, J.; Miloradovic, Z.; Radovanovic, M.; Sredovic Ignjatovic, I.; Radulovic, A.; Nastaj, M.; Sołowiej, B.G.; Tomasevic, I. Sodium Reduction by Partial and Total Replacement of NaCl with KCl in Serbian White Brined Cheese. *Foods* **2022**, *11*, 374. [CrossRef]
11. Raffo, A.; Carcea, M.; Moneta, E.; Narducci, V.; Nicoli, S.; Peparaio, M.; Sinesio, F.; Turfani, V. Influence of different levels of sodium chloride and of a reduced-sodium salt substitute on volatiles formation and sensory quality of wheat bread. *J. Cereal Sci.* **2018**, *79*, 518–526. [CrossRef]

12. Fanari, F.; Desogus, F.; Scano, E.A.; Carboni, G.; Grosso, M. The Effect of the Relative Amount of Ingredients on the Rheological Properties of Semolina Doughs. *Sustainability* **2020**, *12*, 2705. [CrossRef]
13. Fanari, F.; Iacob, C.; Carboni, G.; Desogus, F.; Grosso, M.; Wilhelm, M. Broadband Dielectric Spectroscopy (BDS) Investigation of Molecular Relaxations in Durum Wheat Dough at Low Temperatures and Their Relationship with Rheological Properties. *LWT* **2022**, *161*, 113345. [CrossRef]
14. Angioloni, A.; Dalla Rosa, M. Dough thermo-mechanical properties: Influence of sodium chloride, mixing time and equipment. *J. Cereal Sci.* **2005**, *41*, 327–331. [CrossRef]
15. Beck, M.; Jekle, M.; Becker, T. Impact of sodium chloride on wheat flour dough for yeast-leavened products. I. Rheological attributes. *J. Sci. Food Agric.* **2012**, *92*, 585–592. [CrossRef]
16. Noort, M.W.J.; Bult, J.H.F.; Stieger, M.; Hamer, R.J. Saltiness enhancement in bread by inhomogeneous spatial distribution of sodium chloride. *J. Cereal Sci.* **2010**, *55*, 378–386. [CrossRef]
17. Lynch, E.J.; Dal Bello, F.; Sheehan, E.M.; Cashman, K.D.; Arendt, E.K. Fundamental studies on the reduction of salt on dough and bread characteristics. *Int. Food Res. J.* **2009**, *42*, 885–891. [CrossRef]
18. Jessen, N.; Damasceno, A.; Padrão, P.; Lunet, N. Levels of Salt Reduction in Bread, Acceptability and Purchase Intention by Urban Mozambican Consumers. *Foods* **2022**, *11*, 454. [CrossRef]
19. Voinea, A.; Stroe, S.-G.; Codină, G.G. Use of Response Surface Methodology to Investigate the Effects of Sodium Chloride Substitution with Potassium Chloride on Dough's Rheological Properties. *Appl. Sci.* **2020**, *10*, 4039. [CrossRef]
20. Doyle, M.E.; Glass, K.A. Sodium Reduction and Its Effect on Food Safety, Food Quality, and Human Health. *Compr. Rev. Food Sci. Food Saf.* **2010**, *9*, 44–56. [CrossRef]
21. Voinea, A.; Stroe, S.-G.; Codină, G.G. The Effect of Sea Salt, Dry Sourdough and Fermented Sugar as Sodium Chloride Replacers on Rheological Behavior of Wheat Flour Dough. *Foods* **2020**, *9*, 1465. [CrossRef]
22. Voinea, A.; Stroe, S.-G.; Codină, G.G. The Effect of Sodium Reduction by Sea Salt and Dry Sourdough Addition on the Wheat Flour Dough Rheological Properties. *Foods* **2020**, *9*, 610. [CrossRef] [PubMed]
23. Arena, E.; Muccilli, S.; Mazzaglia, A.; Giannone, V.; Brighina, S.; Rapisarda, P.; Fallico, B.; Allegra, M.; Spina, A. Development of Durum Wheat Breads Low in Sodium Using a Natural Low-Sodium Sea Salt. *Foods* **2020**, *9*, 752. [CrossRef] [PubMed]
24. ICC. Standard Methods of the International Association for Cereal Chemistry. In *Methods 104/1, 110/1, 105/2, 106/1, 107/1, 114/1, 115/1, 126/1*; International Association for Cereal Chemistry: Vienna, Austria, 2010.
25. *Romanian Standard SR 90:2007*; Wheat Flour. Analysis Method. Romanian Standards Association: Bucharest, Romania, 2007.
26. AACC. Approved Methods of Analysis, 11th Edition—AACC Method 89-01.01. Cereals & Grains Association. Yeast Activity, Gas Production. 2000. Available online: https://methods.aaccnet.org/methods/89-01.pdf (accessed on 12 April 2016).
27. *Romanian Standard SR 91:2007*; Bread and Fresh Pastries. Analysis Method. Romanian Standards Association: Bucharest, Romania, 2007.
28. *ISO 3972:2011*; Sensory Analysis—Methodology—Method of Investigating Sensitivity of Taste. 3rd ed. International Organization for Standardization: Geneva, Switzerland, 2011.
29. *ISO8586-1:1993*; Sensory Analysis—General Guidance for Selection, Training and Monitoring of Assessor. Part 1: Selected Assessors. International Organisation for Standardization: Geneva, Switzerland, 1993.
30. *ISO8586-2:1994*; Sensory analysis—General Guidance for the Selection, Training and Monitoring of Assessors—Part 2: Experts. International Organisation for Standardization: Geneva, Switzerland, 1994.
31. Uthayakumaran, S.; Batey, I.L.; Day, L.; Wrigley, C.W. Salt reduction in wheat-based foods-technical challenges and opportunities. *Food Aust.* **2011**, *63*, 137–140.
32. Costa, R.G.B.; Sobrala, D.; Aglaê, V.; Teodoro, M.; Costa Junior, L.C.G.; de Paula, J.C.J.; Landin, T.B.; Oliveira, M.B. Sodium substitutes in Prato cheese: Impact on the physicochemical parameters, rheology aspects and sensory acceptance. *LWT-Food Sci. Technol.* **2018**, *90*, 643–649. [CrossRef]
33. Wellner, N.; Bianchini, D.; Mills, E.N.C.; Belton, P.S. Effect of selected hofmeister anions on the secondary structure and dynamics of wheat prolamins in gluten. *Cereal Chem.* **2003**, *80*, 596–600. [CrossRef]
34. Tuhumury, H.C.D.; Small, D.M.; Day, L. Effects of Hofmeister salt series on gluten network formation: Part, I. Cation series. *Food Chem.* **2016**, *212*, 789–797. [CrossRef]
35. Bernklau, I.; Neußer, C.; Moroni, A.V.; Gysler, C.; Spagnolello, A.; Chung, W.; Becker, T. Structural, textural and sensory impact of sodium reduction on long fermented pizza. *Food Chem.* **2017**, *234*, 398–407. [CrossRef]
36. Nogueira, A.C.; Kussano, J.T.; Steel, C.J. Sourdough reduces sodium in wheat flour doughs. *Int. J. Food Sci. Technol.* **2015**, *50*, 2621–2629. [CrossRef]
37. Codină, G.G.; Dabija, A.; Oroian, M. Prediction of Pasting Properties of Dough from Mixolab Measurements Using Artificial Neuronal Networks. *Foods* **2019**, *8*, 447. [CrossRef]
38. Salvador, A.; Sanz, T.; Fiszman, S.M. Dynamic rheological characteristics of wheat flour-water doughs. Effect of adding NaCl, sucrose and yeast. *Food Hydrocoll.* **2006**, *20*, 780–786. [CrossRef]
39. Codină, G.G.; Voica, D. The influence of different forms of bakery yeast Saccharomyces cerevisie type strain on the concentration of individual sugars and their utilization during fermentation. *Rom. Biotech. Lett.* **2010**, *15*, 5417–5422.
40. Jekle, M.; Necula, A.; Jekle, M.; Becker, T. Concentration dependent rate constants of sodium substitute functionalities during wheat dough development. *Food Res. Int.* **2019**, *116*, 346–353. [CrossRef] [PubMed]

41. Pasqualone, A.; Caponio, F.; Pagani, M.A.; Paradiso, V.M.; Paradiso, V.M. Effect of salt reduction on quality and acceptability of durum wheat bread. *Food Chem.* **2019**, *289*, 575–581. [CrossRef]
42. Codină, G.G.; Zaharia, D.; Ropciuc, S.; Dabija, A. Influence of Magnesium Gluconate Salt Addition on Mixing, Pasting and Fermentation Properties of Dough. *EuroBiotech J.* **2017**, *3*, 222–225. [CrossRef]
43. Belz, M.C.E.; Ryan, L.A.M.; Arendt, E.K. The Impact of Salt Reduction in Bread: A Review. *Crit. Rev. Food Sci. Nutr.* **2012**, *52*, 514–524. [CrossRef] [PubMed]
44. Beck, M.; Jekle, M.; Becker, T. Impact of sodium chloride on wheat flour dough for yeast-leavened products. II. Baking quality parameters and their relationship. *J. Sci. Food Agric.* **2012**, *92*, 299–306. [CrossRef]
45. Paravisini, L.; Sneddon, K.A.; Peterson, D.G. Comparison of the aroma profiles of intermediate wheatgrass and wheat bread crusts. *Molecules* **2019**, *24*, 2484. [CrossRef]
46. Hartman-Petrycka, M.; Klimacka-Nawrot, E.; Ziora, K.; Suchecka, W.; Gorczyca, P.; Rojewska, K.; Błońska-Fajfrowska, B. Sweet, Salty, and Umami Taste Sensitivity and the Hedonic Perception of Taste Sensations in Adolescent Females with Anorexia Nervosa. *Nutrients* **2022**, *14*, 1042. [CrossRef]

applied sciences

MDPI

Article

Study of the Utilization of Spent Grain from Malt Whisky on the Quality of Wafers

Ancuța Chetrariu and Adriana Dabija *

Faculty of Food Engineering, Stefan cel Mare University of Suceava, 720229 Suceava, Romania; ancuta.chetrariu@fia.usv.ro
* Correspondence: adriana.dabija@fia.usv.ro; Tel.: +40-748-845-567

Abstract: This study aimed at determining the quality parameters of the wafer formulated with the addition of grain spent (SG), resulting from the obtainment of whisky. In this sense, wafers were formulated from chickpea flour, spent grain, wild garlic paste, golden flaxseed, and hemp seeds. These food products were analyzed in terms of texture, density, and pH of the batter, but also of the final product for proximate analysis, baking loss, texture, water activity, color, antioxidant capacity, water holding capacity and oil holding capacity, microstructure, and sensorial analysis. The addition of spent grain in the wafer formulation led to products with a high acceptability, the texture of the batter underwent changes due to the addition of spent grain, all parameters increased, and only adhesiveness decreased. The density and pH of the samples with SG decreased. The fracturability of the products with SG decreased with the addition of SG compared to the control sample, and the color becomes darker, influenced by the specific color of the SG. With the addition of spent grain, it increases the fiber and protein content, the antioxidant capacity, but also the baking loss due to the fibers contained in it. The microstructure of samples with the addition of SG shows a heterogeneous distribution of pores on the cross section of the samples, with larger pores in the center of the wafer samples.

Keywords: by-product; valuable resources; valorization; nutritional value; sensorial characteristics

Citation: Chetrariu, A.; Dabija, A. Study of the Utilization of Spent Grain from Malt Whisky on the Quality of Wafers. *Appl. Sci.* **2022**, *12*, 7163. https://doi.org/10.3390/app12147163

Academic Editor: Alessandra Biancolillo

Received: 1 July 2022
Accepted: 14 July 2022
Published: 15 July 2022

1. Introduction

Wafers are food products obtained from thin batter, baked on metal molds, and characterized by the fact that they are found in the form of sheets, empty round figures or covered with chocolate, ice cream cones, and wafer sticks, but also dub other shapes [1]. Baked wafers are used for the manufacture of several confectionery products by spraying a liquid batter between two heated closed plates to allow the dough to spread evenly [2]. Along with crackers, pretzels, and cookies, wafers belong to the category of bakery products with a long shelf life [3]. Flour products in this category are known in a wide variety of assortments, obtained by baking in special forms a fluid batter prepared from wheat flour, water, salt, leaving agents, and other materials used for taste and aroma, which are presented in the form of sheets or different alveoli formats with a high porosity and without filling [4]. The disadvantages lie in the fact that part of these products may have a low nutritional value and poorly outlined sensory characteristics, with a low level of essential amino acids and dietary fiber [5]. Although the basic ingredient in the formulation of waffles is wheat flour, lately, the partial or total replacement of wheat flour has become increasingly used. Crispiness is one of the defining parameters for wafers and reflects the density of the cellular structure [6]. The crispiness of the waffles is given by their low moisture of 1–2%, the crispy texture losing approximately6–7% moisture. The term wafer is often confused with waffle, but between the two products there are major differences: waffles have a soft texture with a moisture of 10% or more, while wafers have a crispy texture, are thin, and with a moisture content of 1–2% [7]. Non-creamed wafers refer to crispy wafers without typical cream fillings, an option being the use of other flours besides wheat, such as chickpea flour or corn flour, or the use of fermented batters to give it a special flavor.

Spent grain is the main by-product resulting in huge quantities in the beer industry and whisky production, with a rich content of fiber and protein, but also minerals, vitamins, and phenolic compounds. Even if lately, it is trying to reuse it, much of the spent grain remains to be managed efficiently still. There is a growing focus on exploring sustainable recovery processes for active compounds released from spent grain [8,9]. In food products, spent grain tends to improve the nutritional value of spent grain-added food products. Due to the biologically active compounds contained, spent grain shows antioxidant activity, antimicrobial, and anti-inflammatory activity, but also potential as a functional ingredient in food or as a nutraceutical ingredient [10,11].

Chickpea (*Cicer arietinum* L.) beans are known all over the world, being used in a wide range of foods as such or used for the fortification of some foods. Chickpea flour can improve the quality of products and increase their nutritional value [12]. Chickpea flour is an excellent source of proteins, carbohydrates with a low and easily digestible glycemic index, dietary fiber, mineral substances: potassium, phosphorus, calcium, manganese, copper, iron and zinc, and vitamins of the B complex, provitamin A, vitamins C, E, and K [12]. Wild garlic (*Alliumursinum*) contains allyl sulfide, vitamin A, vitamin C, carotenoids, ethereal oil, volatile oils, allicin, adenosine, and mineral substances [13]. The seeds of golden flax (*Linumusitatissimum*) and hemp (*Cannabis sativa L.*) are rich in fatty acids, easily digestible proteins, and rich in essential amino acids, carbohydrates, soluble and insoluble fiber, B-complex vitamins, vitamin E, β-carotene, calcium, magnesium, iron, and zinc [14]. Frequent consumption of products based on chickpea flour is effective in reducing cardiovascular risk, type 2 diabetes, some types of cancer and obesity. Wild garlic has depurative, detoxifying, antiseptic, antiviral, antimicrobial, blood fluidifying, expectorant, and antitumor action. Spent grain flour, through fiber intake, has major implications in digestion by slowing down gastric emptying, prolonging the intestinal transit time, and reducing the speed of absorption of nutrients in the small intestine. Golden flax and hemp seeds have potentially beneficial effects on immune function, chronic-degenerative inflammatory diseases, obesity, cancer prevention, and glucose metabolism regulation.

The aim of this study is to develop wafers with sensorial characteristics and improved nutritional value by using the addition of spent grain from malt whisky. The finished product was analyzed in terms of texture, water activity, color, antioxidant capacity, water holding capacity and oil holding capacity, microstructure, and sensorial analysis.

2. Materials and Methods

2.1. Materials

Spent grain from distillery industry was provided from a local factory, Alexandrion Group (Ploiesti, Romania); wet spent grain was stored at −18 °C, dried at 50 °C for 24 h, ground by a mill, and then sieved. Spent grain flour was obtained from a fraction of less than 200 μm and was stored in paper bags at room temperature until further use. Spent grain used in this study has the following characteristics: moisture—5.04 ± 0.42, ash—3.47 ± 0.02, fiber—22.67 ± 0.42, protein—18.88 ± 0.37, and lipids—7.11 ± 0.39, as described in our previous study [15]. Chickpea flour, wild garlic, golden flax, and hemp seeds were purchased from a local market.

All chemicals used in this paper were of analytical grade and were purchased from Sigma Aldrich (St. Louis, MO, USA).

2.2. Wafer Making

The realization of the product provides for the qualitative and quantitative reception of raw and auxiliary materials, dosing and preparation of raw and auxiliary materials, mixing components with the obtainment of fluid batter, filtering the batter for the separation of possible agglomerations, pouring into molds, baking in electric oven, and cooling and packaging of the finished product. In Table 1, the recipe and quantity percentage of wafers used in this study is presented.

Table 1. Wafers recipe.

Ingredients	Control Sample Wafer (kg)	Spent Grain Wafer (kg)
Chickpea flour	60	60
Spent grain flour	-	10
Wild garlic paste	20	20
Golden flax seeds	5	5
Hemp seeds	5	5
Leaving agents	1.5	1.5
Salt	2.5	2.5

2.3. Batter Texture

The density of the wafer can be changed by changing the water–flour ratio used in the formulation of the liquid batter or by using a process trick to release additional moisture. Using a Perten TVT-6700 texturometer (Perten Instruments, Stockholm, Sweden), the batter wafer parameters (batter gumminess, chewiness, springiness, firmness, adhesiveness, and cohesiveness) were analyzed. Texturometer was equipped with a 35 mm compression plate, the flask diameter was 65 mm. Batter was analyzed directly from the glass jar (bowl height = 60 mm; diameter = 50 mm). Two compression cycles were performed for each sample analyzed in the present study at a trigger force of 7 g, test speed 1 mm/s, and retract speed 5 mm/s, method adapted by Oroian et al., 2022 [16] and Iuga and Mironeasa, 2021 [17]. All the measurements have been done in triplicate.

2.4. Batter Density and pH

The density of the waffle batter was determined by the method of Ekramian et al., 2021 [3], with a measuring cylinder by means of the relationship between the weight of the batter and the same volume of distilled water. The density of the dough batter gives us information about how fluffy (aerated) a batter is: the lower its density, the more aerated the batter [18]. The pH of the batter was measured using a Mettler Toledo pH-meter (Columbos, OH, USA) and the optimum pH-value for wafer batter is between pH 5.5 and 7.0 [19].

2.5. Proximate Analysis of Wafers

The moisture of wafers was determined by the method of drying in the air oven at 105 °C until the constant weight. The ash content was determined in an oven at 550 °C for 8 h, until a white or light grey residue is obtained [20]. The protein content was determined by standard Kjeldhal method and lipids content was determined by standard Soxhlet method [20]. Total dietary fibre was assessed by enzymatic method using the AOAC Method 2011.25 with a Megazyme total dietary fiber assay kit (Megazyme, Ireland), according to Lee et al., method [21]. All samples were measured in triplicate.

2.6. Baking Loss

Baking loss was calculated to the following Equation (1) and gives information on the moisture loss during baking [18]:

$$\text{Baking loss (\%)} = [1 - (\text{wafer weight [g]}/\text{batter weight [g]})] \times 100 \quad (1)$$

2.7. Wafers Texture

A very important sensory property is the texture that affects consumer choices and food stability [22]. Dry wafer fracturability was determined using the maximum force F (N) needed to break a wafer piece. A Perten TVT-6700 device (Perten Instruments, Stockholm, Sweden) equipped with an aluminum break rig set adjusted to 50 mm width was used. The test speed was 2 mm/s and the trigger force 0.0005 N [23]. Instrument settings were pulled to break mode: test speed 2 mm/s [24,25]. Three replicates of wafer from each formulation were determined.

2.8. Water Activity

Water activity was measured 2 h after baking, at 25 °C, using the water activity meter AquaLab 4TE (Decagon Devices Inc., Pullman, WA, USA) [3,15].

2.9. Wafers Color

The color of the dried pasta was evaluated with a Konica Minolta CR-400 colorimeter (Tokyo, Japan), using the coordinates of the CIELab color space. The samples were measured in triplicate and the chromameter was calibrated with a whiteboard of the device [20,26,27].

2.10. Antioxidant Capacity of Wafers

The phenolic content was determined by the Folin–Ciocalteau method: 2 g of wafers were ground and mixed with 20 mL 80% methanol (*v/v*) and sonicated for 40 min in a sonication bath at 37 °C and 45 Hz, then the mixture was centrifuged for 5 min at 4000 rpm. A volume of 0.2 mL of the extract was mixed with 2 mL of 1:10 Folin–Ciocalteau reagent and 1.8 mL of 7.5% (*w/v*) sodium carbonate in a container. The mixture was left for 30 min at room temperature in the dark [23]. The total polyphenol content was determined at 750 nm using a UV–VIS–NIR spectrophotometer (Shimadzu Corporation, Kyoto, Japan). The calibration curve of polyphenols was achieved using gallic acid at concentrations of 10–200 mg/L with regression coefficient R^2 = 0.99872. The samples were analyzed in triplicate.

Total antioxidant activity was determined by DPPH assay scavenging activity by adding 2 mL of extract described above with 2 mL of DPPH solution 0.1 mM in methanol [28]. The mixture was vortexed for 2 min and kept for 30 min at room temperature in a dark place, and then the absorbance was determined at 517 nm using a UV–VIS–NIR spectrophotometer (Schimadzu Corporation Kyoto, Japan). The antioxidant capacity was measured in triplicate, using distilled water as a blank sample, as described in Equation (2).

$$\% \text{ Inhibition of DPPH} = [(1 - \text{As}/\text{Ab})] \times 100 \quad (2)$$

where As= absorbance of sample, Ab = absorbance of blank sample.

2.11. Water Holding Capacity and Oil Holding Capacity

To determine the water holding capacity (WHC), 1 g of sample and 10 mL of distilled water are placed in a centrifuge tube and vortexed for 30 min, left to stand for 30 min at room temperature, then centrifuged for 25 min at 3000 rpm [6,24]. Subsequently, the sediment is weighed and water holding capacity is calculated with the Equation:

$$\text{WHC } (\%) = (W_2 - W_1) \times 100/W_0 \quad (3)$$

where W_0 is the weight of the sample, W_1 is the weight of the centrifuge tube plus sample, and W_2 is the weight of the centrifuge tube plus sediments.

Oil holding capacity (OHC)was determined by 0.5 g of sample over which is added 5 mL canola oil placed in a centrifuge tube and vortexed for 30 min, left to stand for 30 min at room temperature, then centrifuged for 25 min at 3000 rpm [24]. Oil holding capacity is determined in the same way as water holding capacity, using the Equation:

$$\text{OHC } (\%) = (W_2 - W_1) \times 100/W_0 \quad (4)$$

where W_0 is the weight of the sample, W_1 is the weight of the centrifuge tube plus sample, and W_2 is the weight of the centrifuge tube plus sediments.

2.12. Microstructure (SEM)

The microstructure was characterized by a scanning electron microscope (SEM) VEGA II LSH (Tescan, Czech Republic) at an acceleration strain of 30 kV and a magnification of 100× and 200× by capturing different images of the cross-section to measure the size of the pores [2,3,29]. The samples were fixed with double adhesive carbon bands.

2.13. Sensory Analysis

For sensory analysis, a trained consumer test panel was recruited by a panel of fifteen experts selected according to their sensorial skills and trained in sensory vocabulary and identification of particular attributes. Panelists were selected based on interest and time availability. The samples were coded and presented randomly. The sensory panel was used to determine the texture rated according to appearance, color, taste, crunchiness, smell, and general acceptance. Panelists used a 5-point scale for sensory evaluation: 5 = appreciated extreme; 4 = pleasant; 3 = I neither liked nor liked; 2 did not like, and 1 =did not liked it extremely [4,30,31]. The acceptability index (AI) was calculated according to the equation

$$AI\ (\%) = Y \times 100/Z \quad (5)$$

where Y =the average score obtained for the product and Z = the maximum score given to the product.

2.14. Statistical Analysis

All of the analyses in the present study were performed in triplicate. Statistical software SPSS 25.0 (trial version) (IBM, New York, NY, USA) was used to calculate the mean values and standard deviations for the quantitative data (https://www.ibm.com/analytics/spss-trials (accessed on 21 June 2022)). A principal component analysis (PCA) was performed to observe the similarities or dissimilarities between the evaluated parameters and formulated samples.

3. Results and Discussion

The textural characteristics of batter wafer are affected by the structural components of the ingredients and manufacturing processes [16]. Gumminess, chewiness, springiness, firmness, and cohesiveness increased in the spent grain sample, while the adhesiveness decreased with the spent grain addition. The wafers batter textural parameters are presented in Table 2. The protein and fibre contents of spent grain affected all those parameters. In comparison, Iuga, 2020 [32] and Chetrariu and Dabija, 2021 [23] obtained the same tendency in their study.

Table 2. Wafer batter textural parameters.

	Gumminess (g)	Chewiness (g)	Springiness (%)	Firmness (g)	Adhesiveness (g × s)	Cohesiveness (Adim.)
Control sample	94.46 ± 3.7 ^{a}	90.20 ± 4.1 ^{a}	0.9995 ± 0.0 ^{a}	64.5 ± 1.5 ^{a}	−359.86 ± 7.0 ^{b}	1.42 ± 0.0 ^{a}
Spent grain wafers	717.85 ± 52.9 ^{a}	725.27 ± 44.0 ^{b}	1.00045 ± 0.0 ^{a}	203.93 ± 24.9 ^{b}	−904.30 ± 33.8 ^{a}	3.61 ± 0.7 ^{b}
p value	n.s.	<0.001	n.s.	0.001	<0.001	0.006

Mean values with different letters in the same column are significantly different ($p < 0.05$), n.s.—not significant.

The batter control sample density was higher (105.26% ± 0.05%) than the batter sample density (89.71% ± 0.03%), which means that the sample density was decreased with adding spent grain into formula wafers; those results are in accord with Ekramian et al., 2021 [3]. The low density indicates that more air is embedded in the batter and is desirable in the wafer batter, being tied to more gas bubbles in the wafers, leading to more bubbles during baking and making a larger volume of the finished product.

A proximate analysis of the wafers is shown in Table 3. The pH value of the samples (6.78 for wafer sample and 6.88 for control sample) are relatively close and fall within the optimal values found in Huber's and Schoenlechner's 2017 study [18]. A slight decrease in pH has been observed with the addition of spent grain to the wafer formulation and these changes can be attributed to the enzymatic reactions between the components of the wafers. A small addition of baking soda gives a slight increase in the pH value of the dough and influences the formation of the more intense Maillard reaction [33].

Table 3. Proximate analysis of spent grain wafer.

Determination	Control Sample(%)	Spent Grain Wafer Sample(%)	p Value
Proteins	9.66 ± 0.35 [a]	11.17 ± 0.21 [a]	0.001
Lipids	16.03 ± 0.02 [a]	17.69 ± 0.03 [b]	<0.001
Fibers	9.82 ± 0.03 [a]	11.64 ± 0.04 [b]	<0.001
Carbohydrates	51.28 ± 0.06 [b]	47.40 ± 0.05 [a]	0.003
Ash	7.35 ± 0.01 [b]	6.22 ± 0.01 [a]	<0.001
Moisture	5.86 ± 0.02 [a]	5.88 ± 0.02 [a]	n.s.

Mean values with different letters in the same column are significantly different ($p < 0.05$), n.s.—not significant.

The moisture content contributes predominantly to the stickiness of the batter, but also to the crispiness of final products and a small moisture extended shelf life of food products [27]. The value for the moisture of the control sample was close to the value of the spent grain wafer sample (5.86% ± 0.02% vs. 5.88% ± 0.02%);the results found in this study were in agreement with Raza et al., 2016 [20]. Matinez-Navarrete et al., 2004, showed a good shelf-life stability and reported that wafers contain 7.00% ± 0.02% moisture and less than 0.52 water activity [34]. Most of the moisture evaporates during the baking process, resulting in a porous foamed cellular structure [29].

The crude proteins content is contributed to by spent grain addition and chickpea flour, with the highest content in the spent grain sample (11.17% ± 0.21%) compared to the control sample (9.66% ± 0.35%), in accordance with Raza et al., 2016 [20].

The lipids content was highest (17.69% ± 0.03%) in the spent grain sample and 16.03% ± 0.02% was the fat content in the control sample, due to the lipid contents of hemp seeds and golden flax seeds. Silva et al., 2022, obtained in their study a lower fat content, between 2.2% and 3.7% [35].

The total dietary fiber content was higher in the wafer sample (11.64% ± 0.04%) compared to the control sample (9.82% ± 0.03%) due to the addition of spent grain that contains a higher amount of fiber; the same trend was observed in study reported by Chetrariu and Dabija, 2022 [36].

The ash content was found to be highest in the control sample (7.35% ± 0.01%) followed by the spent grain sample (6.22% ± 0.01%), in accordance with Tiwari et al., 2020 [26].

The force measured at the first maximum point of the first significant change of the first curve is the force recorded at the breaking or destruction of the sample, i.e., the fracturability. If under stress a material can be fractured then it is considered to be a fragile material [23]. The fracturability of dried wafers decreases with the increase in the addition of spent grain (control sample 5.47 ± 0.09 N, spent grain wafers 3.86 ± 0.07 N), the same result obtained by Iuga, 2020 [32], and with a decrease in pasta fracturability with the increase in the content of whey powder and corn starch and by Chetrariu and Dabija [23] in their study about spent grain pasta.

Baking loss can be influenced by several factors: baking temperature, batter moisture, or leaving agents used. The values for baking loss were 50.15% for the control sample and 56.12% for the spent grain sample, respectively, compared to 19.52–27.48% baking losses in the case of waffle [18], which is to be expected because the wafer humidity is much lower than that of waffles.

Water activity (Aw) is a simple instrumental measure of crispness in wafer, making it an excellent way to monitor crispness over time and from batch to batch [37]. Traditionally, crispness intensity decreases as Aw increases. In their study, Goerlitz et al., 2007, found values for ice cream cones between 0.387 and 0.520, in agreement with this study (0.4302 for the control sample and 0.4350 for the spent grain sample) [37].

The acceptability of a product given by consumers is also influenced by the color of the product. The lightness of the spent grain wafer sample decreased compared to the control sample (Figure 1), a process due to the addition of spent grain, which has a specific dark color. There is a decrease in yellowness (b*) and an increase in redness (a*) in the spent grain wafer sample compared to the control sample (Table 4), in accordance with a study made by Chetrariu and Dabija, 2021 [23].

(A) **(B)**

Figure 1. Control sample (**A**) vs. spent grain sample (**B**).

Table 4. Wafer colors.

	L	a*	b*	ΔL*	ΔE*
Spent grain wafersample	48.86 ± 0.08	2.10 ± 0.03	29.39 ± 0.05	−45.3 ± 0.08	51.93 ± 0.04
Control sample	61.6 ± 0.21	−6.27 ± 0.04	35.65 ± 0.21	−32.57 ± 0.21	45.67 ± 0.01

The total content of polyphenols increased in the spent grain wafer sample (6.46 μg GAE/g) compared to the blank sample (5.29 μg GAE/g), which means that spent grain has a polyphenol intake in the finished product. Spent grain had a contribution of polyphenols also in the case of pasta with the addition of spent grain flour, where the polyphenol content increased with the percentage of added spent grain (17.21 ± 0.19 μg GAE/g—23.06 ± 0.11 μg GAE/g) [23]. The radical scavenging activity on DPPH was higher for the spent grain wafer sample (24.44 ± 0.04) compared to the control sample (1.71 ± 0.01).

The water holding capacity refers to retained water that cannot be removed by ordinary drying processes, bound by electrostatic or chemical interactions. This correlates with the level of protein or ash. The water holding capacity (WHC) is often used to determine the interactions between fiber and water [6]. WHC is closely correlated with the degree of hydration of both proteins and other polar constituents, but also their hydrophilic interactions through hydrogen bonds [38]. The values of the two samples are close (301.88% ± 9.36% for spent grain wafers and 298.22% ± 17.16% for the control sample), with a higher WHC for the sample of spent grain wafers, which is due to the increase in fiber content, in accordance to Dom et al., 2020 [24]. The high protein content could be responsible for the formation of a large number of hydrogen bonds.

The oil holding capacity (OHC) is determined by the size and porosity (density) of the particles, the content of starch, and the concentration and hydrophobic nature of flour proteins [38]. The mechanism of foil absorption is mainly attributed to the physical retention of oils and their binding to the non-polar protein chain by hydrophobic interactions. The retention of oils has significant effects on the rheology of the dough and the quality of the finished product. The values for OHC are close for the two samples (135.28% ± 11.46%

for the spent grain wafers and 136.02% ± 11.88% for the control sample), these functional properties not being influenced by the addition of spent grain.

The wafer sheet was examined with a scanning electron microscope (SEM) to determine the dimensions of the wafer as well as to characterize the internal microstructure, as can be seen in Figure 2.

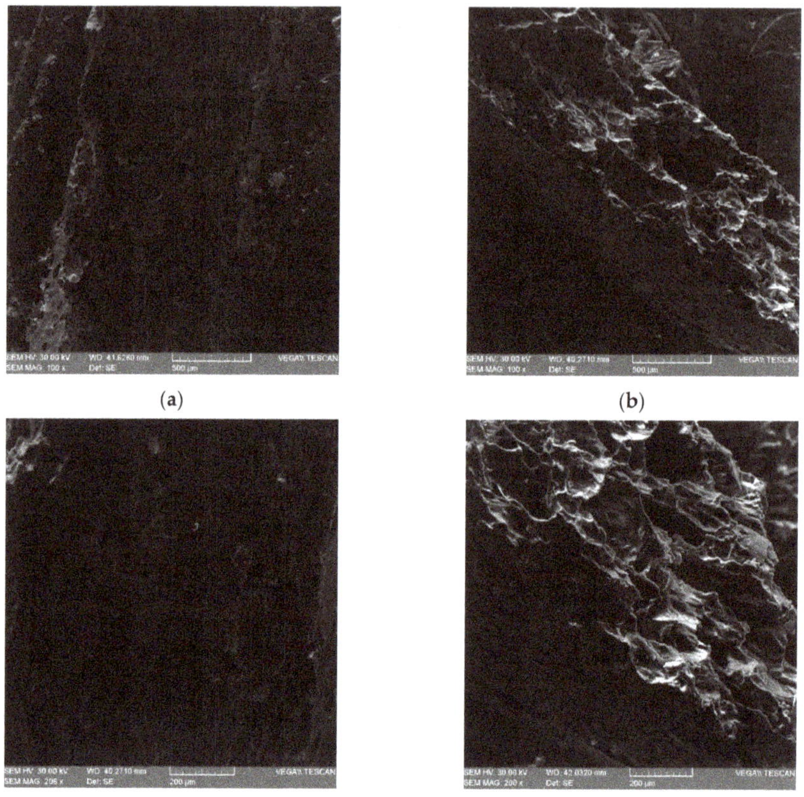

Figure 2. Wafer microstructure by scanning electron microscopy (SEM) of control sample and spent grain sample.(**a**) Control sample at magnification 100×; (**b**) Spent grain sample at magnification 100×; (**c**) Control sample at magnification 200×; (**d**) Spent grain sample at magnification 200×.

The samples were carefully placed inside the SEM so that the cross-section of the sample was directed towards the microscope lens. The characterization of the internal microstructure of the wafers was made using a scanning electron microscope, a suitable choice for imaging studies in that the samples are easy to use, do not require prior sample preparation, and the data acquisition is fast. The microstructure of the wafers is represented by a compact aggregate, giving it a fragile appearance. The presence of a considerable number of air bubbles and surface pores thatpenetrate the matrix is the main textural trait of tenderness and crispness. The microstructure is important because this method can understand why the viscosity and aeration of the batter are important for the quality and resistance of the wafers [39]. SEMs are also an attractive choice for imaging studies because they have an easy-to-use interface, are easy to operate, and require minimal sample preparation. The microstructure of baked wafers is usually associated with the cross-sections of the entire wafer that have dense skin and porous regions of the core, in accordance with Ekramian et al., 2021 [3], and Butt, 2017 [2]. A heterogeneous distribution of pores was observed on the cross section of the samples, with larger pores in the center of the wafer samples, in accordance with the study conducted by Butt, 2017 [2], and Mohammed et al., 2011 [29].

Sensory evaluation, from a statistical perspective, is a scientific method in which experimental results are collected on a set of consumers expressing preferences and reactions to the characteristics of food [2]. The sensory properties of the wafers are shown in Figure 3. The general acceptability showed acceptable values with scores higher than four throughout the analyzed range, indicating that the wafers were pleasant to some extent [35]. The spent grain wafers sample obtained a higher score than the control sample for each of the properties analyzed. The appearance of a food is an important attribute that influences the consumer's perception of a particular product [26]. The biggest difference between the averages of the spent grain wafers sample compared to the control sample was recorded for appearance, and the closest scores were recorded for taste, crunchiness, and smell, which means that the addition of spent grain did not alter these properties, but on the contrary, enriched them. Sensory analysis is useful to see whether or not consumers can detect differences between products or changes in the product's recipe [2]. The acceptability index was higher for spent grain wafers (92%) compared to the control sample (86%).

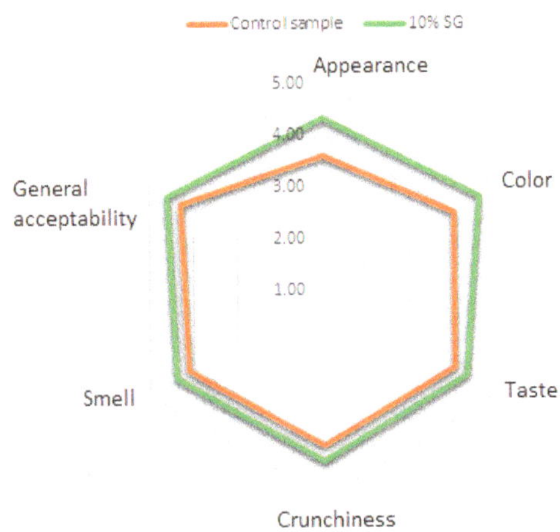

Figure 3. Sensory scores for spent grain wafers.

The principal component analysis (PCA) has been used to highlight similarities or differences between specified characteristics (Figure 4). The loads of the characteristics studied on the first main component, FC1 (70.58%), and the second main component, FC2 (15.22%), described 85.80% of the total variance. The moisture and water holding capacity has a small contribution to the variation of the data, as suggested by their position on the graph, close to the center. In contrast, the proximity of some individual parameters, for example for proteins, firmness, overall acceptability, lipids, color, chewiness, fibers, and total polyphenols content, confirm a close pair correlation, as well as the association between fracturability, adhesiveness, ash, and antioxidant activity (DPPH).

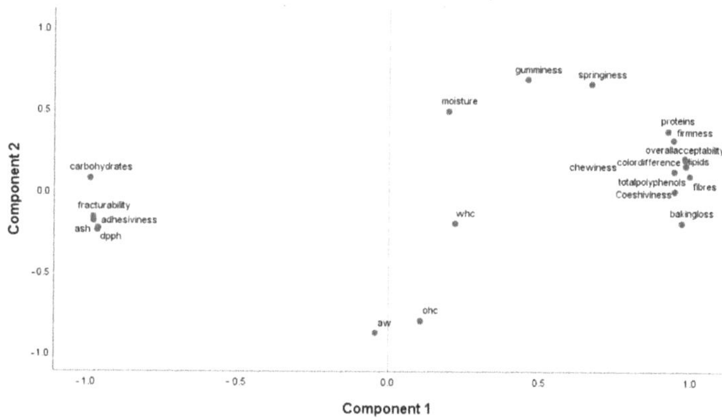

Figure 4. Principal component analysis.

PC1 has been associated with fiber, carbohydrates, fracturability, chewiness, ash, lipids, overall acceptability, color, baking loss, adhesiveness, DPPH, cohesiveness, total polyphenols content, firmness, proteins, and springiness, while PC2 has been associated with water activity, OHC, and gumminess. It may be noted that a high opposition betweenwater holding capacity and moisture, oil holding capacity, and water activity.

4. Conclusions

The study verified that the addition of spent grain can be used to prepare wafers with good sensory and nutritional value, which provide a substantial amount of protein, fat, and dietary fiber. It has been concluded that spent grain wafers are the potential nutritional food that can increase demand and change the consumption pattern of food by-products. The product is economical and can be easily manufactured on a commercial scale compared to other snack products. The addition of spent grain had a positive impact on the formulation of wafers, increasing the protein and fiber content, and antioxidant capacity; wafer samples with the addition of spent grain also led to darker samples. Fracturability decreases with the addition of grain spent. The microstructure of samples with the addition of SG shows a heterogeneous distribution of pores on the cross section of the samples, with larger pores in the center of the wafer samples. The results of the sensory analysis reveal that the addition of spent grain in the wafers recipe leads to valuable value-added products with a high acceptability among consumers. Because of the recent global context, the application of sustainable, clean, and waste-free technologies is no longer just an innovative solution, but an imperative requirement. Therefore, it is very important to adopt a sustainable approach to the management of food by-products, which represent valuable resources from a nutritional and economic or environmental point of view, revealed in the implementation of the system solutions based on the fundamentals of a circular economy.

Author Contributions: Conceptualization, A.C. and A.D.; Methodology, A.C.; Formal Analysis, A.C. and A.D.; Investigation, A.C.; Resources, A.D.; Writing—Original Draft Preparation, A.C.; Writing—Review and Editing, A.C. and A.D. All authors have read and agreed to the published version of the manuscript.

Funding: This research received no external funding.

Institutional Review Board Statement: Not applicable.

Informed Consent Statement: Not applicable.

Data Availability Statement: The data presented in this study are available in this article.

Acknowledgments: We would like to thank Alexandrion Group Romania for providing the spent grain used in this work. The authors acknowledge financial support from Stefan cel Mare University of Suceava, Romania.

Conflicts of Interest: The authors declare no conflict of interest.

Sample Availability: Samples were available from Alexandrion Group Romania.

References

1. Sruthi, V.; Waghray, K.; Narayan Rathod, A. Development of Wafers Incorporated with Pearl Millet Flour & Barnyard Millet Flour. *Int. J. Res. Sci. Technol.* **2018**, *4*, 962–969.
2. Butt, S.S. Geometrical and Material Effects on Sensory Properties of Confectionery Wafers and Similar Extruded Products. Ph.D. Thesis, Mechanics of Materials Department of Mechanical Engineering Imperial College London, London, UK, 2016.
3. Ekramian, H.; Saedi Asl, M.; Karimi, M.; Sheikholeslami, Z.; Pedram Nia, A. Comparison the effect of fruits extract with fungal protease on waffle quality. *J. Food Sci. Technol.* **2021**, *58*, 4766–4774. [CrossRef] [PubMed]
4. Kigozi, J.; Kigozi, J.; Banadda, N.; Byaruhanga, Y.; Kaaya, A.; Lule, F. Sorghum Ice Cream Cone Texture; Correlations Between Instrumental and Sensory Analysis. *J. Adv. Food Sci. Technol.* **2016**, *3*, 134–145.
5. Mahatre, R.; Thankamani, M.; Sonawane, S.K.; Bhushette, P. Comparative Study of Ice-Cream Cones Developed from Refined Wheat, Ragi, Buckwheat, Bajra, Amaranth, and Composite Flour. *Meas. Food* **2022**, *6*, 100033. [CrossRef]
6. Leonard, W.; Zhang, P.; Ying, D.; Fang, Z. Application of extrusion technology in plant food processing byproducts: An overview. *Compr. Rev. Food Sci. Food Saf.* **2020**, *19*, 218–246. [CrossRef]
7. Tiefenbacher Karl, F. Glossary of Terms in Wafers, Waffles and Adjuncts. In *The Technology of Wafers and Waffles II*; Elsevier: Cham, Switzerland, 2019; ISBN 9780128094372.
8. Chetrariu, A.; Dabija, A. Pre-treatments used for the recovery of brewer's spent grain—A minireview. *J. Agroaliment. Process. Technol.* **2020**, *26*, 304–312.
9. Lech, M.; Labus, K. The methods of brewers' spent grain treatment towards the recovery of valuable ingredients contained therein and comprehensive management of its residues. *Chem. Eng. Res. Des.* **2022**, *183*, 494–511. [CrossRef]
10. Naibaho, J.; Butula, N.; Jonuzi, E.; Korzeniowska, M.; Laaksonen, O.; Föste, M.; Kütt, M.L.; Yang, B. Potential of brewers' spent grain in yogurt fermentation and evaluation of its impact in rheological behaviour, consistency, microstructural properties and acidity profile during the refrigerated storage. *Food Hydrocoll.* **2022**, *125*, 107412. [CrossRef]
11. Naibaho, J.; Korzeniowska, M. Brewers' spent grain in food systems: Processing and final products quality as a function of fiber modification treatment. *J. Food Sci.* **2021**, *86*, 1532–1551. [CrossRef]
12. Lu, L.; He, C.; Liu, B.; Wen, Q.; Xia, S. Incorporation of chickpea flour into biscuits improves the physicochemical properties and in vitro starch digestibility. *LWT* **2022**, *159*, 113222. [CrossRef]
13. Khandagale, K.; Roylawar, P.; Randive, P.; Karuppaiah, V. Isolation and Expression Profiling of Insecticidal Lectins from Wild Alliums Against Onion Thrips (*Thrips tabaci* Lindeman). *Proc. Natl. Acad. Sci. India Sect. B Biol. Sci.* **2022**, *92*, 451–459. [CrossRef]
14. Pareek, A.; Singh, N. Seeds as nutraceuticals, their therapeutic potential therapeutic potential and their role in improving sports performance. *J. Phytol. Res.* **2021**, *34*, 127–138.
15. Chetrariu, A.; Dabija, A. Characterization of spent grain from a malt whisky. *Adv. Biotechnol. SGEM Int. Multidiscip. Sci. GeoConf.* **2021**, *21*, 33–40. [CrossRef]
16. Oroian, M.; Codină, G.G.; Dabija, A. Quality Characteristics of Yogurt with Different Levels of Cranberries Powder Addition of Different Particle Sizes. *J. Culin. Sci. Technol.* **2022**, 1–13. [CrossRef]
17. Iuga, M.; Mironeasa, S. Application of heat moisture treatment in wheat pasta production. *Food Control* **2021**, *128*, 108176. [CrossRef]
18. Huber, R.; Kalss, G.; Schoenlechner, R. Waffle Production: Influence of Baking Plate Material on Sticking of Waffles. *J. Food Sci.* **2017**, *82*, 61–68. [CrossRef]
19. Huber, R.; Schoenlechner, R. Waffle production: Influence of batter ingredients on sticking of waffles at baking plates—Part II: Effect of fat, leavening agent, and water. *Food Sci. Nutr.* **2017**, *5*, 513–520. [CrossRef]
20. Raza, K.; Nadeem, M.; Hussain, S.; Jabbar, S.; Din, A.; Qureshi, T.M.; Ainee, A. Development and Physico-Chemical Characterization of Date Wafers. *J. Agric. Res.* **2016**, *54*, 368–1157.
21. Escarnot, E.; Agneessens, R.; Wathelet, B.; Paquot, M. Quantitative and qualitative study of spelt and wheat fibres in varying milling fractions. *Food Chem.* **2010**, *122*, 857–863. [CrossRef]
22. Kigozi, J.; Kigozi, J.; Banadda, N.; Byaruhanga, Y.; Kaaya, A.; Lule, F. Mathematical Modeling of Sorghum Ice-Cream Cone Texture Characteristics Using the Cone Formulation. *J. Adv. Food Sci. Technol.* **2015**, *2*, 71–80.
23. Chetrariu, A.; Dabija, A. Quality Characteristics of Spelt Pasta Enriched with Spent Grain. *Agronomy* **2021**, *11*, 1824. [CrossRef]
24. Dom, Z.M.; Atikah, N.; Zulkeple, M.; Basha, R.K. Sweet Potato Peel Flour Applications in The Textural Quality of Waffle Ice Cream Cone and Other Food Products. *Adv. Agric. Food Res. J.* **2020**, *1*, 1–12.
25. Huber, R.; Schoenlechner, R. Waffle production: Influence of batter ingredients on sticking of fresh egg waffles at baking plates—Part I: Effect of starch and sugar components. *Food Sci. Nutr.* **2017**, *5*, 504–512. [CrossRef]

26. Tiwari, M.; Barooah, M.S.; Bordoloi, P.L. Quality Characterization of Wafers Enriched with Fish Powder Developed from Small Bony Fish. *J. Aquat. Food Prod. Technol.* **2020**, *29*, 775–788. [CrossRef]
27. Schmidt, C.; Geweke, I.; Struck, S.; Zahn, S.; Rohm, H. Blackcurrant pomace from juice processing as partial flour substitute in savoury crackers: Dough characteristics and product properties. *Int. J. Food Sci. Technol.* **2018**, *53*, 237–245. [CrossRef]
28. Ikram, S.; Zhang, H.; Ahmed, M.S.; Wang, J. Ultrasonic pretreatment improved the antioxidant potential of enzymatic protein hydrolysates from highland barley brewer's spent grain (BSG). *J. Food Sci.* **2020**, *85*, 1045–1059. [CrossRef]
29. Mohammed, I.K.; Charalambides, M.N.; Williams, J.G.; Rasburn, J. Modelling deformation and fracture in confectionery wafers. *Procedia Food Sci.* **2011**, *1*, 499–504. [CrossRef]
30. Hardacre, A.K.; Clark, S.M.; Riviere, S.; Monro, J.A.; Hawktns, A.J. Some textural, sensory and nutritional properties of expanded snack food wafers made from corn, lentil and other ingredients. *J. Texture Stud.* **2006**, *37*, 94–111. [CrossRef]
31. Kigozi, J.; Banadda, N.; Byaruhanga, Y.; Kaaya, A.; Musoke, L. Optimization of Texture in Sorghum Ice Cream Cone Production Using Sensory Analysis. *Open Food Sci. J.* **2015**, *8*, 18–21. [CrossRef]
32. Iuga, M. By-products ingredients in corn based pasta: Effects on the technological and quality characteristics. *Sci. Bull. Ser. F Biotechnol.* **2020**, *XXIV*, 97–106.
33. Tiefenbacher, K.F. Recipes for Wafer Disks and Crunchy Waffle Cookies. In *The Technology of Wafers and Waffles II*; Academic Press: Cambridge, MA, USA, 2019; pp. 73–80. [CrossRef]
34. Martínez-Navarrete, N.; Moraga, G.; Talens, P.; Chiralt, A. Water sorption and the plasticization effect in wafers. *Int. J. Food Sci. Technol.* **2004**, *39*, 555–562. [CrossRef]
35. Silva, C.D.S.; Greiner, R.; Queiroz, L.; Sophia, A.; Alves, B.; Almeida, L.; Leonardo, F.; de Almeida, T. Development of a gluten-free ice cream basket alternative using cowpea flour (*Vigna unguiculata*(L.) Walp), rice flour (*Oryza sativa*) and crude palm oil (*Elaeisguineensis* Jacq.). *Int. J. Gastron. Food Sci.* **2022**, *28*, 100431. [CrossRef]
36. Chetrariu, A.; Dabija, A. Valorisation of Spent Grain from Malt Whisky in the Spelt Pasta Formulation: Modelling and Optimization Study. *Appl. Sci.* **2022**, *12*, 1441. [CrossRef]
37. Goerlitz, C.D.; Harper, W.J.; Delwiche, J.F. Relationship of water activity to cone crispness as assessed by positional relative rating. *J. Sens. Stud.* **2007**, *22*, 687–694. [CrossRef]
38. Sathe, S.K.; Deshpande, S.; Salunkhe, D.K. Functional properties of lupin seed (*Lupinus mutabilis*) proteins and protein concentrates. *Food Sci.* **1982**, *47*, 491–497. [CrossRef]
39. Huang, V.T.; Luebbers, S.T.; Lindamood, J.B.; Hansen, P.M.T. Ice cream cone baking: 2. Textural characteristics of rolled sugar cones. *Top. Catal.* **1989**, *3*, 41–55. [CrossRef]

applied sciences

MDPI

Article

Valorization on the Antioxidant Potential of Volatile Oils of *Lavandula angustifolia* Mill., *Mentha piperita* L. and *Foeniculum vulgare* L. in the Production of Kefir

Ovidiu Tița [1,*], Maria Adelina Constantinescu [1], Mihaela Adriana Tița [1], Tiberius Ilie Opruța [1], Adriana Dabija [2] and Cecilia Georgescu [1,*]

[1] Department of Agricultural Sciences and Food Engineering, Lucian Blaga University of Sibiu, Doctor Ion Rațiu No. 7, 550012 Sibiu, Romania
[2] Faculty of Food Engineering, Stefan cel Mare University of Suceava, 720229 Suceava, Romania
* Correspondence: ovidiu.tita@ulbsibiu.ro (O.T.); cecilia.georgescu@ulbsibiu.ro (C.G.); Tel.: +40-745-292-031 (O.T.); +40-742-079-676 (C.G.)

Abstract: (1) Background: Natural antioxidants are health products found in many plants and may have a therapeutic effect on various diseases caused by oxidative stress. The purpose of this research is the antioxidant analysis of some kefir samples enriched with volatile oils extracted from three aromatic plants; (2) Methods: The volatile oils were extracted from lavender, fennel and mint. Four samples of kefir were made: kefir enriched with encapsulated lavender volatile oil, kefir enriched with encapsulated mint volatile oil, kefir enriched with encapsulated fennel volatile oil and a control sample without volatile oils. The analysis took place in three periods of storage: on the first day, on the 10th day and the 20th day; (3) Results: The antioxidant activity of kefir samples had decreased during the storage. The kefir sample with fennel and lavender volatile oil had the highest antioxidant activity, while the control sample had the lowest activity; (4) Conclusions: We can conclude that the volatile oils add value to the finished product.

Keywords: kefir; lavender; mint; fennel; volatile oils; antioxidants; oxidative stress; bioactive compounds

Citation: Tița, O.; Constantinescu, M.A.; Tița, M.A.; Opruța, T.I.; Dabija, A.; Georgescu, C. Valorization on the Antioxidant Potential of Volatile Oils of *Lavandula angustifolia* Mill., *Mentha piperita* L. and *Foeniculum vulgare* L. in the Production of Kefir. *Appl. Sci.* **2022**, *12*, 10287. https://doi.org/10.3390/app122010287

Academic Editor: Ioannis G. Roussis

Received: 9 September 2022
Accepted: 10 October 2022
Published: 13 October 2022

1. Introduction

Stress is a response to environmental stressors that affect the lives of individuals, and relationships with other people. Stress can cause social, physical and mental problems. People may experience many symptoms, such as irritability, difficulty concentrating, sleep disturbances, diarrhea, fatigue, anger, palpitations, frequent urination, constipation and headaches [1]. Long periods of stress lead to reduced cell proliferation and neurogenesis, thus playing a significant role in depression and Alzheimer's disease (AD). Environmental stressors cause immune system symptoms and hormonal effects, neuroplastic changes that result in neurogenesis and impaired neurotransmission. Eventually, this can lead to neurodegenerative changes, dementia, and cognitive decline [2,3].

At high levels, oxidative stress could degrade lipids, proteins and deoxyribonucleic acid (DNA). This leads to inflammation and cell death. It also has an important role in many cardiovascular diseases, such as heart failure, cardiac arrhythmia, atherosclerosis and ischemia-reperfusion injury [4–7]. It can influence the development of leukemia [8] and can harm male fertility [9].

Antioxidants are health care products that can be sold worldwide without a prescription. Many types of research show the benefits of antioxidants, but only a few mention their possible harmful effects. The balance between free radicals and antioxidants in the human body is offset when any of these predominate [10].

Plants activate antioxidant defense mechanisms under abiotic stress, which helps maintain the integrity of cellular components and attenuates oxidative damage [11]. Important compounds with antioxidant properties found in medicinal and aromatic plants include polyphenols, stilbene, flavonoids, chalcone, capsaicinoids and casinoids, lignans and curcuminoids, carotenoids, isothiocyanates and catechins. Thus, natural antioxidants have therapeutic potential for many diseases caused by stress [12].

The antioxidant activity of plants is the result of their phenolic compounds. Phenolic compounds are functional compounds synthesized by plants. They play an important role in human nutrition for a healthy life, and mint is an important source [13]. Mint (*Mentha piperita* L.) is a medicinal and aromatic plant used in traditional medicine and drugs for its antimicrobial and antioxidant properties [14]. The antioxidant activity of mint prevents oxidative stress at the cellular level by its chemical composition [15]. Mint-derived secondary metabolites are antioxidants that have immunostimulatory, cardio-tonic and antiviral properties [16]. Mint contains volatile components, flavonoids, organic acids, and quinones, necessary for the digestive system, central nervous system, and respiratory system. It has antioxidant, antimicrobial, anti-inflammatory and anesthetic properties [14,17–19]. The most important components of volatile oil are menthol, menthone, menthofuran, isomenthone, caryophyllene, eucalyptol, linalool, limonene, carvone, pulegone and α-terpinol [20]. In the study conducted by Bleiziffer et al., in 2017, it was observed that mint and sage oil are the freest amino acids (>4 $mg \cdot g^{-1}$) [21]. In 2013, Tsai et al., demonstrated that the most important compound of mint essential oil is menthol and it has a high antimicrobial activity against *Pseudomonas aeruginosa*, *Staphylococcus aureus* and *Escherichia coli*. Antioxidant activity has been demonstrated using the 2,2-Diphenyl-1-picrylhydrazyl (DPPH) method and the β-Carovtene-linoleic acid test [14]. Brahmi et al., in 2018 demonstrated that mint essential oil is a good source of compounds with antioxidant, cytoprotective and immunomodulatory properties [22]. In 2019, Park et al., using high-performance liquid chromatography (HPLC) analysis demonstrated that the mint flower has a higher level of phenolic compounds, flavonoids and anthocyanins that offer antimicrobial and antioxidant effects compared to the stem and leaves [23]. Two other studies conducted in 2019 and 2021 by Wu et al., and Hejna et al., respectively, demonstrated the antioxidant activity using the Trolox equivalent antioxidant capacity test and the DPPH method [24,25].

Fennel (*Foeniculum vulgare* L.) is part of the family *Apiaceae* Lindl. [26]. It is a medicinal plant used as an analgesic, diuretic, anti-inflammatory and antispasmodic [27–29]. The essential oil is used as a component in cosmetics, pharmaceuticals and flavoring agents in different food products. Fennel essential oils have hepatoprotective and antispasmodic effects. It is also known for its diuretic, analgesic, anti-inflammatory and antioxidant properties [26,27,30]. The most important components of the essential oil are α-pinene [31], estragol, trans-anethole, a-phellandrene and fenchone [27]. Numerous studies have shown its antioxidant, anti-inflammatory and antimicrobial action [28,30,32–36]. In 2021, Korinek et al., showed that volatile fennel oil had a positive effect on human neutrophils [37] and Mazandrani et al., in 2015 studied the antioxidant and antimicrobial effects during the storage of silver carp fillet [38].

Lavender (*Lavandula angustifolia* Mill.) belongs to the *Lamiaceae* Martinov family [39] and is used as a culinary herb and medicine for burns, skin wounds, headaches, digestive problems, insect bites [40], analgesic, antiseptic, sedative and urinary tract improvement [41,42]. Lavender is rich in essential oil, and studies have concluded that it has antimicrobial, antioxidant, insecticidal and anticholinesterase inhibitory activities [39,40]. The essential oil contains many compounds such as borneol, fenchon cineole, terpineol, camphor [41], linalool, linalyl acetate and β-ocimene [43]. Research has concluded that the essential oil has antimicrobial, antioxidant, anti-inflammatory and antiseptic properties due to its high content of flavonoids and polyphenols [39,40,44–48]. In 2017, Küçükyilmaz et al., concluded that supplementing levels of 24 to 48 mg/kg of lavender essential oil is effective in exerting antioxidant and growth-promoting activities [49]. In 2021, Adaszynska-Skwirzynska et al., showed that by adding it as an additive to the drinking water of broilers it can become an excellent alternative to banned antibiotics to stimulate growth [50]. In a study performed on mice by Kozics et al., in 2017, it was concluded that cell death mediated by oxidative stress in liver tissue can be prevented by treatment with lavender essential oil [51].

Kefir is an acidic fermented milk drink with a creamy consistency produced by bacteria through the alcoholic and lactic fermentation of kefir grains. Kefir is a milk-based drink with antioxidant properties and benefits, such as reduced symptoms of lactose intolerance, immune system stimulation, lower cholesterol, and antimutagenic and anticancer activity. The health benefits have been attributed to various bioactive compounds such as vitamins, minerals, lipids, proteins, amino acids and trace elements [52–57]. The antioxidant and anti-inflammatory properties of kefir have been demonstrated in numerous studies [58–63]. Sabokbar et al., in 2014 and 2015 made drinks based on kefir and added apple and pomegranate juice, and the results obtained showed a high antioxidant activity [64–66].

Current studies targeting the use of encapsulated volatile oils in foods are limited [67]. The encapsulation of oils is used especially for their antimicrobial effect to inhibit certain pathogenic microorganisms that can develop in food products [68–71].

In the present research, we propose the antioxidant analysis of some kefir samples enriched with volatile oils extracted from aromatic plants. The volatile oils were extracted from lavender, fennel and mint, and due to their sensitivity to external factors, it was decided to encapsulate them. Four types of kefir were made: kefir enriched with encapsulated mint volatile oil, kefir enriched with encapsulated lavender volatile oil and kefir enriched with encapsulated fennel volatile oil and a control sample in which no volatile oils were added. The samples were made in the laboratories of the Faculty of Agricultural Sciences, Food Industry and Environmental Protection in Sibiu, Romania. To ensure that these products are accepted by consumers, we initially tested the sensory acceptance and texture of these kefir samples [72].

2. Materials and Methods

2.1. Extraction of Volatile Oils

For the extraction were used dried and crushed lavender (*Lavandula angustifolia* Mill.), mint (*Mentha piperita* L.) and fennel (*Foeniculum vulgare* L.) seeds. For the extraction, the Neo-Clevenger apparatus modified by Moritz with water vapor was used. This method was used according to the Romanian Pharmacopoeia edition X [73]. Table 1 presents the volatile oil extraction efficiency. The highest extraction efficiency was obtained in the case of fennel due to the use of seeds. For volatile oils extracted from the herb, the mint had the highest efficiency and lavender had the lowest.

Table 1. Volatile oil extraction efficiency.

Aromatic Plant	Efficiency [%]
Mint (herb)	2.93
Fennel (seeds)	5.17
Lavender (herb)	1.21

The volatile oils obtained had a pale yellow to greenish yellow color—for mint—and a characteristic odor, the strongest being that of fennel, followed by lavender and mint. All the plants used came from authorized plantations in Sibiu, Romania [72].

2.2. Encapsulation of Volatile Oils

To protect the volatile oils from environmental factors, we encapsulated them in sodium alginate. Three types of oils were used to make the capsules with volatile oils: mint, fennel and lavender. 30 μL of each volatile oil in 10 mL of 2% sodium alginate solution was added [72].

The 2% sodium alginate solution was prepared the day before forming the capsules, this was necessary for it to clear.

After mixing the volatile oil with the sodium alginate, we proceeded to the formation of the capsules. The solution was pipetted dropwise into 250 mL of 2% calcium chloride in an ultrasonic water bath (manufacturer Bandelin Sonorex) at 25 °C. After pipetting the entire solution of sodium alginate with volatile oil, solution was held in the water bath for another 10 min. A solution of calcium chloride and capsules with volatile oils was filtered with the help of a muslin cloth. The capsules were washed with distilled water, after which they were left to dry at room temperature for 10 min. The procedure was identical for each type of volatile oil.

The structure of a capsule was gelatinous and the size was 240 μm. They had an opalescent white color and a characteristic odor of the plant [72].

2.3. Obtaining Samples of Kefir with Volatile Oils Encapsulated

The pasteurization of raw milk was made for 25 min at 85–90 °C. The starter culture of 0.15 g:2 L of milk and pre-mixed powder milk of 150 g:2 L of milk were added after cooling the milk to 20 °C. The capsules were added at 1 g for every 100 g of kefir. Afterwards the milk was placed into plastic containers to a final weight of 250 g. The thermostat had two stages: the first thermostat at 18 °C for 10 h, and the second thermostat at 10 °C for 8 h. The glasses with samples were covered with cling film and stored at 4 °C. Raw milk came from healthy animals from an authorized farm in Vulpăr village, Sibiu, Romania. The starter culture was a mix of LYOFAST cultures, MS 059 DT, manufacturer SACCO [72].

2.4. Extraction of Compounds to Determine Antioxidant Capacity

For the extraction of compounds the method adapted from Patel et al. (2016) was used [74]. 0.5 g of the sample was extracted with 10 mL solvent containing methanol, water and hydrochloric acid 0.12 M = 70:29:1 ($v/v/v$) for 24 h at room temperature. For 30 min the mixture was kept in the ultrasonic bath (manufacturer Bandelin Sonorex) at 25 °C, then it was centrifuged at 8000 rpm for 10 min, and the supernatant was collected. The second extraction was performed exactly like the first, except the time spent in the ultrasonic bath was 15 min. The total quantity of supernatant was evaporated using the rotary evaporator (manufacturer Hei-VAP Precision). After evaporation the residue was taken up with 10 mL of methanol and then the mixture was filtered and filled with methanol to a total volume of 10 mL [75].

2.5. Determination of Antioxidant Activity

To determine the antioxidant activity a method adapted from Tylkowski et al. (2011) was used [76]. A DPPH solution with a concentration of 25 μg/mL was prepared by solubilizing a quantity of DPPH in absolute methanol. To complete the solubilization, the mixture was prepared in advance for 2 h. Seven DPPH solutions made from methanol in the range of 0.1–1 μg/mL were prepared for plotting the calibration curve, starting from a stock DPPH solution. 970 μL of a 25 μg/mL DPPH solution was added into 30 μL od methanol extract, obtained using the extraction described above. Absorbance was measured at 515 nm using a spectrophotometer (manufacturer Shimadzu 1900 UV-VIS), and an absorbance graph was plotted as a function of concentration [75].

The concentration of the solutions was calculated with Formula (1):

$$C = \frac{A - i.c.c}{s.c.c.} \quad (1)$$

where:

C—the concentration of the DPPH solution (μg/mL);

A—sample absorbance;

$i.c.c.$—the intercept of the calibration curve;

$s.c.c.$—the slope of the calibration curve.

The antioxidant activity expressed as a DPPH radical inhibition process was calculated according to Formula (2):

$$AA = \frac{C_0 - C_1}{C_0} \cdot 100 \quad (2)$$

where:

AA—antioxidant activity (%);

C_0—the concentration of the blank solution (DPPH solution in methanol without sample);

C_1—the remaining concentration of DPPH in the sample [76].

For each sample, three repeats were performed, and the analysis took place at three periods of storage: on the first day, on the 10th day and the 20th day.

2.6. Statistical Analysis

All results obtained from the determination of the antioxidant activity were expressed according to the following statistical indicators: mean value, standard error of the mean (SEM), median, standard deviation (SD), maximum value, minimum value, skewness. Statistical significance was determined using correlation analysis of data obtained using the Pearson correlation. Thus, we correlated the data obtained for the samples with volatile oils with those of the control sample [77]. All statistical analyses were performed using Minitab version 14 and $p < 0.05$ was considered significant.

3. Results

3.1. Realization of the Calibration Curve to Determine the Antioxidant Activity

To achieve the calibration curve, we prepared seven DPPH solutions made from methanol in the range of 0.1–1.0 μg/mL starting from a stock solution of DPPH.

According to Figure 1, the slope of the calibration curve is 0.0411, and the intercept of the calibration curve is 0.0349. The absorbance at which the measurements were made is equal to 515 nm. Depending on the absorbance of each sample, we determined the concentration of the solutions, and then the antioxidant activity.

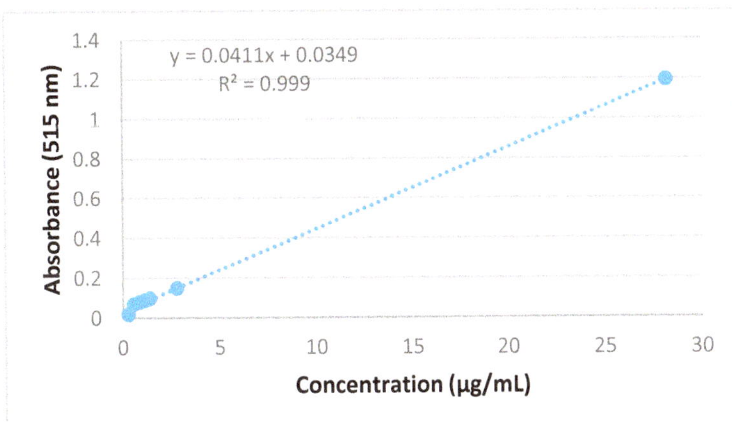

Figure 1. Calibration curve with DPPH to determine antioxidant activity.

3.2. Determination of Antioxidant Activity for Volatile Oil Samples

Figure 2 shows the antioxidant activity for the three assortments of volatile oils used in encapsulation. Fennel volatile oil has a mean value of 30.623 ± 0.012. The median is 30.63 and the histogram is tilted to the left, the skewness being −1.73. The mean value for the antioxidant activity of lavender volatile oil is 12,757 and the standard deviation is 0.015. The skewness is −0.9, the histogram is tilted to the left, and the median value is 12.76. The mean antioxidant activity for mint volatile oil is 25.62 and the standard deviation is 0.01. The median is 25.62 and the histogram is perfectly symmetric because the skewness is 0.

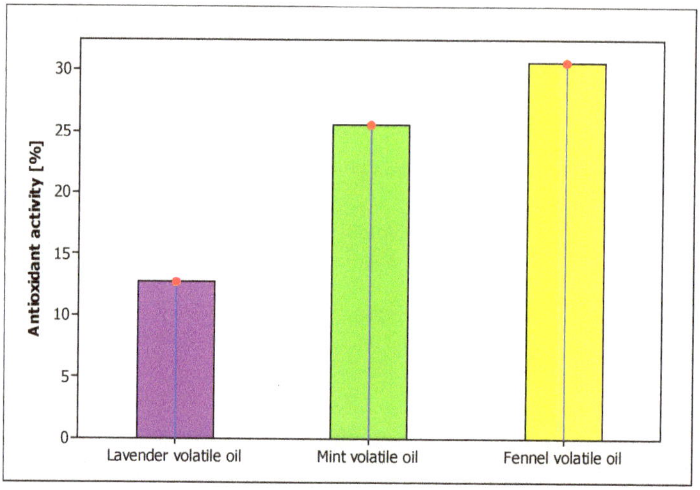

Figure 2. Antioxidant activity for the three assortments of volatile oils used in encapsulation.

3.3. Determination of Antioxidant Activity for Kefir Samples Enriched with Encapsulated Volatile Oils

Figure 3 shows the comparative variation of antioxidant activity in the kefir sample with lavender volatile oil and the control sample. On the first day of storage, the antioxidant activity mean value for the sample of kefir with lavender volatile oil is 65.39, and the standard deviation is 0.021. The median value is 65.38, and the skewness is 1.29, so the histogram is oriented to the right. In the case of the control sample, the mean value of the antioxidant activity is 59.96 and the standard deviation is 0.006. The median value is 59.96, and the skewness is −1.73, the histogram is oriented to the left. On the 10th day, the kefir with lavender volatile oil has an antioxidant activity mean value of 59.78, and the standard deviation is 0.006. The median value is 59.78, and the skewness is 1.73, with the histogram being oriented to the right. For the control sample, the mean value of the antioxidant activity is 35.05 and the standard deviation is 0.01, the median value is 35.05 and the histogram is perfectly symmetrical because the skewness is 0. On the 20th day, the kefir with lavender volatile oil has an antioxidant activity mean value of 13.71 ± 0.01, the median value is 13.71 and the histogram is perfectly symmetrical because the skewness is 0. In the case of the control sample, the mean value of the antioxidant activity is 10.95 and the standard deviation is 0.012. The median value is 10.94 and the skewness is 1.73, the histogram is oriented to the right. On the first day, the Pearson correlation coefficient between the control sample and the kefir with lavender volatile oil is 0.693, which indicates a strong positive association between the two variables. On the 10th and 20th days, the correlation coefficient between the two samples is 0.177, the association being weakly positive.

Figure 4 shows the comparative variation of the antioxidant activity in the kefir sample with mint volatile oil and the control sample. On day 1, the kefir with mint volatile oil has an average antioxidant activity value of 72.47 and the standard deviation is 0.006. The median value is 72.47 and the skewness is −1.73, the histogram is oriented to the left. The mean value of the antioxidant activity for the control sample is 59.96 and the standard deviation is 0.006. The median value is 59.96, and the skewness is −1.73, the histogram is oriented to the left. On day 10 of storage, the antioxidant activity mean value for the kefir sample with mint volatile oil is 41.15 and the standard deviation is 0.01. The value of the median is 41.15, and the histogram is perfectly symmetrical because the skewness is 0. For the control sample, the mean value of the antioxidant activity is 35.05 and the standard deviation is 0.01, the median value is 35.05 and the histogram is perfectly symmetrical because the skewness is 0. On the 20th day of storage, the antioxidant activity mean value for the kefir sample with volatile mint oil is 11.64 ± 0.012. The median value is 11.65 and the skewness is −1.73, the histogram is oriented to the left. The mean value of the antioxidant activity for the control sample is 10.95 and the standard deviation is 0.012. The median value is 10.94 and the skewness is 1.73, the histogram is oriented to the right. For the correlation between the kefir sample with mint volatile oil and the control sample, the Pearson coefficient from day 1 is 1, the association being strongly positive. On day 10 the coefficient is equal to −0.5, the association between the two variables being moderately negative, and on day 20 the coefficient is equal to 0.5, the association between the two variables being moderately positive.

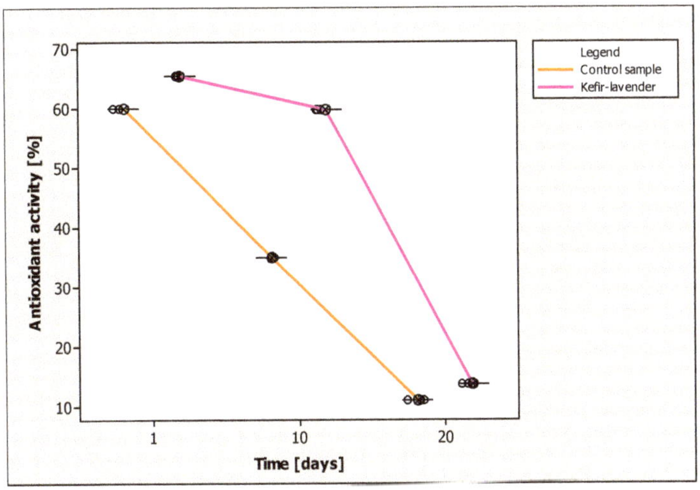

Figure 3. Comparative variation of antioxidant activity in kefir sample enriched with encapsulated lavender volatile oil and control sample.

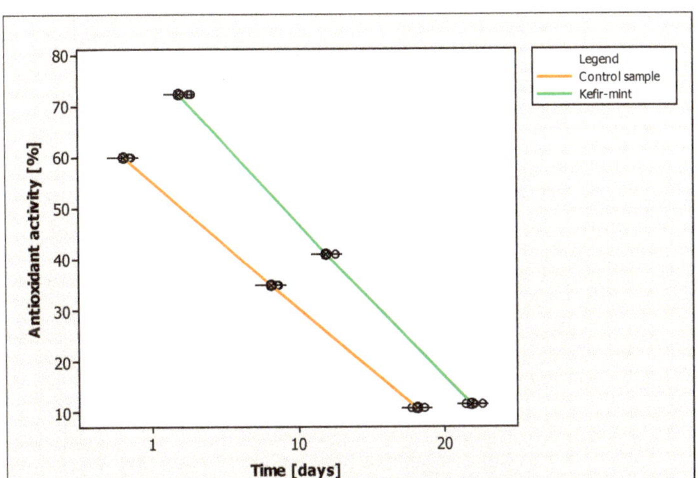

Figure 4. Comparative variation of antioxidant activity in kefir sample enriched with encapsulated mint volatile oil and control sample.

Figure 5 shows the comparative variation of antioxidant activity in the kefir with fennel volatile oil and the control sample. On the first day of storage, the kefir sample with fennel volatile oil has an antioxidant activity mean value of 78.59 ± 0.006. The median is 78.59, and the skewness is −1.73, the histogram being oriented to the left. The mean value of the antioxidant activity for the control sample is 59.96 and the standard deviation is 0.006. The median value is 59.96, and the skewness is −1.73, the histogram being oriented to the left. On the 10th day, the kefir with fennel volatile oil has an antioxidant activity mean value of 41.32 ± 0.017. The median is 41.33, and the skewness is −1.73, the histogram being oriented to the left. For the control sample, the mean value of the antioxidant activity is 35.05 and the standard deviation is 0.01, the median value is 35.05 and the histogram is perfectly symmetrical because the skewness is 0. On the 20th day, the mean value of the antioxidant activity for the kefir sample with fennel volatile oil is 12.59 ± 0.012, and the skewness is −1.73, the histogram being oriented to the left. The mean value of the antioxidant activity for the control sample is 10.95 and the standard deviation is 0.012. The median value is 10.94 and the skewness is 1.73, the histogram being oriented to the right. In the case of the kefir with fennel volatile oil and the control sample, the Pearson coefficient is 1 on the first day, the association being strongly positive. On the 10th day, the correlation coefficient is 0.866, the association being strongly positive, and on the 20th day, the correlation coefficient is 0.5, the association being moderately positive between the two variables.

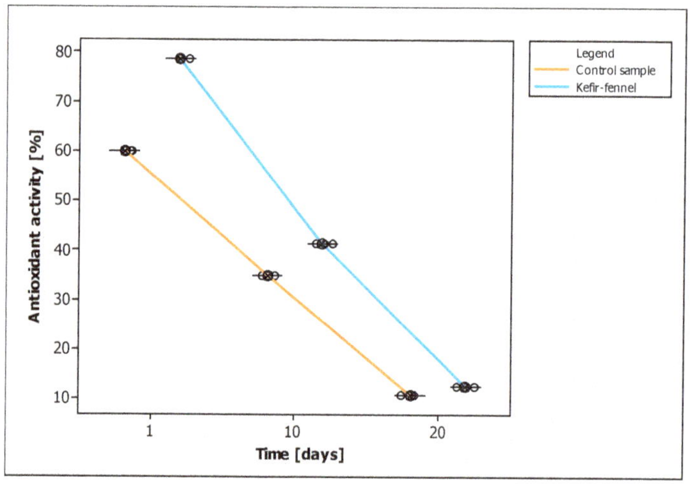

Figure 5. Comparative variation of antioxidant activity in kefir sample enriched with encapsulated fennel volatile oil and control sample.

Figure 6 shows the decline in antioxidant activity on day 10 and day 20 of storage compared to day 1 of storage for kefir samples. For the kefir sample with fennel volatile oil, the decline on day 10 compared to day 1 has a mean value of 8.75%, and that on day 20 is 79.03% compared to day 1. For the kefir sample with lavender volatile oil, the decline on day 10 compared to day 1 is 43.22%, and that on day 20 is 83.93%. For the kefir sample with mint volatile oil, the decline on day 10 is 47.42% and on day 20 it is 83.98%. The control sample has a decline in antioxidant activity on day 10 compared to day 1 of 41.54%, and on day 20 has a decline of 81.74%. We can conclude that the kefir with fennel volatile oil is the most stable in terms of antioxidant activity because it has the lowest value of decline compared to day 1. In the case of the other three kefir samples, the value of the decline on day 10 is approximately equal. The largest decrease is recorded between day 10 and day 20 of storage, this being visible by the high value of the decline on day 20 compared to day 1 for all four kefir samples.

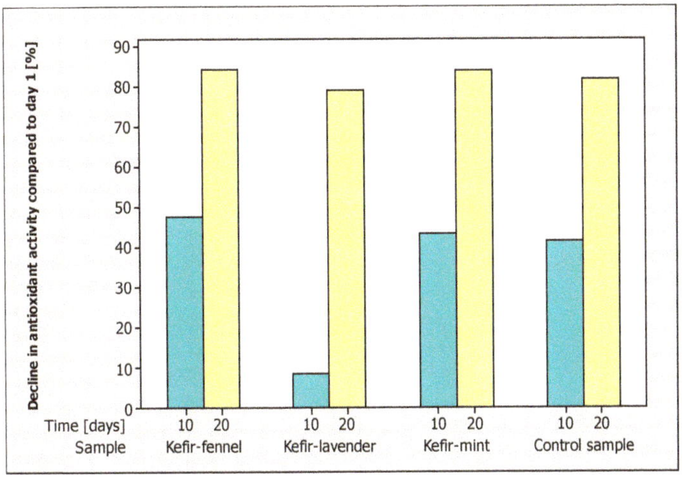

Figure 6. The decline in antioxidant activity on day 10 and day 20 of storage compared to day 1 of storage for kefir samples.

4. Discussion

Stress is one of the factors that can affect people's mental health. Long-term stress leads to various cellular and neuronal disorders, playing a significant role in the onset of depression and Alzheimer's disease [1–3].

Oxidative stress influences physiological and pathological conditions leading to the appearance of various cardiovascular and neuronal diseases. To avoid and reduce the effects of stress on the human body, studies recommend the external intake of antioxidants. Medicinal and aromatic plants have been excellent sources of antioxidants. Creating food products that meet certain food needs, but also contribute to health [78] is a growing interest for many processors in the food industry.

The antioxidant character of the three plants used in this study is given by their different compounds. The main compound in fennel volatile oil is α-pinene [31]. α-pinene is a natural and active monoterpene that is very often used as a flavoring or pharmaceutical agent [79]. This compound is known to have notable antioxidant activity [80]. The antioxidant character of the volatile mint oil is given by menthone, caryophyllene and linalool [20]. Menthone is generally used in the flavoring industry. It is very often found in combination with menthol in various types of volatile oil. This compound has antimicrobial and antioxidant properties, its insecticidal potential has been proven in rice control [81]. Caryophyllene is a sesquiterpene found in reasonable amounts in the volatile oils of many plants [82]. Following studies, it has been shown that caryophyllene gives volatile oil antioxidant and antimicrobial properties [83]. Linalool is a linear monoterpene alcohol [81] that has been shown to have antimicrobial and antioxidant activity [84]. The compounds present in lavender volatile oil that give it antioxidant properties are linalool [43] and terpineol [41]. Terpineol is a monoterpene that has a wide range of properties: antimicrobial [85], antioxidant, anti-inflammatory, anticancer, and insecticidal [81]. The mechanism of action may be due to cytoplasmic membrane compromise or organic membrane penetration that induces deformation, damage and ultimately death of microbial cells [86].

In 2019 we started the study of the antioxidant activity of various yoghurt samples enriched with encapsulated volatile oils, and due to the positive results obtained we decided to test these effects on other dairy products. We chose kefir because of its notoriety and the benefits it brings to consumers' health. Before the antioxidant analysis, we analyzed consumer acceptance of the new kefir assortments through sensory and textural analysis.

The purpose of this study was to produce a dairy product with high nutritional properties. Thus, three types of kefir with volatile oils were made: lavender, mint and fennel. The volatile oils are sensitive to various external factors, and they were encapsulated and introduced as spherical capsules into the dairy product. According to studies, these volatile oils have a high antimicrobial and antioxidant capacity.

Initially, we tested the antioxidant activity of the volatile oils obtained from their extraction from plants harvested from authorized crops in Sibiu, Romania. The highest value of the antioxidant activity of pure volatile oils belongs to the volatile oil of fennel, followed by the volatile oil of mint, and the lowest value belongs to the volatile oil of lavender. Three spectrophotometer readings were taken for each volatile oil.

The antioxidant activity of kefir samples decreased during the twenty days of storage. The highest values of antioxidant activity were obtained on the first day of storage, and the lowest values were obtained on the 20th day of storage. The largest decrease was recorded between day 10 and day 20 of storage for all kefir samples. The highest antioxidant activity for the entire storage period was the kefir sample with fennel volatile oil. This was followed by the kefir with mint volatile oil, and then the lavender volatile oil sample. The control sample showed the lowest values over the entire storage period.

Compared to the study carried out in 2019, the antioxidant activity of kefir samples was higher. The fennel volatile oil brought a significant contribution to both types of dairy products; both the kefir and yogurt samples obtained the highest values during the entire analysis period. In the current study, kefir samples with lavender volatile oil had the lowest antioxidant activity, and in the 2019 study, the yogurt sample with mint volatile oil had the lowest antioxidant activity. In both cases, the volatile oils used added value to the dairy product, giving it a superior antioxidant capacity.

5. Conclusions

We can conclude that volatile oils add value to the finished product. All three samples with volatile oils have good results compared to the control sample for the entire analysis period. The use of medicinal and aromatic plants in the preparation of food is an excellent alternative due to their antioxidant properties.

All these aspects show that the obtained products are an important source of antioxidant compounds that can bring benefits to the health of consumers and can also increase the shelf life of the product due to the incorporation of bioactive compounds.

Author Contributions: Conceptualization, O.T., M.A.C. and A.D.; methodology, M.A.T., T.I.O. and C.G.; software, M.A.C.; validation, O.T., M.A.T. and C.G.; formal analysis, O.T. and T.I.O.; investigation, M.A.C. and A.D.; resources, C.G. and A.D.; data curation, M.A.C. and T.I.O.; writing—original draft preparation, O.T., M.A.C. and A.D.; writing—review and editing, M.A.T., T.I.O. and C.G.; visualization, O.T., M.A.T. and C.G.; supervision, O.T.; project administration, O.T.; funding acquisition, O.T. All authors have read and agreed to the published version of the manuscript.

Funding: This research was funded by the Lucian Blaga University of Sibiu and Hasso Plattner Foundation, grant number LBUS-IRG-2021-07/No. 2946—26 July 2021.

Institutional Review Board Statement: Not applicable.

Informed Consent Statement: Not applicable.

Data Availability Statement: Not applicable.

Acknowledgments: We would like to express our sincere gratitude to the Research Center in Biotechnology and Food Engineering (C.C.B.I.A.), the Lucian Blaga University of Sibiu and the Hasso Plattner Foundation for their support provided throughout the research period. We also appreciate the editor and the anonymous reviewers for their constructive comments and insightful suggestions on the manuscript.

Conflicts of Interest: The authors declare no conflict of interest.

References

1. Atalıkoğlu Başkan, S.; Güneş, D. The Effect of Stress Perceived by Adults Due to the COVID-19 Pandemic on Their Sleep Quality. *J. Turk. Sleep Med.* **2021**, *8*, 57–66. [CrossRef]
2. Juszczyk, G.; Mikulska, J.; Kasperek, K.; Pietrzak, D.; Mrozek, W.; Herbet, M. Chronic stress and oxidative stress as common factors of the pathogenesis of depression and alzheimer's disease; the role of antioxidants in prevention and treatment. *Antioxidants* **2021**, *10*, 1439. [CrossRef] [PubMed]
3. Bukhari, S.N.A. Dietary Polyphenols as Therapeutic Intervention for Alzheimer's Disease: A Mechanistic Insight. *Antioxidants* **2022**, *11*, 554. [CrossRef] [PubMed]
4. Hasanuzzaman, M.; Borhannuddin Bhuyan, M.H.M.; Zulfiqar, F.; Raza, A.; Mohsin, S.M.; Al Mahmud, J.; Fujita, M.; Fotopoulos, V. Reactive oxygen species and antioxidant defense in plants under abiotic stress: Revisiting the crucial role of a universal defense regulator. *Antioxidants* **2020**, *9*, 681. [CrossRef]
5. Wang, W.; Kang, P.M. Oxidative Stress and Antioxidant Treatments in Cardiovascular Diseases. *Antioxidants* **2020**, *9*, 1292. [CrossRef]
6. Zalba, G. Oxidative stress in vascular pathophysiology: Still much to learn. *Antioxidants* **2021**, *10*, 673. [CrossRef]
7. Han, M.; Lee, D.; Lee, S.H.; Kim, T.H. Oxidative stress and antioxidant pathway in allergic rhinitis. *Antioxidants* **2021**, *10*, 1266. [CrossRef]
8. Dong, C.; Zhang, N.J.; Zhang, L.J. Oxidative stress in leukemia and antioxidant treatment. *Chin. Med. J.* **2021**, *134*, 1897–1907. [CrossRef]
9. Sengupta, P.; Dutta, S.; Roychoudhury, S.; D'souza, U.J.A.; Govindasamy, K.; Kolesarova, A. COVID-19, Oxidative Stress and Male Reproduction: Possible Role of Antioxidants. *Antioxidants* **2022**, *11*, 548. [CrossRef]
10. Villanueva, C.; Kross, R.D. Antioxidant-induced stress. *Int. J. Mol. Sci.* **2012**, *13*, 2091–2109. [CrossRef]
11. Caverzan, A.; Casassola, A.; Brammer, S.P. Antioxidant responses of wheat plants under stress. *Genet. Mol. Biol.* **2016**, *39*, 1–6. [CrossRef] [PubMed]
12. Pérez-Torres, I.; Castrejón-Téllez, V.; Soto, M.E.; Rubio-Ruiz, M.E.; Manzano-Pech, L.; Guarner-Lans, V. Oxidative Stress, Plant Natural Antioxidants, and Obesity. *Int. J. Mol. Sci.* **2021**, *22*, 1786. [CrossRef] [PubMed]
13. Elmastaş, M.; Telci, İ.; Akşit, H.; Erenler, R. Comparison of total phenolic contents and antioxidant capacities in mint genotypes used as spices. *Turk. J. Biochem.* **2015**, *40*, 456–462.
14. Tsai, M.L.; Wu, C.T.; Lin, T.F.; Lin, W.C.; Huang, Y.C.; Yang, C.H. Chemical composition and biological properties of essential oils of two mint species. *Trop. J. Pharm. Res.* **2013**, *12*, 577–582. [CrossRef]
15. Parham, S.; Kharazi, A.Z.; Bakhsheshi-Rad, H.R.; Nur, H.; Ismail, A.F.; Sharif, S.; Ramakrishna, S.; Berto, F. Antioxidant, antimicrobial and antiviral properties of herbal materials. *Antioxidants* **2020**, *9*, 1309. [CrossRef]
16. Ilić, Z.S.; Milenković, L.; Tmušić, N.; Stanojević, L.; Stanojević, J.; Cvetković, D. Essential oils content, composition and antioxidant activity of lemon balm, mint and sweet basil from Serbia. *LWT* **2022**, *153*, 112210. [CrossRef]
17. Bajaj, S.; Urooj, A.; Prabhasankar, P. Effect of incorporation of mint on texture, colour and sensory parameters of biscuits. *Int. J. Food Prop.* **2006**, *9*, 691–700. [CrossRef]
18. Benito, M.E.G.; Kremer, C.; Ibáñez, M.A.; Martín, C. Effect of antioxidants on the genetic stability of cryopreserved mint shoot tips by encapsulation-dehydration. *Plant Cell Tissue Organ Cult.* **2016**, *127*, 359–368. [CrossRef]
19. Kapp, K.; Orav, A.; Roasto, M.; Raal, A.; Püssa, T.; Vuorela, H.; Tammela, P.; Vuorela, P. Composition and antibacterial effect of mint flavorings in candies and food supplements. *Planta Med.* **2020**, *86*, 1089–1096. [CrossRef]
20. Kennedy, D.; Okello, E.; Chazot, P.; Howes, M.J.; Ohiomokhare, S.; Jackson, P.; Haskell-Ramsay, C.; Khan, J.; Forster, J.; Wightman, E. Volatile Terpenes and Brain Function: Investigation of the Cognitive and Mood Effects of Mentha x Piperita L. Essential Oil with In Vitro Properties Relevant to Central Nervous System Function. *Nutrients* **2018**, *10*, 1029. [CrossRef]
21. Bleiziffer, R.; Mesaros, C.; Suvar, S.; Podea, P.; Iordache, A.; Yudin, F.D.; Culea, M. Comparative characterization of basil, mint and sage extracts. *Stud. Univ. Babes-Bolyai Chem.* **2017**, *62*, 373–385. [CrossRef]
22. Brahmi, F.; Nury, T.; Debbabi, M.; Hadj-Ahmed, S.; Zarrouk, A.; Prost, M.; Madani, K.; Boulekbache-Makhlouf, L.; Lizard, G. Evaluation of antioxidant, anti-inflammatory and cytoprotective properties of ethanolic mint extracts from Algeria on 7-ketocholesterol-treated murine RAW 264.7 macrophages. *Antioxidants* **2018**, *7*, 184. [CrossRef] [PubMed]
23. Park, C.H.; Yeo, H.J.; Baskar, T.B.; Park, Y.E.; Park, J.S.; Lee, S.Y.; Park, S.U. In vitro antioxidant and antimicrobial properties of flower, leaf, and stem extracts of Korean mint. *Antioxidants* **2019**, *8*, 75. [CrossRef] [PubMed]
24. Hejna, M.; Kovanda, L.; Rossi, L.; Liu, Y. Mint oils: In vitro ability to perform anti-inflammatory, antioxidant, and antimicrobial activities and to enhance intestinal barrier integrity. *Antioxidants* **2021**, *10*, 1004. [CrossRef]
25. Wu, Z.; Tan, B.; Liu, Y.; Dunn, J.; Martorell Guerola, P.; Tortajada, M.; Cao, Z.; Ji, P. Chemical Composition and Antioxidant Properties of Essential Oils from Peppermint, Native Spearmint and Scotch Spearmint. *Molecules* **2019**, *24*, 2825. [CrossRef]
26. Sakr, S.A.; Shalaby, S.Y.; Beder, R.H. Ameliorative Effect of Fennel Oil on Cyclophosphamide Induced Hepatotoxicity in Albino Rats. *Br. J. Pharm. Res.* **2017**, *17*, 1–11. [CrossRef]
27. Ahmed, A.F.; Shi, M.; Liu, C.; Kang, W. Comparative analysis of antioxidant activities of essential oils and extracts of fennel (*Foeniculum vulgare* Mill.) seeds from Egypt and China. *Food Sci. Hum. Wellness* **2019**, *8*, 67–72. [CrossRef]
28. Anwar, F.; Ali, M.; Hussain, A.I.; Shahid, M. Antioxidant and antimicrobial activities of essential oil and extracts of fennel (*Foeniculum vulgare* Mill.) seeds from Pakistan. *Flavour Fragr. J.* **2009**, *24*, 170–176. [CrossRef]

29. Ahmad, B.S.; Thierry, T.; Saad, Z.; Hijazi, A.; Cerny, M.; Kanaan, H.; Chokr, A.; Merah, O. Fennel oil and by-products seed characterization and their potential applications. *Ind. Crops Prod.* **2018**, *111*, 92–98. [CrossRef]
30. Shahat, A.A.; Ibrahim, A.Y.; Hendawy, S.F.; Omer, E.A.; Hammouda, F.M.; Abdel-Rahman, F.H.; Saleh, M.A. Chemical composition, antimicrobial and antioxidant activities of essential oils from organically cultivated fennel cultivars. *Molecules* **2011**, *16*, 1366–1377. [CrossRef]
31. Zeng, H.; Chen, X.; Liang, J. In vitro antifungal activity and mechanism of essential oil from fennel (*Foeniculum vulgare* L.) on dermatophyte species. *J. Med. Microbiol.* **2015**, *64*, 93–103. [CrossRef] [PubMed]
32. Kara, M.; Soylu, S.; Türkmen, M.; Kaya, D.A. Determination and antifungal activities of laurel and fennel essential oils against fungal disease agents of cypress seedlings. *J. Tekirdag Agric. Fac.* **2020**, *17*, 264–275.
33. Angelov, G.; Boyadzhieva, S. Extraction of fennel (*Foeniculum vulgare*) seeds: Process optimization and antioxidant capacity of the extracts. *Chem. Biochem. Eng. Q.* **2016**, *30*, 245–253. [CrossRef]
34. Imbabi, T.; Sabeq, I.; Osman, A.; Mahmoud, K.; Amer, S.A.; Hassan, A.M.; Kostomakhin, N.; Habashy, W.; Easa, A.A. Impact of fennel essential oil as an antibiotic alternative in rabbit diet on antioxidant enzymes levels, growth performance, and meat quality. *Antioxidants* **2021**, *10*, 1797. [CrossRef] [PubMed]
35. Marín, I.; Sayas-Barberá, E.; Viuda-Martos, M.; Navarro, C.; Sendra, E. Chemical Composition, Antioxidant and Antimicrobial Activity of Essential Oils from Organic Fennel, Parsley, and Lavender from Spain. *Foods* **2016**, *5*, 18. [CrossRef]
36. Farid, A.; Kamel, D.; Abdelwahab Montaser, S.; Mohamed Ahmed, M.; El Amir, M.; El Amir, A. Synergetic role of senna and fennel extracts as antioxidant, anti-inflammatory and anti-mutagenic agents in irradiated human blood lymphocyte cultures. *J. Radiat. Res. Appl. Sci.* **2020**, *13*, 191–199. [CrossRef]
37. Korinek, M.; Handoussa, H.; Tsai, Y.H.; Chen, Y.Y.; Chen, M.H.; Chiou, Z.W.; Fang, Y.; Chang, F.R.; Yen, C.H.; Hsieh, C.F.; et al. Anti-Inflammatory and Antimicrobial Volatile Oils: Fennel and Cumin Inhibit Neutrophilic Inflammation via Regulating Calcium and MAPKs. *Front. Pharmacol.* **2021**, *12*, 674095. [CrossRef]
38. Mazandrani, H.A.; Javadian, S.R.; Bahram, S. The effect of encapsulated fennel extracts on the quality of silver carp fillets during refrigerated storage. *Food Sci. Nutr.* **2016**, *4*, 298–304. [CrossRef]
39. Radu (Lupoae), D.; Alexe, P.; Stănciuc, N. Overview on the potential role of phytochemicals from lavender as functional ingredients. *Ann. Univ. Dunarea Jos Galati Fascicle VI—Food Technol.* **2020**, *44*, 173–188. [CrossRef]
40. Radu (Lupoae), D.; Râpeanu, G.; Bahrim, G.E.; Stănciuc, N. Investigations on thermal degradation of phytochemicals from lavender extract. *Ann. Univ. Dunarea Jos Galati Fascicle VI—Food Technol.* **2019**, *43*, 33–47. [CrossRef]
41. Taşkaya, L.; Hasanhocaoğlu Yapici, H.; Metin, C.; Alparslan, Y. The effect of lavender (Lavandula stoechas) on the shelf life of a traditional food: Hamsi kaygana. *Food Sci. Technol.* **2018**, *38*, 711–718. [CrossRef]
42. Wang, D.; Yuan, X.; Liu, T.; Liu, L.; Hu, Y.; Wang, Z.; Zheng, Q. Neuroprotective activity of lavender oil on transient focal cerebral ischemia in mice. *Molecules* **2012**, *17*, 9803–9817. [CrossRef] [PubMed]
43. Woronuk, G.; Demissie, Z.; Rheault, M.; Mahmoud, S. Biosynthesis and therapeutic properties of lavandula essential oil constituents. *Planta Med.* **2011**, *77*, 7–15. [CrossRef]
44. Benbrahim, C.; Barka, M.S.; Basile, A.; Maresca, V.; Flamini, G.; Sorbo, S.; Carraturo, F.; Notariale, R.; Piscopo, M.; Khadir, A.; et al. Chemical composition and biological activities of oregano and lavender essential oils. *Appl. Sci.* **2021**, *11*, 5688. [CrossRef]
45. da Silva, G.L.; Luft, C.; Lunardelli, A.; Amaral, R.H.; da Silva Melo, D.A.; Donadio, M.V.F.; Nunes, F.B.; de Azambuja, M.S.; Santana, J.C.; Moraes, C.M.B.; et al. Antioxidant, analgesic and anti-inflammatory effects of lavender essential oil. *An. Acad. Bras. Cienc.* **2015**, *87*, 1397–1408. [CrossRef] [PubMed]
46. Aboutaleb, N.; Jamali, H.; Abolhasani, M.; Pazoki Toroudi, H. Lavender oil (Lavandula angustifolia) attenuates renal ischemia/reperfusion injury in rats through suppression of inflammation, oxidative stress and apoptosis. *Biomed. Pharmacother.* **2019**, *110*, 9–19. [CrossRef]
47. Radu (Lupoae), S.D.; Mihalcea, L.; Aprodu, I.; Socaci, S.A.; Cotârleț, M.; Enachi, E.; Crăciunescu, O.; Barbu, V.; Oancea, A.; Dulf, F.V.; et al. Fostering Lavender as a Source for Valuable Bioactives for Food and Pharmaceutical Applications through Extraction and Microencapsulation. *Molecules* **2020**, *25*, 5001. [CrossRef]
48. Aebisher, D.; Cichonski, J.; Szpyrka, E.; Masjonis, S.; Chrzanowski, G. Essential oils of seven lamiaceae plants and their antioxidant capacity. *Molecules* **2021**, *26*, 3793. [CrossRef]
49. Küçükyilmaz, K.; Kiyma, Z.; Akdağ, A.; Çetinkaya, M.; Atalay, H.; Ateş, A.; Gürsel, F.E.; Bozkurt, M. Effect of lavender (Lavandula Stoechas) essential oil on growth performance, carcass characteristics, meat quality and antioxidant status of broilers. *S. Afr. J. Anim. Sci.* **2017**, *47*, 178–186. [CrossRef]
50. Adaszyńska-Skwirzyńska, M.; Szczerbińska, D.; Zych, S. The use of lavender (Lavandula angustifolia) essential oil as an additive to drinking water for broiler chickens and its in vitro reaction with enrofloxacin. *Animals* **2021**, *11*, 1535. [CrossRef]
51. Kozics, K.; Srancikova, A.; Sedlackova, E.; Horvathova, E.; Melusova, M.; Melus, V.; Krajcovicova, Z.; Sramkova, M. Antioxidant potential of essential oil from Lavandula angustifolia in in vitro and ex vivo cultured liver cells. *Neoplasma* **2017**, *64*, 485–493. [CrossRef] [PubMed]
52. Paredes, J.L.; Escudero-Gilete, M.L.; Vicario, I.M. A new functional kefir fermented beverage obtained from fruit and vegetable juice: Development and characterization. *LWT* **2022**, *154*, 112728. [CrossRef]
53. Kesenkaş, H.; Dinkçi, N.; Seçkin, K.; Kinik, Ö.; Gönç, S. Antioxidant properties of kefir produced from different cow and soy milk mixtures. *J. Agric. Sci.* **2011**, *17*, 253–259.

54. Liu, J.R.; Chen, M.J.; Lin, C.W. Antimutagenic and antioxidant properties of milk-kefir and soymilk-kefir. *J. Agric. Food Chem.* **2005**, *53*, 2467–2474. [CrossRef] [PubMed]
55. Farag, M.A.; Jomaa, S.A.; El-wahed, A.A.; El-seedi, H.R. The many faces of kefir fermented dairy products: Quality characteristics, flavour chemistry, nutritional value, health benefits, and safety. *Nutrients* **2020**, *12*, 346. [CrossRef]
56. Darvishzadeh, P.; Orsat, V.; Martinez, J.L. Process Optimization for Development of a Novel Water Kefir Drink with High Antioxidant Activity and Potential Probiotic Properties from Russian Olive Fruit (Elaeagnus angustifolia). *Food Bioprocess Technol.* **2021**, *14*, 248–260. [CrossRef]
57. Aiello, F.; Restuccia, D.; Spizzirri, U.G.; Carullo, G.; Leporini, M.; Loizzo, M.R. Improving kefir bioactive properties by functional enrichment with plant and agro-food waste extracts. *Fermentation* **2020**, *6*, 83. [CrossRef]
58. Liu, J.R.; Lin, Y.Y.; Chen, M.J.; Chen, L.J.; Lin, C.W. Antioxidative activities of kefir. *Asian-Australas. J. Anim. Sci.* **2005**, *18*, 567–573. [CrossRef]
59. Carullo, G.; Governa, P.; Spizzirri, U.G.; Biagi, M.; Sciubba, F.; Giorgi, G.; Loizzo, M.R.; Di Cocco, M.E.; Aiello, F.; Restuccia, D. Sangiovese cv pomace seeds extract-fortified kefir exerts anti-inflammatory activity in an in vitro model of intestinal epithelium using caco-2 cells. *Antioxidants* **2020**, *9*, 54. [CrossRef]
60. Diken, H.O.; Demirtas, B.; Kaya, H.K. Effects of kefir on paraoxanase activity (PON1), total antioxidant status (TAS), tooxidant status (TOS), and serum lipid profiles in smokers and non-smokers. *Food Sci. Technol.* **2022**, *42*, 42. [CrossRef]
61. Łopusiewicz, Ł.; Drozłowska, E.; Siedlecka, P.; Mężyńska, M.; Bartkowiak, A.; Zielińska-Bliźniewska, H.; Kwiatkowski, P. Development, characterization, and bioactivity of non-dairy kefir-like fermented beverage based on flaxseed oil cake. *Foods* **2019**, *8*, 544. [CrossRef] [PubMed]
62. Can, E.; Kurtoğlu, İ.Z.; Benzer, F.; Erişir, M.; Kocabaş, M.; Kızak, V.; Kayım, M.; Çelik, H.T. The Effects of Different Dosage of Kefir with Different Durations on Growth Performances and Antioxidant System in the Blood and Liver Tissues of Çoruh Trout (*Salmo coruhensis*). *Turk. J. Fish. Aquat. Sci.* **2012**, *12*, 277–283. [CrossRef]
63. Yirmibeşoğlu, S.S.S.; Tefon Öztürk, B.E. Comparing microbiological profiles, bioactivities, and physicochemical and sensory properties of donkey milk kefir and cow milk kefir. *Turk. J. Vet. Anim. Sci.* **2020**, *44*, 774–781. [CrossRef]
64. Sabokbar, N.; Khodaiyan, F.; Moosavi-Nasab, M. Optimization of processing conditions to improve antioxidant activities of apple juice and whey based novel beverage fermented by kefir grains. *J. Food Sci. Technol.* **2015**, *52*, 3422–3432. [CrossRef] [PubMed]
65. Sabokbar, N.; Khodaiyan, F. Characterization of pomegranate juice and whey based novel beverage fermented by kefir grains. *J. Food Sci. Technol.* **2015**, *52*, 3711–3718. [CrossRef]
66. Sabokbar, N.; Khodaiyan, F. Total phenolic content and antioxidant activities of pomegranate juice and whey based novel beverage fermented by kefir grains. *J. Food Sci. Technol.* **2016**, *53*, 739–747. [CrossRef]
67. Reis, D.R.; Ambrosi, A.; Di Luccio, M. Encapsulated essential oils: A perspective in food preservation. *Future Foods* **2022**, *5*, 100126. [CrossRef]
68. Meng, F.B.; Gou, Z.Z.; Li, Y.C.; Zou, L.H.; Chen, W.J.; Liu, D.Y. The Efficiency of Lemon Essential Oil-Based Nanoemulsions on the Inhibition of Phomopsis sp. and Reduction of Postharvest Decay of Kiwifruit. *Foods* **2022**, *11*, 1510. [CrossRef]
69. Gulin-Sarfraz, T.; Kalantzopoulos, G.N.; Haugen, J.E.; Axelsson, L.; Kolstad, H.R.; Sarfraz, J. Controlled Release of Volatile Antimicrobial Compounds from Mesoporous Silica Nanocarriers for Active Food Packaging Applications. *Int. J. Mol. Sci.* **2022**, *23*, 7032. [CrossRef]
70. Ghorbanzade, T.; Akhavan-Mahdavi, S.; Kharazmi, M.S.; Ibrahim, S.A.; Jafari, S.M. Loading of fish oil into β-cyclodextrin nanocomplexes for the production of a functional yogurt. *Food Chem. X* **2022**, *15*, 100406. [CrossRef]
71. Elhama, A.; Farid, M. Encapsulation of thyme essential oil using electrospun zein fiber for strawberry preservation. *Chem. Biol. Technol.* **2022**, *9*, 2.
72. Tița, M.A.; Constantinescu, M.A.; Tița, O.; Mathe, E.; Tamošaitienė, L.; Bradauskienė, V. Food Products with High Antioxidant and Antimicrobial Activities and Their Sensory Appreciation. *Appl. Sci.* **2022**, *12*, 790. [CrossRef]
73. Agenţia Naţională a Medicamentului. *Farmacopeea Română*, 10th ed.; Medicală: Bucharest, Romania, 1993.
74. Patel, M.K.; Mishra, A.; Jha, B. Non-targeted metabolite profiling and scavenging activity unveil the nutraceutical potential of psyllium (*Plantago ovata forsk*). *Front. Plant Sci.* **2016**, *7*, 431–432. [CrossRef] [PubMed]
75. Tița, O.; Constantinescu, M.A.; Tița, M.A.; Georgescu, C. Use of yoghurt enhanced with volatile plant oils encapsulated in sodium alginate to increase the human body's immunity in the present fight against stress. *Int. J. Environ. Res. Public Health* **2020**, *17*, 7588. [CrossRef] [PubMed]
76. Tylkowski, B.; Tsibranskaa, I.; Kochanova, R.; Peeva, G.; Giamberini, M. Concentration of biologically active compounds extracted from *Sideritis* ssp. L. by nanofiltration. *Food Bioprod. Process.* **2011**, *89*, 307–314. [CrossRef]
77. Pearson's Correlation Using Minitab. Available online: https://statistics.laerd.com/minitab-tutorials/pearsons-correlation-using-minitab.php (accessed on 26 March 2021).
78. Bătușaru, C.M. Sustainability of the small business environment in Romania in the context of increasing economic competitiveness. *Manag. Sustain. Dev.* **2019**, *11*, 37–41. [CrossRef]
79. Niu, F.X.; Huang, Y.B.; Ji, L.N.; Liu, J.Z. Genomic and transcriptional changes in response to pinene tolerance and overproduction in evolved Escherichia coli. *Synth. Syst. Biotechnol.* **2019**, *4*, 113–119. [CrossRef] [PubMed]

80. Elansary, H.O.; Abdelgaleil, S.A.M.; Mahmoud, E.A.; Yessoufou, K.; Elhindi, K.; El-Hendawy, S. Effective antioxidant, antimicrobial and anticancer activities of essential oils of horticultural aromatic crops in northern Egypt. *BMC Complement. Med. Ther.* **2018**, *18*, 214. [CrossRef]
81. Tahghighi, A.; Maleki-Ravasan, N.; Dinparast Djadid, N.; Alipour, H.; Ahmadvand, R.; Karimian, F.; Yousefinejad, S. GC-MS analysis and anti-mosquito activities of Juniperus virginiana essential oil against *Anopheles stephensi* (Diptera: Culicidae). *Asian Pac. J. Trop. Biomed.* **2019**, *9*, 168–175. [CrossRef]
82. Oppong-Damoah, A.; Blough, B.E.; Makriyannis, A.; Murnane, K.S. The sesquiterpene beta-caryophyllene oxide attenuates ethanol drinking and place conditioning in mice. *Heliyon* **2019**, *5*, e01915. [CrossRef]
83. De Oliveira, J.R.; Camargo, S.E.A.; De Oliveira, L.D. Rosmarinus officinalis L. (rosemary) as therapeutic and prophylactic agent. *J. Biomed. Sci.* **2019**, *26*, 5. [CrossRef] [PubMed]
84. Chua, L.Y.W.; Chua, B.L.; Figiel, A.; Chong, C.H.; Wojdyło, A.; Szumny, A.; Choong, T.S.Y. Antioxidant Activity, and Volatile and Phytosterol Contents of Strobilanthes crispus Dehydrated Using Conventional and Vacuum Microwave Drying Methods. *Molecules* **2019**, *24*, 1397. [CrossRef] [PubMed]
85. Dore, S.; Ferrini, A.M.; Appicciafuoco, B.; Massaro, M.R.; Sotgiu, G.; Liciardi, M.; Cannas, E.A. Efficacy of a terpinen-4-ol based dipping for post-milking teat disinfection in the prevention of mastitis in dairy sheep. *J. Essent. Oil Res.* **2019**, *31*, 19–26. [CrossRef]
86. Fikry, S.; Khalil, N.; Salama, O. Chemical profiling, biostatic and biocidal dynamics of *Origanum vulgare* L. essential oil. *AMB Express* **2019**, *9*, 41. [CrossRef] [PubMed]

applied sciences

MDPI

Article

Hemp Seed Oil Extraction and Stable Emulsion Formulation with Hemp Protein Isolates

Noemi Baldino [1,*], Ilaria Carnevale [1], Olga Mileti [1], Donatella Aiello [2], Francesca R. Lupi [1], Anna Napoli [2] and Domenico Gabriele [1]

[1] Department of Information, Modeling, Electronics and System Engineering, University of Calabria, Via P. Bucci, Cubo 39C, 87036 Rende, Italy
[2] Department of Chemistry and Chemical Technologies, University of Calabria, 87036 Rende, Italy
* Correspondence: noemi.baldino@unical.it; Tel.: +39-0984-494011; Fax: +39-0984-494009

Abstract: Industrial hemp (*Cannabis sativa* L.) is traditionally processed for its high fibre content in the textile industry, but in recent years, it has come to constitute a new raw material in the food industry. Hemp seeds, but also the seed meal, are rich in protein (25%) and oil (30%), the latter consisting of 80–90% of unsaturated fatty acids; they represent a rich and balanced nutritional source to replace the classic animal sources, and they are used in the food industry to supply new food trends. In this work, the hemp oil extraction process was studied, taking advantage of the supercritical CO_2 and ultrasounds, and comparing it with the exhaustive Soxhlet technique. The residual cake from extraction is a protein-rich waste that can be used for food formulations. From this perspective, the hemp oil extracted was used to formulate emulsions with the consistency of vegetable drinks, enriched with standard hemp protein isolate and stabilized with the addition of 0.05% *w/w* of thickening polysaccharides (Gellan gum). The formulated emulsion is stable, and this can encourage the process improvement and the use of the waste from hemp seed extraction for the valorisation of by-products and waste to obtain complete food products with high nutritional value.

Keywords: hemp seed oil; emulsion; gellan gum; assisted extraction; rheology

Citation: Baldino, N.; Carnevale, I.; Mileti, O.; Aiello, D.; Lupi, F.R.; Napoli, A.; Gabriele, D. Hemp Seed Oil Extraction and Stable Emulsion Formulation with Hemp Protein Isolates. *Appl. Sci.* **2022**, *12*, 11921. https://doi.org/10.3390/app122311921

Academic Editor: Adriana Dabija

Received: 25 October 2022
Accepted: 18 November 2022
Published: 22 November 2022

1. Introduction

The hemp plant (*Cannabis sativa* L.) is widely cultivated and used in industrial production because of its multi-functionality. It is possible to obtain isolated proteins, fibres for paper and textile, oil from seeds and flours either for food or for cosmetics and pharmaceutical industries [1–3].

Particularly, the oil from hemp seeds has a high nutritional profile because of polyunsaturated essential fatty acids (EFAs) contained in this oil, which are linoleic acid (C18:2ω6) and α-linolenic acid (C18:2ω3) contained in the hemp seed oil at a ratio of 3:1 [4]. In addition, hemp seed oil has a rare acid, γ-linolenic acid, that is advantageous for hormonal balance. Linolenic acid also a positive impact on rheumatoid arthritis, atopic dermatitis, and allergies and has anti-inflammatory, antihypertension, anti-vasoconstrictive, and anti-cancer effects [5]. Additionally, the oil is suitable for both human nutrition and the preparation of several oils and body creams, thanks to its good absorption through the skin. Tocopherols contained in hemp oil are also used to reduce the risk of cardiovascular diseases, cancers and skin-ageing alterations within the blood vessels [6,7], and the well-balanced quantity of linolenic acid and α-linolenic acid gives hemp seed oil unique healthy properties [5].

Hemp oil is generally produced by solvent extraction or by cold pressing. The first type of extraction uses solvents, and for this reason is dangerous for the environment and can contain toxic residue. Thus, in the last few years, cold pressing has come to be preferred despite the lower yield of extraction [8]. The cold-pressing extraction can recover up to

65% of the oil in the seeds, and 35% of the available oil remains in the seed cake, but a high chlorophyll pigment content is also present in the oil [9]. The pigments modify the oil colour from dark to light green and promote the oxidation of lipids, which leads to the degradation of hemp oil quality, requiring its storage in dark or matt packaging [9,10]. Therefore, other extraction techniques are investigated by researchers such as supercritical extraction with CO_2 (SFE-CO_2), which can give an oil of very high quality also in the presence of an ultrasound pre-treatment [8,10]. SFE-CO_2 is becoming an important technology to obtain oil with a high purification degree for nutritional purposes or pharmaceutical applications. The solvent used is nontoxic and easy to separate from the extract, and moreover, it is tunable; the use of ultrasounds, coupled with this green technique, can improve the yield of oil obtained because they can disrupt the oil bodies [8]. The ultrasonic pretreatment was also evaluated to extract oil with SFE-CO_2 from hemp seeds and other types of seeds [8,11–13]. The use of ultrasonic conditions reduces the extraction time [12–14], but when ultrasounds are used, they generate heat, the amount of which is directly proportional to the ultrasonic power used.

It should be pointed out that In recent years, food industries have also focused their attention on the use of hemp proteins because hemp proteins consist of albumin and edestin, and these are an excellent source of digestible amino acids [15–18]. The amount of protein inside the hemp seeds is around 25%, and it is possible to extract them as hemp protein isolates after extracting the oil from the seeds and after performing particular steps to isolate them [15,17].

This type of protein can be used for different foods, such as baked goods [19], food foams, or emulsions [15,17,20,21] because they are also easily digested [20,22]. Unfortunately, hemp proteins have low solubility at neutral pH, making their use limited in systems such as foams or emulsions [20].

Tang and co-workers [21] studied the stability of hemp protein emulsions to find a trend with pH and compare their functional properties with those of soy protein isolate. Other literature works have looked for a correlation between the ways of isolation of hemp proteins, their solubility and the stability of the emulsions [17,21,23]. It is also possible to stabilize emulsions prepared with hemp protein isolate using nanoparticles [20] in a complex of hemp globulin (HG) with sodium caseinate (SC) via a pH-cycling method, even if the obtained stable emulsion has a solid-like consistency, or by adding pectins [24]. With the aim of obtaining hemp seed oil emulsion from hemp seeds, two nonionic surfactants, tween and span, were also used in the literature [25]. The use of polysaccharides in addition to proteins is suitable for stabilizing emulsions because of the stabilizing effect of the interaction between the two biopolymers but also because of the thickening effect induced by polysaccharides [22,26].

Due to the interesting hemp seed oil properties and high-quality profile of the hemp proteins, which are also characterized by interesting interfacial activity [15], this work aims at preparing stable oil-in-water emulsions based on hemp oil and commercial hemp protein isolates, trying also to investigate the possibility of maximizing the yield of oil extraction from seeds and enhancing the waste. Oil extraction by SFE-CO_2 was also investigated, combining the extraction with ultrasound pretreatment and comparing the lipidic profile with the oil extracted by cold pressing and by classical Soxhlet extraction.

2. Material and Methods

2.1. Raw Material and Samples Pre-Treatment

Hemp seed (*Cannabis sativa* L.) samples with 11.50 ± 0.28% w/w moisture and water activity of 0.385 ± 0.004 were collected from experimental cultivation of hemp Futura 75 carried out at Spezzano Sila (Cosenza, Italy).

The hemp seed oil was obtained from grounded seeds. A quantity of 30 g of hemp seeds was ground for 5 min in a stainless-steel blender to obtain hemp seed particles. After this first step, the hemp seed particles were put in a beaker without any solvent according to the literature [8] and pretreated before extraction with an ultrasound bath Bandelin Sonorex

RK 102 H (BANDELIN electronic GmbH & Co., Berlin, Germany) or directly processed by SFE-CO_2 or Soxhlet extractor. The ultrasound bath works at a fixed frequency of 35 kHz and has a power of 120/480 W. The device is equipped with a thermostat suitable to change the bath temperature (range: 20–80 °C).

Different sonication times (10, 20, and 40 min) were used to investigate the effect of the ultrasounds pretreatment on the hemp oil obtained by SFE-CO_2 to increase the yield of extraction without altering the lipid profile. The temperature during ultrasound pretreatment increases, and for this reason during sonication, the temperature was maintained at the desired level, between 20 and 30 °C, by controlling the ultrasound bath. The sample temperature was also monitored by a thermocouple during the sonication time. Table 1 shows the samples investigated and the sample identifiers.

Table 1. Hempseed samples processed by Soxhlet and SFE-CO_2 at different conditions.

Sample ID	Pre-Treatment Type	Type of Extraction
HS_SFE	-	SFE-CO_2
HS_SFE _10U	10 min ultrasounds	SFE-CO_2
HS_SFE _20U	20 min ultrasounds	SFE-CO_2
HS _SFE_40U	40 min ultrasounds	SFE-CO_2
HS_S	-	Soxhlet extraction

2.2. Soxhlet Extraction

30 ± 1 g of hemp seeds and 250 mL of n-hexane of analytical grade (Sigma-Aldrich Co., Milano, Italy) were used. Ground hemp seeds were placed into a thimble filter and located in the Soxhlet extractor. N-hexane was added, and the system was heated until boiling. Reflux was kept for 8 h. A mixture of n-hexane rich in hemp seed oil was collected [8].

To remove water, anhydrous sodium sulphate was added to the mixture of oil and n-hexane before evaporation [27]. After the water removal, the oil was isolated thanks to a Rotavapor at 50 °C (Heidolph G3, Hei-VAP Value) [8]. In the end, the yield was calculated after weighting. Extractions were done in triplicate and the maximum yield was computed according to the following equation:

$$Yield_{max}(\%) = \frac{mass\ of\ oil\ extracted\ [g]}{mass\ of\ hemp\ seed\ feed\ [g]} \times 100 \quad (1)$$

Determination was done in triplicate.

2.3. Supercritical CO_2 Extraction (CO_2-SFE)

25 ± 2 g of grounded hemp seeds were put in the vessel extractor. Supercritical CO_2 extraction experiments were carried out in a laboratory-scale plant (Spe-ed SFE, Applied Separations, Allentown (PA), USA) using a temperature of 40 °C and a pressure of 300 bar according to Da Porto and coworkers [8]. The used carbon dioxide (purity >99.99%) is supplied by SIAD Spa (Bergamo, Italy).

The SFE procedure is composed of alternated static and dynamic extraction steps. After an initial static phase of 30 min, to promote intimate contact between the fluid and the matrix, the extraction was carried out for 4 h and 30 min. During the total extraction time, two steps were alternated: one of 15 min in dynamic and another one of 15 min in static conditions. Extracts obtained from hemp seeds were collected in a volumetric flask and weighed during extraction. The yield max was calculated according to Equation (1).

2.4. Determination of Fatty Acids Compositions

2.4.1. Oil Sample Preparation

The oil samples to be analysed were diluted to a ratio of 1:10 with chloroform ($CHCl_3$) saturated with NaCl. Solvents ($CHCl_3$ HPLC grade) and NaCl were purchased from Sigma Aldrich Fluka (Milano, Italy).

2.4.2. Mass Spectrometry

Each sample was directly spotted three times on 384-well insert Opt-TOFTM stainless steel MALDI plates (AB SCIEX. Darmstadt. Germany). Mass spectrometric analyses were performed using a 5800 MALDI-TOF-TOF Analyzer (AB SCIEX. Darmstadt. Germany) equipped with an Nd: YLF Laser with λ = 345-nm wavelength of <500 ps pulse length and p to 1000 Hz repetition rate, in reflectron positive mode with a mass accuracy of 5 ppm. Mass spectra were acquired automatically in the positive reflector mode between 200 and 2000 with fixed laser intensity. Spectra with signal-to-noise below 200 were discarded automatically by the instrument. The operation parameter was optimized for the mass region of interest. Laser intensity was adjusted manually to avoid detector saturation. At least 4000 laser shots are typically accumulated with a laser pulse rate of 400 Hz in the MS mode. After acquisition, spectra were handled using Data Explorer version 4.11 (AB Sciex). All data presented in this work are averages of three replicates. LD MS spectra were evaluated against a computationally generated database of lipids [28] by entering a list of precursor ion m/z values restricted to 10 ppm of mass tolerance, and to commonly occurring acyl chains in edible oil, Chain positions and double bond regiochemistry and geometry was not specified.

2.5. Emulsion

2.5.1. Emulsions Preparation

In this study, gellan gum (G) (Sigma Aldrich, Hamburg, Germania) was used as a stabilizing agent with hemp protein isolate (H). The H used is a commercial product purchased from Bulk Powders® (Brunel Way, Colchester, UK) and its composition, as described in the datasheet supplied by the manufacturer, is 47% *w*/*w* of protein, 15.3% *w*/*w* of carbohydrate, 12.7% *w*/*w* of fat and 20% *w*/*w* of fibre.

All the emulsions were prepared with distilled MilliQ ultrapure water (W) (Millipore, Burlington, VT, USA), hemp seed oil and H. The hemp oil used is obtained by SFE-CO_2 without ultrasound pretreatment because of the light green colour of the other extracts obtained with sonication. The H was mixed into a beaker at 35 °C with water for 2 h with a magnetic stirrer (AREX Heating Magnetic Stirrer, Velp Scientifica, Usmate (MB), Italia), and then the solution was centrifuged at 2900 rpm (centrifuge 5810, Eppendorf, Hamburg Germany) for 30 min to remove the fiber part. After centrifugation, NaCl and $CaCl_2$ were added and the solution was mixed for 30 min by a magnetic stirrer (AREX Heating Magnetic Stirrer, Velp Scientifica, Italia) [22]. All solutions were prepared by considering the real protein content in raw materials, then calculating the protein content purity, and then using, for each sample, the protein isolate amount suitable for giving the correct protein content in the investigated samples reported in the label sample in Table 2 (i.e., H_2 means that 2% *w*/*w* of hemp protein is present in the formulation).

When emulsions with stabilizing agents were prepared, the right quantity of stabilizing agent was added to bidistilled MilliQ ultrapure water (Millipore, USA), and the solution was kept at 90 °C for 10 min [15] to ensure the solubilization by magnetic stirrer with temperature control (AREX Heating Magnetic Stirrer, Velp Scientifica, Italia); then the H was added.

After obtaining the aqueous solution, the production of emulsions followed a typical direct homogenization method. Two steps of homogenization were carried out [29]: In the first step, the two phases were mixed at high shear rates with a rotor-stator device (UT T50, IKA, Königswinter, Germany, tool S50N G45F) at 4000 rpm for 120 s, whereas the second step was carried out with gentle mixing with a magnetic stirrer (AREX Heating

Magnetic Stirrer, Velp Scientifica, Italia) at room temperature to complete the emulsion stabilization [30]. The pH of samples was not varied, and nor were reverse methods.

Table 2. Emulsions formulation.

Emulsion ID	H (% w/w)	Hemp Seed Oil (% w/w)	W (% w/w)	G (% w/w)	NaCl (% w/w)	$CaCl_2$ (% w/w)
H_1.5	3.20	5	91.69	-	0.1	0.01
H_2	4.26	5	90.63	-	0.1	0.01
H_3	6.38	5	88.51	-	0.1	0.01
H_2_G	4.26	5	90.58	0.05	0.1	0.01

Following [20], the dispersion of one emulsion droplet into the water and hemp oil was analysed, and it was found that the emulsions are of O/W type. After emulsion preparation, the samples were transferred into a beaker and covered with parafilm to prevent drying for 3 h, and after this time, the characterization was performed [31].

2.5.2. Emulsion Morphology and Determination of Droplet Size

The morphology of the emulsion droplets was observed with a contrast phase microscope (MX5000, Meiji, Chikumazawa, Japan) equipped with a phase contrast 20× lens. The samples were prepared according to [31,32].

Particle size distribution (PSD) was obtained with the statistical analysis of data measured thanks to the contrast phase optical microscope (MX5300H, MEIJI, Japan, magnification 20× for all samples). The statistical analysis of the DSD (Droplets size distribution) was obtained with the image processing software Dhs Particle Analysis (DHS image database, Greifenstein, Germany), which by the selective colouring of the particles allows tracing their average diameter (d_s) and the standard deviation of the log-normal distribution (σ_s) [31,32] through an analysis of the number of pixels necessary for the selective colouring of the single particle.

The drop size distribution can be well described by a lognormal model:

$$f(d) = \frac{1}{d \cdot \sigma_{ln} \cdot \sqrt{2\pi}} exp\left[\frac{-(\ln(d) - d_{ln})^2}{2\sigma_{ln}^2}\right] \quad (2)$$

where d_{ln} and σ_{ln} are, respectively, the mean and the standard deviation of the normal model [31,32], which allow the evaluation of the mean diameter d_s and the variance σ_s^2:

$$d_s = e^{d_{ln} + \sigma_{ln}^2/2} \quad (3)$$

$$\sigma_s^2 = e^{2d_{ln} + \sigma_{ln}^2 (e^{\sigma_{ln}^2} - 1)} \quad (4)$$

σ_s, is the standard deviation and it is commonly considered an index of polydispersity [31,32].

2.5.3. ζ-Potential Measurements

The ζ-potential measurements were performed with a zeta potential analyser (Zetasizer Nano ZS, Malvern Instrument, Malvern, UK) using electrophoretic dynamic light scattering. A static electrical field was applied by electrodes to the emulsion sample in a cell in order to move charged oil drops towards the oppositely charged electrode. The ζ-potential was calculated from the Smoluchowski equation. The absolute zeta potential gives information about the stability of the emulsions; in particular, for absolute zeta potential greater than 30 mV, the emulsion can be considered stable [33]. For the testing, the emulsions were diluted to a droplet concentration of about 0.001 wt% using buffer solutions of the appropriate pH according to Noshad and coworkers [34].

2.5.4. Rheological Characterization

The stable emulsion was characterized by rheological investigation with a rotational rheometer ARES (Rheometric Scientific, New Castle, DE 19720, USA) adopting parallel plates geometry (diameter = 50 mm) thermostated with a Peltier system (±0.1 °C). Flow curves were performed at 25 °C in triplicate, and the data are shown as mean and standard deviation.

3. Results and Discussion

3.1. Comparison of SFE-CO_2 Extraction and Solvent Extraction

Before analysing the ultrasound effect on the yield of extraction, the yield max and the composition of hemp seed oil obtained with supercritical CO_2 are compared. It was found that 35.33 ± 2.00 g hemp oil/100 g of hemp seed and 14.77 ± 0.82 g hemp oil/100 g of hemp seed feed of oil was extracted by soxhlet and supercritical CO_2 method, respectively.

Figure 1 shows the yield obtained for the samples processed at 300 bar and 40 °C to evaluate the effect of pretreatment with ultrasounds.

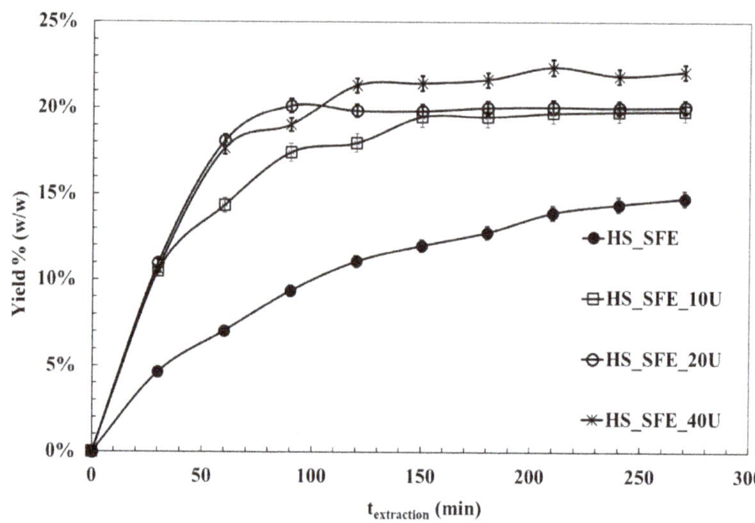

Figure 1. Yield (% w/w) of oil extracted by supercritical CO_2 from untreated and ultrasound pretreated seeds.

From the obtained trend, it is possible to point out that the penetration of CO_2 inside the tissue is favoured by ultrasound pretreatment because it facilitates the oil extraction, as evidenced by the higher slope of the three pretreated samples compared with the untreated one. In particular, the slope increases increase the sonication time even if the amount of the extracted hemp oil does not increase in a relevant way.

Specifically speaking, a longer ultrasound pretreatment from 10 to 40 min causes extraction yields of 19.85 ± 0.02, 20.04 ± 0.03 and 20.92 ± 0.02 (% w/w) compared with the yield obtained without pretreatment, but it is possible to observe a slight greenish colour due to the probable presence of chlorophyll [10].

The increase in the penetration rate of supercritical CO_2 solvent into the tissue is evidenced by the differences in the slopes of the curve with and without ultrasound pretreatment. Moreover, it is possible to observe that the penetration velocity of the supercritical fluid increases with increasing the ultrasound time from 10 to 20 min because of the higher yield at the same extraction time. According to Ivanovs and Blumberga [35], ultrasound pretreatment decreases the extraction time and solvent consumption, also giving a higher penetration of chosen solvent into the cellular material and an enhanced release of cell content, even if it is possible to highlight that the quantity of oil extracted as the pretreatment time increases remains unchanged contrary to Da Porto and coworkers [8].

The quality of the extracted oil is fundamentally related to the extraction technique employed. Hemp seed oil is generally extracted by cold pressing, avoiding the use of organic solvents if intended for human consumption. Solvent extraction takes a long time, and the presence of toxic residues in the final product is a disadvantage, but it is in general used to know the maximum yield of extraction [10]. A green alternative with relatively low extraction times could be optimal because the oil colour is yellow without sonication pretreatment, contrary to cold pressing, and light green using the sonication. Moreover, SFE is considered a clean technology (green process/eco-friendly) that allows the use of low-environmental-impact solvents such as carbon dioxide that allow you to work at low temperatures and extraction pressures [10].

3.2. LD TOF MS Analysis of Hemp Seed Oil

Considering that the high nutritional quality of hempseed oil is essentially determined by its fatty acid composition, the application of MS-based chemical component profiling offers significant opportunities to obtain detailed information that can be directly correlated to oil quality [36].

A typical distribution of the major fatty acids in industrial hemp seed oil is as follows [37]: palmitic acid (C16:0; 6.66–6.98%), stearic acid (C18:0; 2.08–2.82%), oleic acid (C18:1; 9.38–13.00%), linoleic acid (C18:2; 55.56–56.58 %), α-linolenic acid (C18:3; 14.69–17.27%) γ-linolenic acid (C18:3; 2.56–4.49%). The fatty acid composition of the hemp seed oils, determined in accordance with [38,39] obtained by Soxhlet and supercritical CO_2 extraction with and without ultrasound pretreatment is reported in Table 3.

Table 3. Fatty acid composition of hempseed oil, cultivar Futura 75, extracted by supercritical CO_2 from untreated and ultrasound pre-treated hemp seeds and Soxhlet. Each data represents the mean value of three replicates ± standard deviation.

	HS_SFE	**HS_SFE_10U**	**HS_SFE_20U**	**HS_SFE_40U**	**HS_S**
Monounsaturated of	11.74 ± 0.04	11.8 ± 0.2	11.71 ± 0.07	11.54 ± 0.04	11.5 ± 0.2
Oleic acid (C18:1)	11.74 ± 0.04	11.8 ± 0.2	11.71 ± 0.07	11.54 ± 0.04	11.5 ± 0.2
PUFA sum of	**81.71**	**80.67**	**81.46**	**81.29**	**81.77**
Linoleic acid (C18:2)	59.88 ± 0.09	59.37 ± 0.05	59.6 ± 0.3	59.79 ± 0.08	59.86 ± 0.08
alfa-linolenic gamma-linolenic (C18:3)	21.8 ± 0.2	21.3 ± 0.2	21.8 ± 0.2	21.51 ± 0.06	21.91 ± 0.01
ω-6/ω-3 ratio	2.74	2.79	2.73	2.78	2.73
Saturated	7.18 ± 0.03	7.5 ± 0.1	7.18 ± 0.04	7.28 ± 0.07	7.28 ± 0.05
PUFA sum/saturated	11.39	10.7	11.35	11.16	11.24

No significant differences were found when the hemp seed oil was extracted by supercritical fluid extraction using CO_2 as a fluid or by Soxhlet. The main components were oleic acid (O) (18:1), linolenic acid (L) (18:2) and linolenic (Ln) (18:3) (Table 3). The determined polyunsaturated fatty acids (PUFA) were 81% of total fatty acids, while the monounsaturated and saturated fatty acids amounted to 11.4–11.7% and 7–7.5%, respectively. The ratio of polyunsaturated to saturated components was 11–11.5%. The determined omega 6/omega 3 ratio for all samples (2.7–2.9%) is slightly different from the commonly accepted value (3–3.3%) because this approach does not allow for distinguishing α-linolenic acid from γ-linolenic acid. This composition agrees with results reported in Canada [40,41] and for hemp cultivated in Europe [10].

3.3. Droplet Size Distribution of Emulsions

The emulsifying properties of the vegetable proteins were investigated, varying the protein concentration in the emulsions. Drop size distributions (DSD) were obtained by image analysis of emulsion microphotographs; the images were captured for the emulsions after 3 h storage time and are reported in Figure 2.

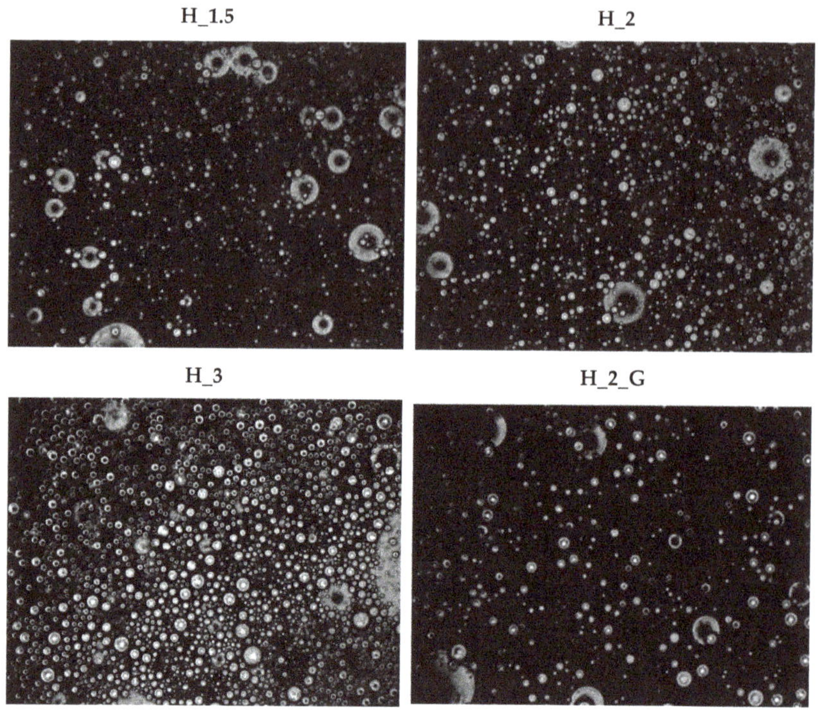

Figure 2. Microphotographs of O/W emulsions (dilution 1:10) after a storage time of 3 h (magnification 20×).

As it is possible to observe in the microphotographs, the emulsions show a broad distribution of oil droplets. The average droplet diameter (d_s) and standard deviation (σ_s) of emulsions were evaluated from the lognormal model to investigate the effects of the increased hemp protein isolate fraction in the emulsions as well as the gellan gum effect, which allows for stabilizing the sample with a final quantity of 2% *w*/*w* of hemp protein.

Experimental data show the mean droplet diameter and the standard deviation for all the tested emulsions. It is possible to point out that both mean diameter and standard deviation increase by increasing the quantity of the protein from 1.5 to higher values in the emulsions. It is possible to observe that emulsions with only protein isolate have an average droplet diameter that goes from 2.1 μm at 1.5% *w*/*w* up to 2.8 μm at the highest protein concentration. The value found at the lowest protein level is compatible with the mean droplet diameter found by Dapčević-Hadnađev and coworkers [17], who prepared emulsions with protein isolates from hemp seed meal obtained using two different isolation techniques [17].

The emulsions with only protein show instability phenomena after about 6 h from the preparation. For this reason, it was necessary to add a stabilizing agent. The gellan gum in the quantity of 0.05% *w*/*w* was found to be suitable for obtaining a stable emulsion, but as seen in Figure 3, the mean diameter and the standard deviation increase after the addition of the gellan gum as compared with the same emulsion without it. The increase in d_s after the addition of a polysaccharide had already been observed in the literature [31] and was attributed to the thickening effect of the polysaccharide and to the variations in the adsorption behaviour of proteins at the interface O/W. These phenomena can make it difficult to form small droplets of oil. The polydispersity has also a similar trend after the gellan gum addition.

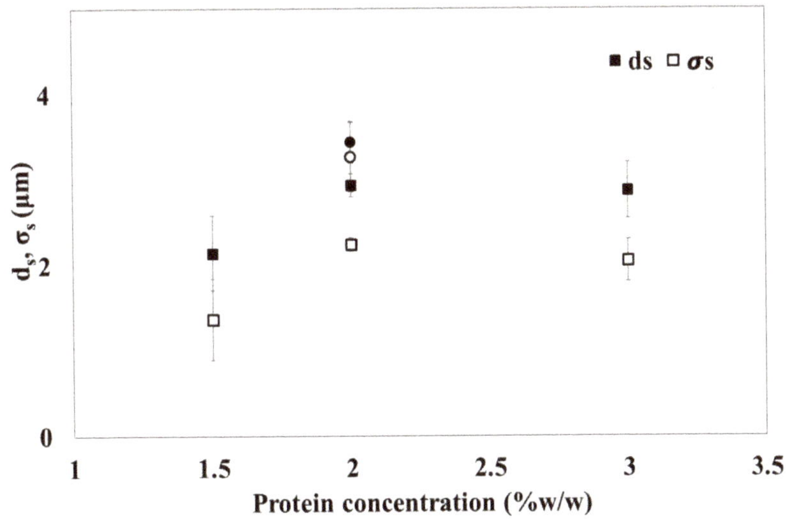

Figure 3. Average droplet diameter, d_s and standard deviation, σ_s, for the different O/W emulsions. The square symbol indicates samples without stabilizing while the circle symbol is for the emulsion with gellan gum (H_2_G).

3.4. Rheology and ζ-Potential Emulsions Measurements

The emulsions prepared using only vegetable proteins as stabilizing agents were all unstable, indicating that the interfacial properties controlled by vegetable proteins are not sufficient to stabilize the whole system. The instability phenomena are not only due to the interface but also to the difference in density between the mixed phases (oily and aqueous), which leads to obvious phenomena of phase separation, such as the creaming of large drops, or sedimentation. On the contrary, the emulsion with gellan gum was also visually monitored to verify the formation of any layers, and it was observed that no stratification had occurred within one week. Therefore, the visual observation evidence that the polysaccharide, in the quantity of 0.05% *w*/*w*, can avoid destabilization phenomena. In light of these results, the viscosity of the sample H_2_G was measured because this is an important measure for understanding the structural organization and the network formation. The viscosity is also an important commercial parameter that influences the texture and the sensory properties [17]. A previous work of literature shows that emulsions prepared with only hemp protein isolate have a Newtonian behaviour [17] while, as Figure 4 shows, the emulsion prepared with gellan shows a non-Newtonian behaviour. The shear thinning behaviour can be due to the stabilizing agent added to control the instability phenomena. In fact, as reported by Herrera [22], emulsions obtained with gellan gum show high viscosity, probably caused by the presence of Ca^{2+} that causes carboxylate–cation2^{+}–carboxylate interactions increasing the gel structure [22]. The same result was found by Feng and coworkers [24] using pectin as a stabilizing agent. Pectin as gellan gum probably forms a three-dimensional network, and this network gives a high-viscosity and shear-thinning behaviour to the emulsion [24].

The result found is also confirmed by the charge analysis of the emulsions carried out at the different protein quantities as shown in Figure 5. Before measuring the zeta-potential, pH was measured, obtaining a pH of 6.8 ± 0.1 for all the emulsions.

As is possible to observe, all the samples have a negative charge in accordance with the H isoelectric point [15]. For the studied emulsions, it is worth noting that the ζ-potential seems to slightly decrease with increasing protein quantity, but only after the addition of gellan gum, the ζ-potential absolute value at 2% *w*/*w* of H increase from −33 ± 2 to -26.5 ± 2.6 mV. As is well known, absolute zeta potential greater than 30 mV indicates a stable system. The electrostatic repulsion between the polysaccharide and protein particles, coupled with the tridimensional network formation, enhances the stability and the result obtained thanks to the polysaccharide addition, which was also obtained for oil-in-water hemp emulsions with another polysaccharide, pectin, as a stabilizing agent by Feng et al. [24].

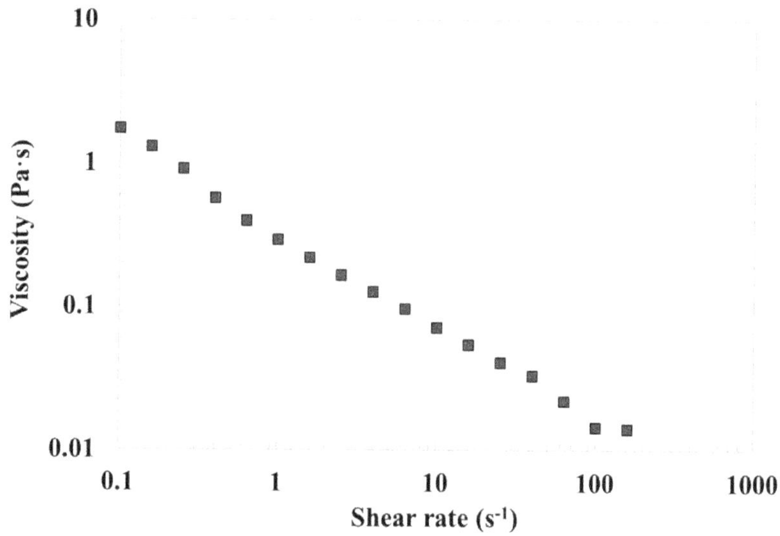

Figure 4. Flow curve of the sample with 2% *w*/*w* of hemp protein and 0.05% of gellan gum (H_2_G) at 25 °C.

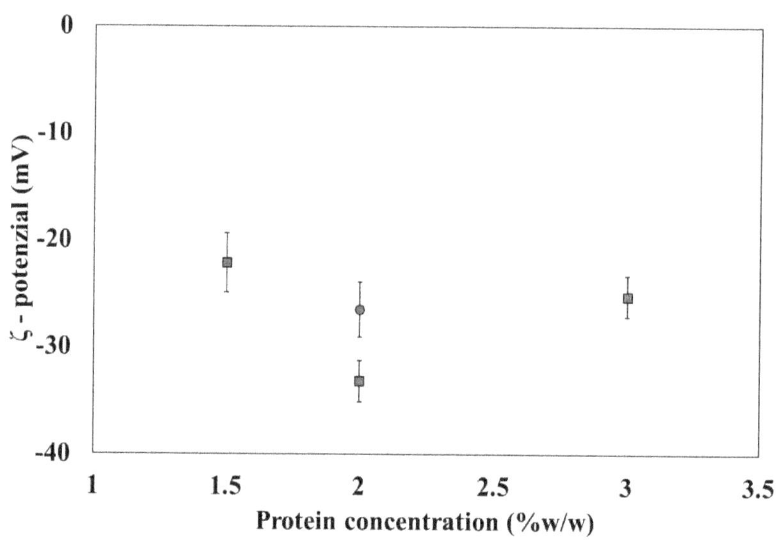

Figure 5. ζ-potential values for investigated samples. The square symbol indicates samples without stabilizing while the circle symbol is for the emulsion with gellan gum (H_2_G).

This formulation can give the possibility of creating food-grade emulsions, capable of valorising both the oil and the hemp protein isolate. Interesting applications could be in creating high nutritional food thanks to the good amino acid profile of the isolate proteins [17,25,42]; even if the sample H_2_G has a non-Newtonian behaviour with high viscosity, which does not yet make it in line with the rheological characteristics of other systems based on vegetable proteins as vegetable drinks, it is possible in any case to use the formulation for other food applications (i.e., desserts or soft foods).

4. Conclusions

In this work, the possibility of extracting oil from hemp seeds using supercritical CO_2 was investigated. It is noted how the lipid profile remains optimal even in the presence of pretreatment with ultrasounds, which increases the extraction yield even if a slight green colouring is observable. The oil extracted by the supercritical process was then used to obtain oil-in-water emulsions, by adding isolated hemp proteins. A commercial isolate was used to give a standard residue for the formulation of the emulsions. The emulsions obtained were characterized, and the 2% *w*/*w* protein emulsions were stabilized by adding 0.05% *w*/*w* of gellan gum. Although the emulsion obtained was stable, the sample showed a relatively high viscosity, which contributed to the stability of the system but which requires optimizing for eventual marketing as a beverage. The product can instead be used as a base for obtaining slightly structured products for people with special dietary needs (e.g., dysphagia).

Author Contributions: N.B.: was responsible for the conceptualization and design of the study, and for the writing of the manuscript and she also partially performed the extraction of the oil; I.C.: performed the supercritical fluid extraction and organized the data; O.M.: performed the emulsions characterization and its formulation and she contributed for the final writing; D.A.: performed the characterization of the hemp seed oil and organized the data; F.R.L.: performed the emulsion microscopic characterization and analysed the data; A.N.: supervised the characterization of the hemp seed oil also writing the results and D.G.: supervised the emulsion characterization also contributing to the writing. All authors have read and agreed to the published version of the manuscript.

Funding: This research was funded by Regione Calabria (Italy) grant number J28C17000210006.

Institutional Review Board Statement: Not applicable.

Informed Consent Statement: Not applicable.

Data Availability Statement: The data presented in this study are available on request from the corresponding author.

Acknowledgments: The work was supported by Regione Calabria (Italy) Azione 1.2.2 POR Calabria FESR FSE 2014–2020—Asse I"Promozione della ricerca e dell'innovazione" "Produzione di latti vegetali a reologia e stabilità controllata" RHEO VEGETABLE MILK CUP J28C17000210006.

Conflicts of Interest: The authors declare that there is no conflict of interest.

References

1. Sorrentino, G. Introduction to Emerging Industrial Applications of Cannabis (*Cannabis Sativa* L.). *Rend. Lincei Sci. Fis. Nat.* **2021**, *32*, 233–243. [CrossRef] [PubMed]
2. Burton, R.A.; Andres, M.; Cole, M.; Cowley, J.M.; Augustin, M.A. Industrial Hemp Seed: From the Field to Value-Added Food Ingredients. *J. Cannabis Res.* **2022**, *4*, 45. [CrossRef] [PubMed]
3. Nissen, L.; Casciano, F.; Babini, E.; Gianotti, A. Chapter 10—Industrial Hemp Foods and Beverages and Product Properties. In *Industrial Hemp*; Pojić, M., Tiwari, B.K., Eds.; Academic Press: Cambridge, MA, USA, 2022; pp. 219–246. ISBN 978-0-323-90910-5.
4. Devi, V.; Khanam, S. Comparative Study of Different Extraction Processes for Hemp (*Cannabis Sativa*) Seed Oil Considering Physical, Chemical and Industrial-Scale Economic Aspects. *J. Clean. Prod.* **2019**, *207*, 645–657. [CrossRef]
5. Teh, S.-S.; Birch, J. Physicochemical and Quality Characteristics of Cold-Pressed Hemp, Flax and Canola Seed Oils. *J. Food Compos. Anal.* **2013**, *30*, 26–31. [CrossRef]
6. Laskoś, K.; Pisulewska, E.; Waligórski, P.; Janowiak, F.; Janeczko, A.; Sadura, I.; Polaszczyk, S.; Czyczyło-Mysza, I.M. Herbal Additives Substantially Modify Antioxidant Properties and Tocopherol Content of Cold-Pressed Oils. *Antioxidants* **2021**, *10*, 781. [CrossRef] [PubMed]
7. Atalay, S.; Atalay, S.; Jarocka-Karpowicz, I.; Skrzydlewska, E. Antioxidative and Anti-Inflammatory Properties of Cannabidiol. *Antioxidants* **2019**, *9*, 21. [CrossRef]

8. da Porto, C.; Natolino, A.; Decorti, D. Effect of Ultrasound Pre-Treatment of Hemp (*Cannabis Sativa* L.) Seed on Supercritical CO_2 Extraction of Oil. *J. Food Sci. Technol* **2015**, *52*, 1748–1753. [CrossRef]
9. Aachary, A.A.; Liang, J.; Hydamaka, A.; Eskin, N.A.M.; Thiyam-Holländer, U. A New Ultrasound-Assisted Bleaching Technique for Impacting Chlorophyll Content of Cold-Pressed Hempseed Oil. *LWT Food Sci. Technol.* **2016**, *72*, 439–446. [CrossRef]
10. da Porto, C.; Decorti, D.; Tubaro, F. Fatty Acid Composition and Oxidation Stability of Hemp (*Cannabis Sativa* L.) Seed Oil Extracted by Supercritical Carbon Dioxide. *Ind. Crops Prod.* **2012**, *36*, 401–404. [CrossRef]
11. da Porto, C.; Decorti, D.; Natolino, A. Separation of Aroma Compounds from Industrial Hemp Inflorescences (*Cannabis Sativa* L.) by Supercritical CO_2 Extraction and on-Line Fractionation. *Ind. Crops Prod.* **2014**, *58*, 99–103. [CrossRef]
12. Liu, X.; Ou, H.; Xiang, Z.; Gregersen, H. Ultrasound Pretreatment Combined with Supercritical CO_2 Extraction of Iberis Amara Seed Oil. *J. Appl. Res. Med. Aromat. Plants* **2020**, *18*, 100265. [CrossRef]
13. Li, X.; Shi, J.; Scanlon, M.; Xue, S.J.; Lu, J. Effects of Pretreatments on Physicochemical and Structural Properties of Proteins Isolated from Canola Seeds after Oil Extraction by Supercritical-CO_2 Process. *LWT* **2021**, *137*, 110415. [CrossRef]
14. Lin, J.-Y.; Zeng, Q.-X.; An, Q.I.; Zeng, Q.-Z.; Jian, L.-X.; Zhu, Z.-W. Ultrasonic Extraction of Hempseed Oil. *J. Food Process. Eng.* **2012**, *35*, 76–90. [CrossRef]
15. Mileti, O.; Baldino, N.; Carmona, J.A.; Lupi, F.R.; Muñoz, J.; Gabriele, D. Shear and Dilatational Rheological Properties of Vegetable Proteins at the Air/Water Interface. *Food Hydrocoll.* **2022**, *126*, 107472. [CrossRef]
16. Kahraman, O.; Kahraman, O.; Petersen, G.E.; Fields, C. Physicochemical and Functional Modifications of Hemp Protein Concentrate by the Application of Ultrasonication and PH Shifting Treatments. *Foods* **2022**, *11*, 587. [CrossRef]
17. Dapčević-Hadnađev, T.; Dizdar, M.; Pojić, M.; Krstonošić, V.; Zychowski, L.M.; Hadnađev, M. Emulsifying Properties of Hemp Proteins: Effect of Isolation Technique. *Food Hydrocoll.* **2019**, *89*, 912–920. [CrossRef]
18. Gorissen, S.H.M.; Gorissen, S.H.M.; Crombag, J.J.R.; Senden, J.M.G.; Waterval, W.A.H.; Bierau, J.; Verdijk, L.B.; van Loon, L.J.C. Protein Content and Amino Acid Composition of Commercially Available Plant-Based Protein Isolates. *Amino Acids* **2018**, *50*, 1685–1695. [CrossRef]
19. Baldino, N.; Carnevale, I.; Laitano, F.; Lupi, F.; Curcio, S.; Gabriele, D. Formulation of Bread Model Doughs with Resistant Starch, Vegetable Proteins and Transglutaminase. *Eur. Food Res. Technol.* **2020**, *246*, 397–408. [CrossRef]
20. Chuang, C.-C.; Ye, A.; Anema, S.G.; Loveday, S.M. Concentrated Pickering Emulsions Stabilised by Hemp Globulin–Caseinate Nanoparticles: Tuning the Rheological Properties by Adjusting the Hemp Globulin : Caseinate Ratio. *Food Funct.* **2020**, *11*, 10193–10204. [CrossRef]
21. Tang, C.H.; Ten, Z.; Wang, X.S.; Yang, X.Q. Physicochemical and Functional Properties of Hemp (*Cannabis Sativa* L.) Protein Isolate. *J. Agric. Food Chem.* **2006**, *54*, 8945–8950. [CrossRef]
22. Sosa-Herrera, M.G.; Berli, C.L.A.; Martínez-Padilla, L.P. Physicochemical and Rheological Properties of Oil-in-Water Emulsions Prepared with Sodium Caseinate/Gellan Gum Mixtures. *Food Hydrocoll.* **2008**, *22*, 934–942. [CrossRef]
23. Wang, X.S.; Tang, C.H.; Yang, X.Q.; Gao, W.R. Characterization, Amino Acid Composition and in Vitro Digestibility of Hemp (*Cannabis Sativa* L.) Proteins. *Food Chem.* **2008**, *107*, 11–18. [CrossRef]
24. Feng, Y.; Yu, D.; Lin, T.; Jin, Q.; Wu, J.; Chen, C.; Huang, H. Complexing Hemp Seed Protein with Pectin for Improved Emulsion Stability. *J. Food Sci.* **2021**, *86*, 3137–3147. [CrossRef] [PubMed]
25. Mikulcová, V.; Kašpárková, V.; Humpolíček, P.; Buňková, L. Formulation, Characterization and Properties of Hemp Seed Oil and Its Emulsions. *Molecules* **2017**, *22*, 700. [CrossRef]
26. Ganzevles, R.A.; Cohen Stuart, M.A.; van Vliet, T.; de Jongh, H.H.J. Use of Polysaccharides to Control Protein Adsorption to the Air-Water Interface. *Food Hydrocoll.* **2006**, *20*, 872–878. [CrossRef]
27. Steiger, M.; Asmussen, S. Crystallization of Sodium Sulfate Phases in Porous Materials: The Phase Diagram Na_2SO_4–H_2O and the Generation of Stress. *Geochim. Cosmochim. Acta* **2008**, *72*, 4291–4306. [CrossRef]
28. Sud, M.; Fahy, E.; Cotter, D.; Brown, A.; Dennis, E.A.; Glass, C.K.; Merrill, A.H., Jr.; Murphy, R.C.; Raetz, C.R.H.; Russell, D.W.; et al. LMSD: LIPID MAPS Structure Database. *Nucleic Acids Res.* **2007**, *35*, D527–D532. [CrossRef]
29. McClements, D.J. *Food Emulsions: Principles, Practices, and Techniques*, 2nd ed.; Taylor & Francis: Boca Raton, FL, USA, 2004.
30. McClements, D.J. Chapter 4—Biopolymers in Food Emulsions. In *Modern Biopolymer Science;* Kasapis, S., Norton, I.T., Ubbink, J.B., Eds.; Academic Press: San Diego, CA, USA, 2009; pp. 129–166. ISBN 978-0-12-374195-0.
31. Seta, L.; Baldino, N.; Gabriele, D.; Lupi, F.R.; de Cindio, B. The Influence of Carrageenan on Interfacial Properties and Short-Term Stability of Milk Whey Proteins Emulsions. *Food Hydrocoll.* **2013**, *32*, 373–382. [CrossRef]
32. Lupi, F.; de Santo, M.; Ciuchi, F.; Baldino, N.; Gabriele, D. A Rheological Modelling and Microscopic Analysis of Bigels. *Rheol. Acta* **2017**, *56*, 753–763. [CrossRef]
33. Cano-Sarmiento, C.; Téllez-Medina, D.I.; Viveros-Contreras, R.; Cornejo-Mazón, M.; Figueroa-Hernández, C.Y.; García-Armenta, E.; Alamilla-Beltrán, L.; García, H.S.; Gutiérrez-López, G.F. Zeta Potential of Food Matrices. *Food Eng. Rev.* **2018**, *10*, 113–138. [CrossRef]
34. Noshad, M.; Mohebbi, M.; Koocheki, A.; Shahidi, F. Influence of Interfacial Engineering on Stability of Emulsions Stabilized with Soy Protein Isolate. *J. Dispers. Sci. Technol* **2016**, *37*, 56–65. [CrossRef]
35. Ivanovs, K.; Blumberga, D. Extraction of Fish Oil Using Green Extraction Methods: A Short Review. *Energy Procedia* **2017**, *128*, 477–483. [CrossRef]

36. Aiello, A.; Pizzolongo, F.; Scognamiglio, G.; Romano, A.; Masi, P.; Romano, R. Effects of Supercritical and Liquid Carbon Dioxide Extraction on Hemp (*Cannabis Sativa* L.) Seed Oil. *Int. J. Food Sci. Technol.* **2020**, *55*, 2472–2480. [CrossRef]
37. Vonapartis, E.; Aubin, M.-P.; Seguin, P.; Mustafa, A.F.; Charron, J.-B. Seed Composition of Ten Industrial Hemp Cultivars Approved for Production in Canada. *J. Food Compos. Anal.* **2015**, *39*, 8–12. [CrossRef]
38. Indelicato, S.; Bongiorno, D.; Ceraulo, L.; Emmanuello, C.; Mazzotti, F.; Siciliano, C.; Piazzese, D. One-Pot Analysis: A New Integrated Methodology for Determination of TAG and FA Determination through LC/MS and in-Silico Saponification. *Food Anal. Methods* **2018**, *11*, 873–882. [CrossRef]
39. Calvano, C.D.; Palmisano, F.; Zambonin, C.G. Laser Desorption/Ionization Time-of-Flight Mass Spectrometry of Triacylglycerols in Oils. *Rapid Commun. Mass Spectrom.* **2005**, *19*, 1315–1320. [CrossRef]
40. Oomah, B.D.; Busson, M.; Godfrey, D.V.; Drover, J.C.G. Characteristics of Hemp (*Cannabis Sativa* L.) Seed Oil. *Food Chem.* **2002**, *76*, 33–43. [CrossRef]
41. Alonso-Esteban, J.I.; González-Fernández, M.J.; Fabrikov, D.; de Cortes Sánchez-Mata, M.; Torija-Isasa, E.; Guil-Guerrero, J.L. Fatty Acids and Minor Functional Compounds of Hemp (*Cannabis Sativa* L.) Seeds and Other Cannabaceae Species. *J. Food Compos. Anal.* **2023**, *115*, 104962. [CrossRef]
42. Baldino, N.; Mileti, O.; Lupi, F.; Gabriele, D. Chapter 1—Interfacial Rheology of Food: Protein as a Model Food. In *Advances in Food Rheology and Its Applications*, 2nd ed.; Ahmed, J., Basu, S., Eds.; Woodhead Publishing: Cambridge, UK, 2023; pp. 3–26. ISBN 978-0-12-823983-4.

applied sciences

MDPI

Article

Optimization of Experimental Parameters in the Solvent Extraction of Trans-Resveratrol from Pruning Waste of *Vitis vinifera*, Fetească Neagră Variety

Alina Lenuța Crăciun and Gheorghe Gutt *

Faculty of Food Engineering, Stefan cel Mare University of Suceava, 720225 Suceava, Romania
* Correspondence: g.gutt@fia.usv.ro

Abstract: The past few decades have seen a marked expansion in market demand for food supplements with therapeutic value. Due to this demand, the recovery of vine waste for obtaining certain phytochemicals or plant synthesized compounds with health-promoting activities can be an important economic component, principally with the agreement of the European Union for resveratrol as a new food ingredient. For the sake of economic capitalization, it is necessary to determine optimum extraction parameters for maximum recovery. In this paper, we have determined the optimum parameters for the solvent extraction of trans-resveratrol from vine prunings. We tested different extraction conditions: 35 different types of solvents, 10 types of solid-to-liquid ratios, 10 extraction times, 10 types of granulosity of the ground material and 7 consecutive extractions on the same material. The optimal parameters determined were: solvent ethanol:diethyl ether 4:1 ratio, 1:35 solid liquid ratio g/mL, 4 days for extraction time, 500 µm–350 µm granulosity of powdered material and one extraction on the material. These findings are confirmed by optimization of extracting parameters according to Box–Behnken design.

Keywords: trans-resveratrol; pruning wastes; food supplements; HPLC; phytoalexin

Citation: Crăciun, A.L.; Gutt, G. Optimization of Experimental Parameters in the Solvent Extraction of Trans-Resveratrol from Pruning Waste of *Vitis vinifera*, Fetească Neagră Variety. *Appl. Sci.* **2023**, *13*, 823. https://doi.org/10.3390/app13020823

Academic Editor: Antonio Valero

Received: 29 November 2022
Revised: 28 December 2022
Accepted: 4 January 2023
Published: 6 January 2023

1. Introduction

Recently, there has been a rising interest in the identification, quantification and extraction of plant compounds with a beneficial effect on human health that can be utilized in the food, cosmetic or para-pharmaceutical industry. The identification of these compounds can generate new food supplements or food products [1].

Resveratrol (3,5,4′-trihydroxystilbene) is a natural phenol compound which is a part of the stilbene family [2] and it is synthesized by plants as a result of exposure to various stressful factors, such as exposure to ultraviolet radiation [3], damage and fungal parasites, etc. [4]. This compound offers potential positive therapeutic effects on slowing down the aging process [5], antioxidant effect against free radicals [6], anti-aging effects on skin by suppressing the responsible enzymes, antibacterial effects [7], antifungal effects [8], anti-inflammatory activity [9], anti-carcinogenic effect, cardio protective effects [10], protection against diseases of the nervous system [11] and beneficial effects on the symptoms of diabetes [7].

Due to the positive impact of resveratrol intake to the human body confirmed by research, resveratrol supplements have been commercialized since 2012 [12]. This antioxidant is further used in the cosmetic [13] and food industries, with the permission of the Commission Implementing Decision (EU) 2016/1190 for the use of resveratrol as a new food ingredient in the European Union [14]. In the food industry, resveratrol started to be used in many areas, such as in manufacturing edible films enriched with this compound [15,16] or in the chocolate industry as an antioxidant [17].

Although resveratrol has been successfully extracted from several types of waste from the wine industry (seeds [18], grape skins [19], grape marc [4,20–23], stems [24,25], leaves [26]), the highest level of resveratrol was determined in the prunings resulting from the annual winter pruning of the vineyard [27].

A wide range of research has shown that vine punings contain a large amount of trans-resveratrol [28]. In pruning waste, the predominant stilbenoids identified were trans-ε-viniferin and trans-resveratrol [24].

Moreover, after the pruning, vine wastes still accumulate this compound [29], due to structural genes, 4CL, PAL, C4H, and STS (forming trans-resveratrol), which are still active, the content of trans-resveratrol rising by up to 40 times post-harvest [30]. Ji et al. (2014) analyzed the content of some parts of vine and grapes and reported that trans-resveratrol content starts rising in the spring months, meeting a maximum level in autumn, and then declines with the reduction in temperatures [27].

Both the plant growth conditions (climatic conditions and plant management) as well as the type and variety of *Vitis genus* significantly influence the composition and concentration of stilbenes in the shoots. The content of stilbene can fluctuate depending on the variety, although the climatic conditions of agricultural cultivation of vines are the same. This is because stilbenes are a part of the plant's protection to pathogen infecion, and varieties resistant to these infections will synthesize a greater amount of stilbenes [31].

Grapes (*Vitis* spp.) are one of the world's largest fruit crops, with 74.5 million tons cultivated in 2014 [16]. The cultivation of vines annually generates significant amounts of viticulture waste, such as shoots, tendrils, leaves and wood, with an estimated amount of 1–5 t per hectare per year [30]. These wastes can represent a source of bioactive substances, especially due to the fact that they are usually composted or burned for disposal [1]. Polyphenols were extracted often from pruning wastes using different techniques, supercritical fluid extraction method being used lately [32,33].

Lately, due to a rising widespread consumption of grapes and the masive waste generation, multiple uses were found for this waste. Vine prunings, pomace and grape seeds are valuable wastes, rich in polyunsaturated fatty acids, phenolics, vitamin E and phytosterols, with uses in the food industry for preservation or nutritional value, for the production of bio-fuel, aclohols, animal feed and fertilizers [34]. Extraction of resveratrol from puning wastes is one of the uses among many others.

Due to the promising health benefits of this antioxidant and the high amount contained by pruning wastes, an optimization of extraction parameters is necessary due to economic reasons and industrialization practice.

The abundance of trans-resveratrol in vine shoots such as routine pruning, has already been demonstrated in numerous studies over the last decade [31].

Statistical analysis was made with Box–Behnken design, which is a proficient and effective statistical method to obtain the optimized experimental conditions, despite the effect of various process variables such as solid–liquid ratio, type of solvent, maceration time, granulosity and number of extractions [35]. The aim of this study was to determine the optimal experimental parameters for the extraction of trans-resveratrol in order to obtain a higher yield from the material of the vine prunings due to the growing demand on the commercial preparation of compounds with therapeutic effects.

2. Materials and Methods

2.1. Vine Pruning Samples

Vine pruning samples Fetească Neagră variety were collected from region Cotnari, Iași on 7 November 2020.

2.2. Reagent and Chemicals

Diethyl ether and the trans-resveratrol standard (99% GC) was bought from Sigma-Aldrich (Hamburg, Germany). Acetonitrile, methanol and ethanol (LiChrosolv for HPLC) were purchased from Merk (Darmstadt, Germany). The PTFE membrane filters with 0.22 μm dimension were obtained from Phenomenex (Torrance, CA, USA).

2.3. Standard Solutions

A set of 11 standards diluted in methanol of different concentrations were prepared to obtain the calibration curve. Resveratrol standards were kept in brown glass jars at −20 °C, away from exposure to light to prevent the phenomenon of photochemical isomerization, which resveratrol is susceptible of.

2.4. Sample Collection and Processing

The prunings samples collected in November 2020 were stored for 12 weeks at 20 °C until they were dry. After that, the samples were dried in a thermostat at 45 °C for 24 h. Then, the samples were grinded in a laboratory grinder and the powder obtained was kept in a plastic container until maceration, away from direct sunlight and moisture.

2.4.1. Solvent Selection

The powder obtained was macerated with 35 different solvents, 5 mL/g of the powdered material for 72 h at room temperature in darkness. The selected solvents are as follows: 10% ethanol (Et), 20% Et, 30% Et, 40% Et, 50% Et, 60% Et, 70% Et, 80% Et, 90% Et, 96.6% Et, 10% methanol (Met), 20% Met, 30% Met, 40% Met, 50% Met, 60% Met, 70% Met, 80% Met, 90% Met, 99.9% Met, 10% acetone (Ac), 20% Ac, 30% Ac, 40% Ac, 50% Ac, 60% Ac, 70% Ac, 80% Ac, 90% Ac, 99.9% Ac, Met-HCl 99:1, Met-HCl-H_2O 1:88:19, ethanol-diethyl ether (Et-Diet) 1:4, Et-Diet 1:1 and Et-Diet 4:1. The extract obtained was then filtered through filter paper and kept at 4 °C until use in analysis.

2.4.2. Solid–Liquid Ratio

The solid-to-liquid ratio was varied to determine the optimal parameters of extraction. The powdered material was macerated with ethanol:diethyl ether 4:1 ratio for 72 h at room temperature in darkness with the ratio of solvent as follows: 1:5 (5 mL solvent/g of the powdered material), 1:10, 1:15, 1:20, 1:25, 1:30, 1:35, 1:40, 1:45 and 1:50. The extract obtained was then filtered through filter paper and kept in a refrigerator at 4 °C until use.

2.4.3. Maceration Time

The powdered material was macerated with ethanol:diethyl ether 4:1 ratio with 35 mL/g of the powdered material for 1, 2, 3, 4, 5, 6 and 7 days at room temperature in darkness. The extracts were collected every day, filtered through filter paper and stored at refrigeration temperature.

2.4.4. Granulosity Selection

The powdered material was sieved in stirring sieve system (Sieve shaker Retsch, Haan, Germany) and then was macerated with ethanol:diethyl ether 4:1 ratio with 35 mL/g of the powdered material for 4 days. The granulosities used were particles larger than 1000 μm, particles between 1000 μm and 710 μm, 710 μm–630 μm, 630 μm–500 μm, 500 μm–350 μm, 350 μm–250 μm, 250 μm–200 μm, 200 μm–180 μm, 180 μm–125 μm and smaller than 125 μm. A sample was taken from each type of powder, depending on the granulosity, and was macerated with ethanol:diethyl ether 4:1 ratio with 35 mL/g for 4 days. The extracts obtained were then filtered through filter paper and kept at 4 °C until use.

2.4.5. Number of Extractions on the Same Material

The powdered material with a granulosity of particles between 500 μm–350 μm was macerated with ethanol:diethyl ether 4:1 ratio with 35 mL/g of the powdered material for 4 days. After 4 days, the extract was filtered through filter paper and the remaining powder was then extracted in the same conditions 9 times. All the 10 extracts were kept at 4 °C until use in analyses.

2.5. Analytical HPLC Procedure

Prior to HPLC analysis, the extracts were partially purified. Using a rotary evaporator (Rotary evaporator with water bath HS-2005S-N HAHNSHIN-Korea) the extracts were evaporated closer to dryness and then were resolved in diethyl ether. The diethyl ether layer was extracted three times with 5% sodium bicarbonate solution in a separatory funnel and then was recovered, rotary evaporated closer to dryness and resolved in methanol. The final extract obtained was filtered through 0.22 μm PTFE membrane filters.

The analysis of the resveratrol content was carried out with the HPLC chromatograph (Shimadzu, Kyoto, Japan), with SPD-M-20A photodiode array detector equipped with a Phenomenex Kinetex 2.6 μm Biphenyl 100 Å 150 × 4.6 mm HPLC column, thermostated at 20 °C. The injection volume was 8 μL.

Elution was executed by a method described by Marshall et al. (2012) with some modifications, with a solvent system composed of two solvents: pure water (solvent A) and acetonitrile (solvent B). A binary gradient was used as follows: linear gradient from 0% to 10% B in 42 min, 10–40% B in 42.6 min, 40–90% B in 46.5 min. The analysis conditions were a total running time of 49.5 min and a solvent flow rate of 0.5 mL/min [36].

The standard calibration curves were obtained from peak area and had a high degree of linearity ($R^2 > 0.99$). For data collection and processing, the version 1.21 version of LC solution software was used (Shimadzu, Kyoto, Japan). Analyses were performed in duplicate.

2.6. Statistical Analysis

The results with trans-resveratrol content of the extracts were entered to analysis of variance (ANOVA) using XLSTAT (trial version, Addinsoft Inc., New York, NY, USA) using the Fisher's test at the 95% confidence level ($p < 0.05$).

The second experiment was conducted in a three-factor full factorial design according to Box–Behnken design. Each independent variable (solid–liquid ratio, time, granulosity) had at least 3 levels, as follows: ratio (1:20, 1:35 and 1:50 g/mL), time (2, 4 and 6 days) and granulosity (>1000 μm, 500–350 μm and <125 μm). Design Expert 12 (trial version, Stat-Ease Inc., Minneapolis, MN, USA) was used for the full factorial design.

3. Results

3.1. Statistical Analysis of Variance

The analysis of variance (ANOVA) for trans resveratrol content of the extracts are presented in Table 1.

Table 1. Analysis of variance for trans-resveratrol content for the samples extracted with different solvents.

Solvent Type	Resveratrol, mg/kg D.W.	F-Value
Et 10%	0.29 (0.007) ^{af}	
Et 20%	0.46 (0.024) ^{ae}	
Et 30%	1.27 (0.017) ^{z}	
Et 40%	1.58 (0.018) ^{w}	
Et 50%	3.11 (0.039) ^{t}	
Et 60%	3.28 (0.033) ^{s}	
Et 70%	18.51 (0.012) ^{o}	
Et 80%	66.86 (0.012) ^{h}	1×10^7 ***
Et 90%	39.00 (0.014) ^{l}	
Et 96%	55.11 (0.014) ^{k}	
Met 10%	0.44 (0.027) ^{ae}	
Met 20%	1.27 (0.028) ^{z}	
Met 30%	1.13 (0.024) ^{aa}	
Met 40%	0.95 (0.012) ^{ad}	
Met 50%	1.00 (0.010) ^{ac}	
Met 60%	1.93 (0.012) ^{v}	

Table 1. *Cont.*

Solvent Type	Resveratrol, mg/kg D.W.	F-Value
Met 70%	5.73 (0.031) ^{p}	
Met 80%	33.86 (0.012) ^{m}	
Met 90%	75.30 (0.020) ^{f}	
Met 99.9%	84.06 (0.015) ^{d}	
Acet 10%	1.34 (0.014) ^{y}	
Acet 20%	1.48 (0.012) ^{x}	
Acet 30%	1.05 (0.011) ^{ab}	
Acet 40%	2.02 (0.012) ^{u}	
Acet 50%	1.94 (0.011) ^{v}	
Acet 60%	4.72 (0.018) ^{r}	
Acet 70%	4.90 (0.025) ^{q}	
Acet 80%	18.69 (0.018) ^{n}	
Acet 90%	66.46 (0.020) ^{i}	
Acet 99.9%	61.29 (0.012) ^{j}	
Met-HCl 99:1	92.06 (0.013) ^{c}	
Met-HCl-H_2O 1:88:19	67.84 (0.012) ^{g}	
Et-Diet 1:4	76.80 (0.009) ^{e}	
Et-Diet 1:1	92.12 (0.012) ^{b}	
Et-Diet 4:1	147.14 (0.011) ^{a}	

(Mean values ± standard deviation), a–z—different letters in the same column indicate significant differences between samples ($p < 0.0001$) according to the LSD test with $\alpha = 0.05$. *** $p < 0.001$. (Et = ethanol; Met = methanol; Acet = acetone; Et-Diet = ethanol-diethyl ether).

3.2. Solvent Selection

For appraising the efficacy of solvents to extract trans-resveratrol, the requirement was that the solvent efficiently extracts the intended compound from vine shoots material. The extraction yield of trans-resveratrol from pruning waste for 35 different solvents is presented in Figure 1.

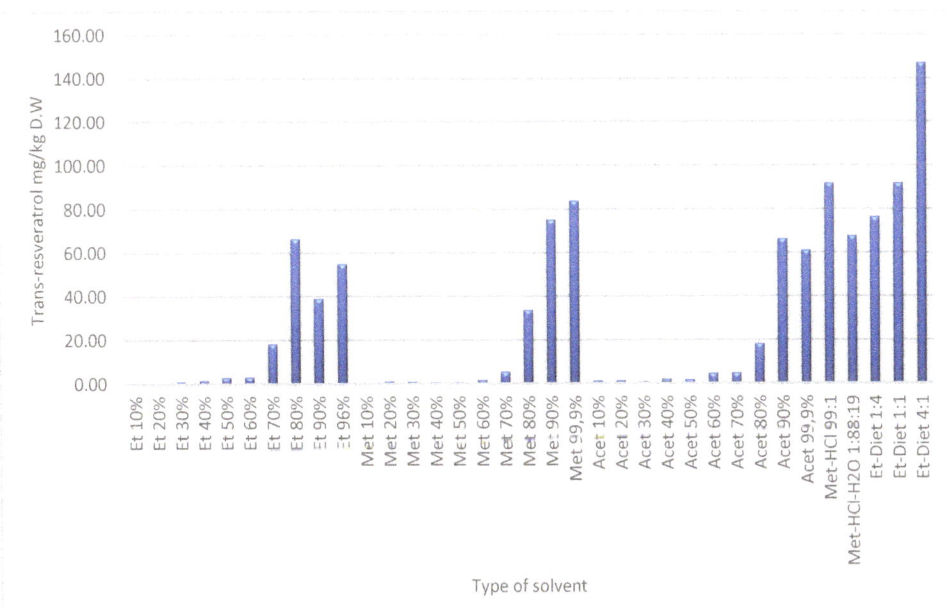

Figure 1. Extraction yield of resveratrol of different solvents from vine shoots of Fetească Neagră variety.

3.3. *Solid–Liquid Ratio*

Ten different solid–liquid ratio of maceration solvents was tested in order to identify the one who will offer a better extraction yield of resveratrol from vine shoots and the results are presented in Figure 2.

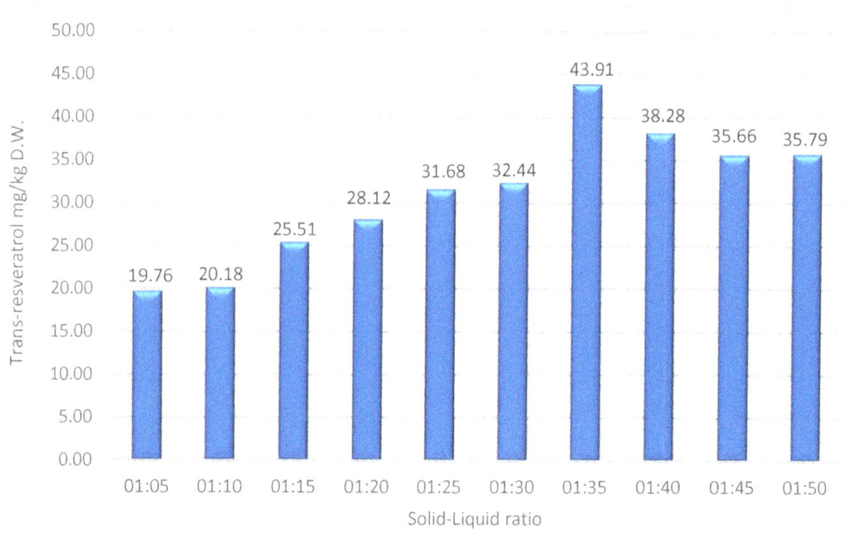

Figure 2. Extraction yield of resveratrol from vine shoots for different solid–liquid ratio.

3.4. *Time of Maceration*

Figure 3 shows the high difference in extraction yield of resveratrol depending on the extraction time.

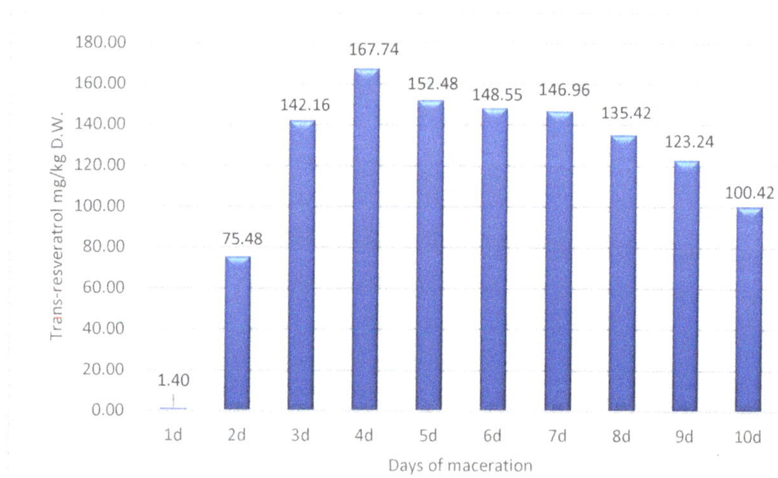

Figure 3. Extraction yield of resveratrol from vine shoots for different time of maceration.

3.5. Granulosity Selection

The extraction yield should increase by the reduction of the particle size because in that case, the superficial area available is larger and facilitates the mass transfer [37]. The results of resveratrol extraction are presented in Figure 4. and the grinded pruning waste were divided into 10 categories according to the size of the particles, as shown in Figure 5.

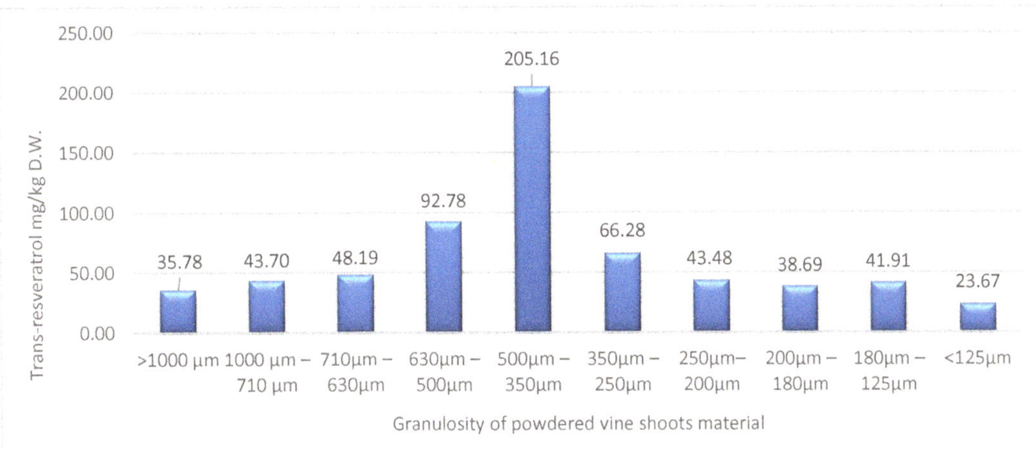

Figure 4. Extraction yield of resveratrol for different granulosity of the powdered material of vine shoots.

3.6. Number of Extractions on the Same Material

Seven extractions were performed on the same material in an attempt to improve extraction yield and to reduce the loss by recovering any resveratrol that could not be extracted on the first extraction and the results are shown in Figure 6.

3.7. Optimization of Extracting Parameters

The experimental and coded values presented in Table 2. were used to analyze the variance with Box–Behnken design. The measured response and the predicted response obtained are shown in Table 3 and the 3D graphs of trans-resveratrol content evolution after extraction are shown in Figure 7.

Figure 5. *Cont.*

<125 μm

Figure 5. Types of granulosity.

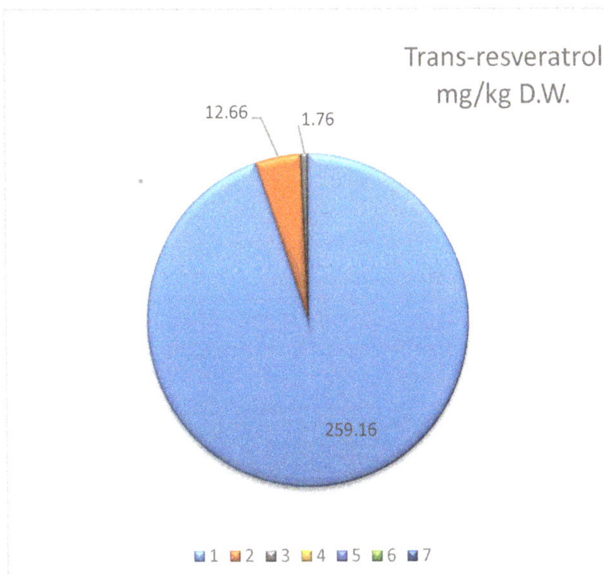

Figure 6. Extraction yield of resveratrol after 7 extractions on the same material.

Table 2. Experimental and coded values according to Box–Behnken design.

Factor	−1	0	1
Ratio S/L (g/mL) A	1:20	1:35	1:50
Time (days) B	2	4	6
Particle size intervals (μm) C	<125	500–350	>1000

Table 3. Measured response and predicted response according to Box–Behnken design.

No.	Ratio S/L (g/mL)	Time (Days)	Particle Size Intervals (μm)	Measured Response	Predicted Response
1	1:35	6	1 mm	148.55	147.43
2	1:20	2	500	67.13	91.65
3	1:20	4	1	43.91	42.95
4	1:20	6	500	43.70	45.78
5	1:50	2	500	128.12	126.04
6	1:50	4	125	89.17	90.13
7	1:35	2	125	108.64	109.76
8	1:20	4	125	92.35	66.71
9	1:50	4	1	158.98	184.62
10	1:20	2	1	184.39	160.83
11	1:20	6	125	104.2	127.76
12	1:50	6	500	201.02	176.50
13	1:35	4	500	205.16	205.16

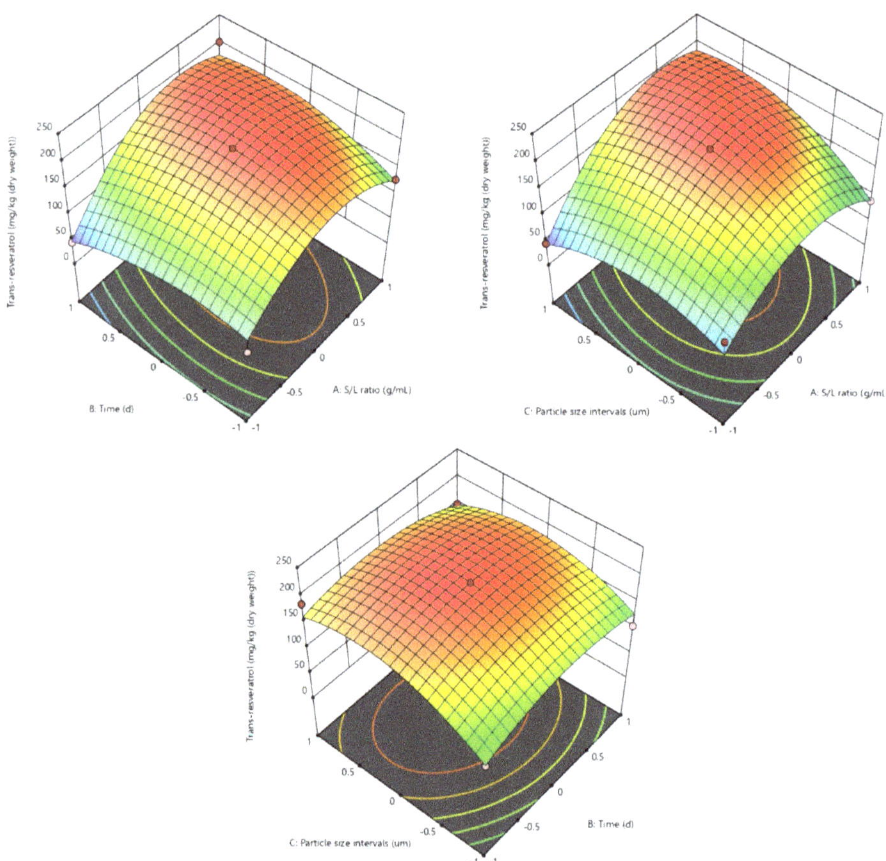

Figure 7. Three-dimensional graphs of trans-resveratrol content evolution after extraction with ethanol:diethyl ether 4:1 ratio.

4. Discussion

4.1. Solvent Selection

Figure 1 presents the result of the extraction of resveratrol by different solvents. Among the extractions with different concentrations of ethanol, the 80% ethanol solvent stood out with the highest content of resveratrol extracted (66.86 mg/kg dry weight (D.W.)). Of the methanol solvents, 99.9% methanol extracted the highest concentration of resveratrol (84.06 mg/kg D.W.), exceeding 80% ethanol solvent. Series 10–99.9% acetone solvents had the lowest concentration of resveratrol after extraction. By adding 1% HCl to the 99.9% methanol solvent positively improved the extraction yield, reaching 92.06 mg/kg D.W. By adding 19% water to this solvent, the extraction yield dropped. Among the solvents with ethanol–diethyl ether in different ratio, the ethanol–diethyl ether 4:1 ratio stood out, obtaining the highest concentration of resveratrol (147.14 mg/kg D.W.), exceeding all solvents.

In the work of Rayne et al. (2008) a broad spectrum of protic and aprotic solvents were examined for extracting trans-resveratrol. Aprotic solvents gave optimum yields at solvent dielectric constants ($\varepsilon/\varepsilon_0$) between 36 and 42. That indicates that any solvent mixture giving this volumetrically averaged ($\varepsilon/\varepsilon_0$) range would give an optimum analyte yield. This optimal polarity range matches up with solutions of 70:30 and 80:20 ethanol:water (v/v) for protic systems. Changes in the protic solvent polarity from this optimum quickly reduced trans-resveratrol yields [38]. This could explain the higher yields of trans-resveratrol for 80:20 ethanol:water among the ethanolic solvents.

Soural et al. (2015) compared methanol and acetone as extraction solvents for resveratrol from vine shoots. Their findings are in agreement with the results of this study, as they conclude that the extraction yield is decreasing when acetone is used. The highest difference between methanol and acetone resulted when extractions were performed on cut vine shoots at laboratory temperature, in which case the extracts obtained with acetone solvents had a very low content of trans-resveratrol in comparison with the extracts obtained with methanol [39].

A few studies are focused on establishing the optimal solvents for trans-resveratrol extraction, and the majority tend to search for the optimal solvents for extracting all polyphenols or stilbenes from grape cane [40] and grape pomace [41]. Therefore, all of them recommend ethanol and methanol as extracting solvents in different concentrations (60–80%). Romero-Perez et al. (2001) established in their study that the optimal solvent for extracting trans-resveratrol from grape berry skins is ethanol 80% and the lowest concentration was obtained using acetone 99.9% [42].

The result that shows resveratrol is more soluble in solutions with a higher content of alcohol were confirmed by Angelov et al. (2016), who separated the vine stems extract in two fractions, containing either ethanol or water-soluble compounds. The aqueous fraction was essentially free of resveratrol and the ethanol fraction had 284.6 mg/kg trans-resveratrol [2].

The most frequent solvents used for the extraction of resveratrol from pruning wastes were ethanol 80% [1,42,43] and methanol 80% [39]. Although methanol offers high extraction yields due to the fact that it is a polar organic solvent, it is not a selective solvent for polyphenols [37]. Resveratrol has also been extracted from the bark of *Pinus sibirica* [44] and from the bark of spruce *Picea mariana* [45] by using diethyl ether with great results.

Furthermore, despite the fact that it is a fairly nonpolar solvent, diethyl ether is more recommended for extraction due to the selectivity for polyphenols, including trans-resveratrol. In addition, due to the selectivity for polyphenols, this solvent eases the further steps of the purification methodology [46].

In the attempt to partially purify the dry extracts of vine shoots obtained by maceration with 80% ethanol solvent, Aavikasaar et al. (2003) reported that these were insoluble in diethyl ether. To overcome this issue, diethyl ether was added to the extraction solvent in a 1:4 ratio, resulting in the appearance of a precipitate and facilitating the decrease in brown colored impurities, without the loss of stilbenoids [28]. Although the solvent diethyl ether was used in the prepurification procedure to precipitate the impurities, in this study it was used as a solvent for maceration to establish the selectivity for trans-resveratrol. Surprisingly, this solvent managed to extract the largest amount of resveratrol when it was used together with ethanol in a ratio of 1:4. The solvent used, 80/20 (*v*/*v*) ethanol/diethyl, is in the range of 70–80% ethanol, a range that has been reported as having a high extraction affinity to resveratrol [38].

4.2. Solid–Liquid Ratio

As it can be seen in Figure 2, 1:35 liquid–solid ratio seemed to suit the resveratrol extraction perfectly, providing the highest concentration. This fact is confirmed by the results obtained for the other extractions, who showed an increase for resveratrol content from 1:5 liquid–solid ratio to 1:35 liquid–solid ratio, and decreased after that, until 1:50 liquid–solid ratio.

Preliminary studies have established that, to provide an effective interphase contact for an optimal wetting of raw material, a ratio of 5:1 solvent/solid is necessary (hydro module). From this value upwards the solubility constraints because of insufficient quantity of solvent are prevented [2].

Garcia et al. (2016) reported an optimal solid–liquid ratio of 1:40 in their attempt to extract resveratrol from peanut grass *Aracis repens* with the highest resveratrol content of 3.93 ± 0.35 mg/L. This result is near to the result obtained for vine shoots [47].

4.3. Time of Maceration

After one day, the degree of resveratrol extraction was the lowest. The resveratrol content of the extract obtained on day 3 was double that of the previous day. The highest concentration of resveratrol was obtained on day 4—167.74 mg/kg D.W. After the 4th day, the content of resveratrol extracted decreased, continuing to decline in the other days.

Soural et al. (2015) tested different temperatures and different maceration times for the extraction of resveratrol from vine shoots with different solvents. The time intervals tested for maceration at laboratory temperature were 8 h, 2.4 and 7 days. The results obtained after this test showed an increasing slope for the concentration of resveratrol until day 4, the day when the maximum amount of resveratrol extracted was reached, after which it started to decrease slightly. This was observed for both solvents used in the extraction, methanol and acetone [39]. These results are in agreement with the result of this study.

The effect of the contact time (respectively 1, 4, 7 and 10 days) and variation of ethanol concentration of the solvent (0, 4, 7.5 and 13%) on extraction of the phenolics was tested by Gambuti et al. (2009) on the skins and seeds of the grapes of Uva di Troia and Aglianico variety (*Vitis vinifera*). The results indicate that many compounds reach a maximum concentration on 4th day of maceration, compounds such as anthocyanins, vanillin reactive flavones, phenolics and proanthocyanidins [48]. These results validate the findings of this study.

4.4. Granulosity Selection

The material of the grounded vine shoots was sieved in 12 separate sieves and was macerated under the same conditions and with the same solvent to determine whether the granulosity of the material has an effect on the extraction content of trans-resveratrol (Figure 5).

Figure 4 shows that there were significant disparities in yield depending on the granulosity of the material, highlighting the material with 500 µm–350 µm granulosity, with a yield of 205.16 mg/kg D.W. Unexpectedly, the lowest yield (23.67 mg/kg D.W.) was recorded for the least granular material—125 µm. This is unexpected, because the surface area for mass transfer is larger than that of other materials. An amount of 55.1% of the chemical composition of vine prunings is represented by holocellulose (31.9% α-cellulose and 23.2% hemicellulose), followed by lignin, with 38.5% [49]. These differences in the extraction yield can be attributed to the non-uniformity of the chemical composition in the composition of the shoots. This result highlights the fact that it is not necessary for the solid material to be grinded to fine particles in order to obtain a superior extraction yield. Thus, in the process of obtaining trans-resveratrol from vine shoots, the grinding of the material can be carried out through an energy-saving process by grinding the material up to 350 μm.

Soural et al. (2015) compared the resveratrol content of extracts for cut material (~1 cm) and powdered material (>1 mm) from vine shoots using methanol and preceding the extraction at laboratory temperature. Although the dimensions of the extraction material used in this study are different, a slight positive difference was recorded (4189.0 μg/g D.W. for the cut material and 4109.0 μg/g D.W. for the powdered material) [39].

4.5. Number of Extractions on the Same Material

The second extraction managed to extract another 5% resveratrol content compared to the first extraction, and the third only 0.67%. The next four extracts did not contain resveratrol (Figure 6).

From an economical point of view, further extractions on the same material are not advantageous. The use of solvents, time and manufacture practice are not economical for the second and third extraction, which together add up to only 5.67% out of the total content of trans-resveratrol from vine shoots. This attempt to increase the amount of trans-resveratrol extracted from the same material highlights the fact that the optimal parameters of the first extraction are well chosen, thus extracting 94.33% of the amount of resveratrol available.

Soural et al. (2015) conducted multiple Soxhlet extractions with methanol on vine shoots and compared the extraction yield of stilbene for each extraction. The yield of stilbene for the second extraction is under 5%, and the yields of the third and further extractions were under 1% [39]. These results are the same with the result obtained by conventional maceration with diethyl ether–ethanol 1:4 ratio.

4.6. Optimization of Extracting Parameters

The experimental results of trans-resveratrol content at different solid–liquid ratios (1:20, 1:35 and 1:50 g/mL), time (2, 4 and 6 days) and granulosity (<125, 500–350 μm and >1000 μm) were placed to the quadratic equation using the response surface analysis. The equation for trans-resveratrol content is displayed in Equation (1):

$$\begin{aligned}\text{Trans} - \text{resveratrol} \\ &= 205.16 + 41.27 \times A + 1.15 \times B + 17.68 \times C + 24.08 \times AB + 29.56 \times AC - 7.85 \times BC \\ &-67.75 \times A^2 - 27.41 \times B^2 - 41.30 \times C^2\end{aligned} \quad (1)$$

where A—solid/liquid ratio g/mL, B—Time, C—particle size interval.

The analysis of variance indicated that the quadratic regression model is statistically significant ($p < 0.05$) with a coefficient of regression $R^2 = 0.90$. Figure 7 displays the 3D evolution of trans-resveratrol content in the function of the parameters employed (Table 2).

According to Figure 7, the highest trans-resveratrol content is recorded at the correlation of the following parameter values 1:35 g/mL solid:liquid ratio, 4 days with 500–350 μm (148.55 mg/kg D.W.), having as extraction solvent ethanol:diethyl ether in a ratio of 4:1. At these conditions, the predicted response for the optimized value is the same as the measured response, namely a resveratrol content of 205.16 mg/kg dry weight. The lowest value was observed at 1:20 g/mL solid–liquid ratio, 6 days with 500–350 μm (43.70 mg/kg D.W.). These results are in compliance with those reported by Soural et al. (2015) on the extraction of resveratrol from vine shoots [39] and Gambuti et al. (2009) on the extraction of resveratrol from skin and seeds of grape [48], regarding obtaining a maximum yield on day 4 of extraction. Additionally, Garcia et al. (2014) reported an optimum solid–liquid of sample/solvent ratio of 1:40 for extracting resveratrol from peanut grass *Aracis repens* [47], results that are near the results obtained, namely an optimal ratio of 1:35 solid/liquid.

The predicted optimum conditions for trans resveratrol extraction (Table 3) were very similar to the experimental concentration achieved under these conditions. Any increase or decrease in the value of the extraction parameters negatively affected the extraction yield.

5. Conclusions

Numerous studies have highlighted the evidence of the therapeutic effect of trans-resveratrol on the human body. Due to the fact that vine wastes are recently considered a valuable source of this compound through the high amount that they contain, the optimization of the extracting parameter is needed. Various extracting conditions were applied to determine the optimal method.

The results of the present study showed that the solvent ethanol:diethyl ether in a ratio of 4:1 is recommended for the extraction of trans-resveratrol from vine shoots, obtaining by its use the highest yield compared to other solvents (147.14 mg/kg D.W.). By increasing the diethyl ether quantity, the extraction yield was negatively affected. Acetone was a weak solvent for extracting trans-resveratrol, resulting in lower quantities extracted. Among the ethanol and methanol series, Et 80% (66.86 mg/kg D.W.) and Met 99.9% (84.06 mg/kg D.W.) had a good yield, but it was almost half that of the solvent with diethyl ether. Adding HCl to Met 99.9% positively improved the extraction yield, being the second solvent by the quantity extracted (92.06 mg/kg D.W.).

Among the different reports of the solid liquid, 1:35 liquid:solid ratio had extracted the highest content of trans-resveratrol (43.91 mg/kg D.W.). The optimal time of extraction was 4 days (167.74 mg/kg D.W.) and the optimal granulosity of the grinded vine shoots was 500 μm–350 μm (205.16 mg/kg D.W.).

With the extraction method described above, the yield of trans-resveratrol extraction was 94.33%. Economically, another extraction on the same material is not recommended.

The result obtained was confirmed by optimization of extracting parameters according to Box–Behnken design, in which the predicted response was similar to the obtained response. The highest trans-resveratrol content was recorded at the association of the following parameter values: 1:35 g/mL solid:liquid ratio, 4 days with 500–350 μm (148.55 mg/kg D.W.).

Author Contributions: Conceptualization, A.L.C. and G.G.; methodology, A.L.C.; validation, A.L.C. and G.G.; formal analysis, A.L.C.; investigation, A.L.C.; resources, A.L.C. and G.G.; data curation, A.L.C.; writing—original draft preparation, A.L.C.; writing—review and editing, A.L.C.; visualization, G.G.; supervision, G.G.; project administration, G.G.; funding acquisition, G.G. All authors have read and agreed to the published version of the manuscript.

Funding: This research received no external funding.

Institutional Review Board Statement: Not applicable.

Informed Consent Statement: Not applicable.

Data Availability Statement: All data related to this study are explicitly plotted in the figures in this article.

Conflicts of Interest: The authors declare no conflict of interest.

References

1. Piñeiro, Z.; Marrufo-Curtido, A.; Vela, C.; Palma, M. Microwave-assisted extraction of stilbenes from woody vine material. *Food Bioprod. Process.* **2017**, *103*, 18–26. [CrossRef]
2. Angelov, G.; Boyadzhiev, L.; Georgieva, S. Useful Bioactive Substances from Wastes: Recovery of Trans-Resveratrol from Grapevine Stems. *Open Chem. Eng. J.* **2016**, *10*, 4–9. [CrossRef]
3. Langcake, P.; Pryce, R.J. The production of resveratrol by *Vitis vinifera* and other members of the Vitaceae as a response to infection or injury. *Physiol. Plant Pathol.* **1976**, *9*, 77–86. [CrossRef]
4. Billet, K.; Houillé, B.; Besseau, S.; Mélin, C.; Oudin, A.; Papon, N.; Courdavault, V.; Clastre, M.; Giglioli-Guivarc'h, N.; Lanoue, A. Mechanical stress rapidly induces E-resveratrol and E-piceatannol biosynthesis in grape canes stored as a freshly-pruned byproduct. *Food Chem.* **2018**, *240*, 1022–1027. [CrossRef]

5. Howitz, K.; Bietterman, K.J.; Cohen, Y.H.; Lamming, D.W.; Wood, J.; Zipkin, R.E.; Chimg, P.; Kisielewski, A.; Zhang, L.L.; Brandy, S.; et al. Small molecule activators of sirutuins extend Saccharomyces cerevisiae lifespan. *Nature* **2003**, *425*, 191–196. [CrossRef]
6. Murcia, M.A.; Marti'Nez-Tome, M. Antioxidant activity of resveratrol compared with common food additives. *J. Food Prot.* **2001**, *64*, 379–384. [CrossRef] [PubMed]
7. Oryza Oil & Fat Chemical Co., Tokyo, Japan, Resveratrol for Anti-Ageing, Anti-Oxidation, Neuron Protection & Metabolic Syndrome. 2008. Available online: http://www.oryza.co.jp/ (accessed on 3 January 2023).
8. Adrian, M.; Jeandet, P.; Veneau, J.; Weston, L.; Bessis, R. Biological Activity of Resveratrol, a Stilbenic Compound from Grapevines, Against Botrytis cinerea, the Causal Agent for Gray Mold. *J. Chem. Ecol.* **1997**, *23*, 1689–1702. [CrossRef]
9. Yang, C.J.; Lin, C.Y.; Hsieh, T.; Olson, S.C.; Wu, J.M. Control of eotaxin-1 expression and release by resveratrol and its metabolites in culture human pulmonary artery endothelial cells. *Am. J. Cardiovasc. Dis.* **2011**, *1*, 16–30.
10. Vidavalur, R.; Otani, H.; Singal, P.; Máulik, N. Significance of wine and resveratrol in cardiovascular disease: French paradox revisited. *Exp. Clin. Cordial.* **2006**, *11*, 217–225.
11. Kumar, V.; Pandy, A.; Jahan, S.; Shukla, R.K.; Kumar, D.; Srivastava, A.; Singh, S.; Rajpurohit, C.; Yadav, S.; Khanna, Z.K. Differential responses of trans-resveratrol on proliferation of neuronal progenitor cells and aged rat hippocampal neuro-genesis. *Sci. Rep.* **2016**, *6*, 28142. [CrossRef]
12. Heger, A.; Ferk, F.; Nerresyam, A. Intake of a resveratrol-containing dietary supplement has no impact on DNA stability. *Mutat. Res.* **2012**, *749*, 82–86. [CrossRef] [PubMed]
13. Jeandet, P.; Clément, C.; Tisserant, L.-P.; Crouzet, J.; Courot, E. Use of grapevine cell cultures for the production of phytostilbenes of cosmetic interest. *Comptes Rendus Chim.* **2016**, *19*, 1062–1070. [CrossRef]
14. Available online: https://eur-lex.europa.eu/legal-content/GA/TXT/?uri=CELEX:32016D1190 (accessed on 20 May 2022).
15. Ansarian, E.; Aminzare, M.; Azar, H.H.; Mehrasbi, M.R.; Bimakr, M. Nanoemulsion-based basil seed gum edible film containing resveratrol and clove essential oil: In vitro antioxidant properties and its effect on oxidative stability and sensory characteristic of camel meat during refrigeration storage. *Meat Sci.* **2021**, *185*, 108716. [CrossRef] [PubMed]
16. Pastor, C.; Sánchez-González, L.; Chiralt, A.; Cháfer, M.; González-Martínez, C. Physical and antioxidant properties of chitosan and methylcellulose based films containing resveratrol. *Food Hydrocoll.* **2013**, *30*, 272–280. [CrossRef]
17. Mendez, J.F.; Flórez, J.Y.; Pérez, E.; Lares, M. Effect of the consumption of chocolate enriched with tryptophan and resveratrol on biochemical markers and oxidative stress in a healthy population. *Vitae* **2019**, *26*, 8–16. [CrossRef]
18. Farhadi, K.; Esmaeilzadeh, F.; Hatami, M.; Forough, M.; Molaie, R. Determination of phenolic compounds content and antioxidant activity in skin, pulp, seed, cane and leaf of five native grape cultivars in West Azerbaijan province, Iran. *Food Chem.* **2016**, *199*, 847–855. [CrossRef]
19. Pugajeva, I.; Pērkons, I.; Górnaś, P. Identification and determination of stilbenes by Q-TOF in grape skins, seeds, juice and stems. *J. Food Compos. Anal.* **2018**, *74*, 44–52. [CrossRef]
20. Zagklis, D.; Paraskeva, C.A. Purification of grape marc phenolic compounds through solvent extraction, membrane filtration and resin adsorption/desorption. *Sep. Purif. Technol.* **2015**, *156*, 328–335. [CrossRef]
21. Brezoiu, A.M.; Matei, C.; Deaconu, M.; Stanciuc, A.-M.; Trifan, A.; Gaspar, A.; Berger, D. Polyphenols extract from grape pomace. Characterization and valorisation through encapsulation into mesoporous silica-type matrices. *Food Chem. Toxicol.* **2019**, *133*, 110787. [CrossRef]
22. Casas, L.; Mantell, C.; Rodríguez, M.; de la Ossa, E.M.; Roldán, A.; De Ory, I.; Caro, I.; Blandino, A. Extraction of resveratrol from the pomace of Palomino fino grapes by supercritical carbon dioxide. *J. Food Eng.* **2010**, *96*, 304–308. [CrossRef]
23. Aliakbarian, B.; Fathi, A.; Perego, P.; Dehghani, F. Extraction of antioxidants from winery wastes using sub-critical water. *J. Supercrit. Fluids* **2012**, *65*, 18–24. [CrossRef]
24. Ewald, P.; Delker, U.; Winterhalter, P. Quantification of stilbenoids in grapevine canes and grape cluster stems with a focus on long-term storage effects on stilbenoid concentration in grapevine canes. *Food Res. Int.* **2017**, *100*, 326–331. [CrossRef] [PubMed]
25. Barros, A.; Gironés-Vilaplana, A.; Teixeira, A.; Collado-González, J.; Moreno, D.A.; Gil-Izquierdo, A.; Rosa, E.; Domínguez-Perles, R. Evaluation of grape (*Vitis vinifera* L.) stems from Portuguese varieties as a resource of (poly) phenolic compounds: A comparative study. *Food Res. Int.* **2014**, *65*, 375–384. [CrossRef]
26. Silva, V.; Igrejas, G.; Falco, V.; Santos, T.P.; Torres, C.; Oliveira, A.M.; Poeta, P. Chemical composition, an-tioxidant and antimicrobial activity of phenolic compounds extracted from wine industry by-products. *Food Control* **2018**, *92*, 516–522. [CrossRef]
27. Ji, M.; Li, Q.; Ji, H.; Lou, H. Investigation of the distribution and season regularity of resveratrol in *Vitis amurensis* via HPLC–DAD–MS/MS. *Food Chem.* **2014**, *142*, 61–65. [CrossRef]
28. Aaviksaar, A.; Haga, M.; Pussa, T.; Roasto, M.; Tsoupras, G. Purification of resveratrol from vine stems. *Proc. Est. Acad. Sci. Chem.* **2003**, *52*, 155–164. [CrossRef]
29. Crăciun, A.L.; Gutt, G. Study on Kinetics of Trans-Resveratrol, Total Phenolic Content, and Antioxidant Activity Increase in Vine Waste during Post-Pruning Storage. *Appl. Sci.* **2022**, *12*, 1450. [CrossRef]
30. Houillé, B.; Besseau, S.; Courdavault, V.; Oudin, A.; Glévarec, G.; Delanoue, G.; Guérin, L.; Simkin, A.J.; Papon, N.; Clastre, M.; et al. Biosynthetic origin of resveratrol acummulation in grape canes during pastharvest storage. *J. Agric. Food Chem.* **2015**, *63*, 1631–1638. [CrossRef]
31. Guerrero, R.F.; Biais, B.; Richard, T.; Puertas, B.; Waffo-Teguo, P.; Merillon, J.-M.; Cantos-Villar, E. Grapevine cane's waste is a source of bioactive stilbenes. *Ind. Crops Prod.* **2016**, *94*, 884–892. [CrossRef]

32. Aresta, A.; Cotugno, P.; De Vietro, N.; Massari, F.; Zambonin, C. Determination of Polyphenols and Vitamins in Wine-Making by-Products by Supercritical Fluid Extraction (SFE). *Anal. Lett.* **2020**, *53*, 2585–2595. [CrossRef]
33. Barba, F.J.; Zhu, Z.; Koubaa, M.; Sant'Ana, A.S.; Orlien, V. Green alternative methods for the extraction of antioxidant bioactive compounds from winery wastes and by-products: A review. *Trends Food Sci. Technol.* **2016**, *49*, 96–109. [CrossRef]
34. Maamoun, M.A.I. An Insight into the Brilliant Benefits of Grape Waste. In *Mediterranean Fruits Bio-Wastes;* Springer: Cham, Switzerland, 2022; pp. 433–465.
35. Bimakr, M.; Rahman, R.A.; Taip, F.S.; Adzahan, N.M.; Sarker, M.Z.I.; Ganjloo, A. Optimization of ultra-sound-assisted extraction of crude oil from winter melon (*Benincasa hispida*) seed using response surface methodology and evaluation of its antioxidant activity, total phenolic content and fatty acid composition. *Molecules* **2012**, *17*, 11748–11762. [CrossRef] [PubMed]
36. Marshall, D.A.; Stringer, S.J.; Spiers, J.D. Stilbene, ellagic acid, flavonol, and phenolic content of muscadine grape (*Vitis rotundifolia* Michx.) Cultivars. *J. Pharm. Crops* **2012**, *3*, 69–77. [CrossRef]
37. Spigno, G.; Tramelli, L.; De Faveri, D.M. Effects of extraction time, temperature and solvent on concentration and antioxidant activity of grape marc phenolics. *J. Food Eng.* **2007**, *81*, 200–208. [CrossRef]
38. Rayne, S.; Karacabey, E.; Mazza, G. Grape cane waste as a source of trans-resveratrol and trans-viniferin: High-value phytochemicals with medicinal and anti-phytopathogenic applications. *Ind. Crop. Prod.* **2008**, *27*, 335–340. [CrossRef]
39. Soural, I.; Vrchotová, N.; Tříska, J.; Balík, J.; Horník, Š.; Cuřínová, P.; Sýkora, J. Various Extraction Methods for Obtaining Stilbenes from Grape Cane of *Vitis vinifera* L. *Molecules* **2015**, *20*, 6093–6112. [CrossRef] [PubMed]
40. Piñeiro, Z.; Marrufo-Curtido, A.; Serrano, M.J.; Palma, M. Ultrasound-Assisted Extraction of Stilbenes from Grape Canes. *Molecules* **2016**, *21*, 784. [CrossRef] [PubMed]
41. Pintać, D.; Majkić, T.; Torović, L.; Orčić, D.; Beara, I.; Simin, N.; Lesjak, M. Solvent selection for efficient ex-traction of bioactive compounds from grape pomace. *Ind. Crops Prod.* **2018**, *111*, 379–390. [CrossRef]
42. Romero-Pérez, A.I.; Lamuela-Raventós, R.M.; Andrés-Lacueva, C.; de la Torre-Boronat, M.C. Method for the Quantitative Extraction of Resveratrol and Piceid Isomers in Grape Berry Skins. Effect of Powdery Mildew on the Stilbene Content. *J. Agric. Food Chem.* **2000**, *49*, 210–215. [CrossRef]
43. Jesus, M.S.; Genisheva, Z.; Romaní, A.; Pereira, R.N.; Teixeira, J.A.; Domingues, L. Bioactive compounds recovery optimization from vine pruning residues using conventional heating and microwave-assisted extraction methods. *Ind. Crops Prod.* **2019**, *132*, 99–110. [CrossRef]
44. Gromova, A.S.; Tyukavkina, N.A.; Luckiy, V.I.; Kalabin, G.A.; Kushnarev, D.F. Oxystilbenes in the bark of *Pinus sibirica*. *Khim. Prir. Soedin.* **1975**, 677–682. (In Russian)
45. Piyaratne, P.S. Extraction and Purification of (E)-Resveratrol from the Bark of Maine's Native Spruces. Ph.D. Thesis, University of Maine, Orono, ME, USA, 2018.
46. Pajović-Šćepanović, R.; Wendelin, S.; Eder, R. Phenolic composition and varietal discrimination of Montenegrin red wines (*Vitis vinifera* var. Vranac, Kratošija, and Cabernet Sauvignon). *Eur. Food Res. Technol.* **2018**, *244*, 2243–2254. [CrossRef]
47. Garcia, L.; Garcia, R.; Pacheco, G.; Sutili, F.; De Souza, R.; Mansur, E.; Leal, I. Optimized Extraction of Resveratrol from *Arachis repens* Handro by Ultrasound and Microwave: A Correlation Study with the Antioxidant Properties and Phenol Contents. *Sci. World J.* **2016**, *2016*, 1–10. [CrossRef] [PubMed]
48. Gambuti, A.; Capuano, R.; Lecce, L.; Fragasso, M.G.; Moio, L. Extraction of phenolic compounds from 'Aglianico'and 'Uva di Troia'grape skins and seeds in model solutions: Influence of ethanol and maceration time. *Vitis* **2009**, *48*, 193–200.
49. Cebrián, C.; Sánchez-Gómez, R.; Salinas, M.R.; Alonso, G.L.; Zalacain, A. Effect of post-pruning vine-shoots storage on the evolution of high-value compounds. *Ind. Crops Prod.* **2017**, *109*, 730–736. [CrossRef]

applied sciences

MDPI

Article

The Use of Vacuum Impregnation of Barley Grain in the Production of Malt for Wort

Leszek Rydzak [1], Renata Różyło [2,*], Tomasz Guz [2], Marcin Natoniewski [1] and Kostiantyn Vasiukov [1]

1 Department of Biological Bases of Food and Feed Technologies, University of Life Sciences in Lublin, Głęboka 28, 20-612 Lublin, Poland
2 Department of Food Engineering and Machines, University of Life Sciences in Lublin, Głęboka 28, 20-612 Lublin, Poland
* Correspondence: renata.rozylo@up.lublin.pl

Featured Application: This study shows that the use of the vacuum impregnation process of barley grain for the production of light malt significantly reduces the soaking time of the grain, which speeds up the malt production process without causing significant negative changes in the parameters of wort. Vacuum impregnation is not used in industrial beer production. However, it can be quickly implemented with a small investment, i.e., the purchase of a vacuum pump and vacuum chamber.

Abstract: In this study, the process of the vacuum impregnation of Kangoo barley grain, at the stage of soaking, was used in the production of light malt. The influence of vacuum impregnation on the speed of the water uptake by the barley, at temperatures of 12, 14, 16 and 18 °C, was also analyzed. At this stage of the research, the grain was soaked in water to obtain a moisture content of approximately 42%. The samples for the moisture content tests were taken every 2 h. The grain intended for malt was soaked in an air–water system and was kept submerged in water for 6 h. It was then removed from the water and kept for 18 h. The grain was aerated during the soaking process. The malting and soaking lasted eight days at temperatures of 12, 14, 16 and 18 °C. The samples for further testing were taken daily. Then, each of the samples was dried, following the same procedure. The sprouts were removed immediately after the dried samples contained approximately 4% moisture. Following a 3-month maturation process, the congress wort was produced from the malt. The pH and the extract content in the wort were tested. It was found that the process of vacuum impregnation significantly accelerates the uptake of water by the grain. In almost all cases, the influence of the tested factors on the pH of the wort and the extract content was also observed.

Keywords: vacuum impregnation; malt; barley; wort

Citation: Rydzak, L.; Różyło, R.; Guz, T.; Natoniewski, M.; Vasiukov, K. The Use of Vacuum Impregnation of Barley Grain in the Production of Malt for Wort. *Appl. Sci.* **2023**, *13*, 1628. https://doi.org/10.3390/app13031628

Academic Editors: Adriana Dabija, Georgiana Gabriela Codină and Lăcrămioara Rusu

Received: 14 December 2022
Revised: 24 January 2023
Accepted: 25 January 2023
Published: 27 January 2023

1. Introduction

Impregnation, i.e., filling materials with additional liquid components, has been a known and widely used process for over 100 years. Since the end of the 1960s, the appearance of and the expansion in the use of the impregnation process under reduced pressure conditions can be observed on an industrial scale. Such impregnation is called vacuum impregnation, in both the scientific and trade literature. Its importance can be proven by the fact that it is among the processes covered by the standards of the American defense industry (MIL-STD-276A) [1].

In traditional applications, the purpose of impregnation is to ensure tightness and to eliminate the porosity of various materials. It is mostly used in the production of engine blocks and cylinder heads in the automotive industry, as well as in power systems, brake systems and landing gear components in the aviation industry. The knowledge already gathered allows us to state that impregnation is a process that modifies the structure and the sensory and functional properties of products and increases their physicochemical stability by creating a coating of liquid material on the surface of a solid body [2,3]. In recent years, there has been a rapid development of a new branch of food engineering, referred to as the food matrix engineering (spatial architecture) of materials [4,5].

In comparison to vacuum impregnation, it was demonstrated that freezing/thawing and compression caused more changes in the internal microstructure of apple tissue, resulting in the formation of cell cavities [6]. Regardless of the changes in the microstructure caused by vacuum impregnation, the integrity of the plant tissue cells is unaffected [7,8].

Vacuum impregnation can be carried out in many ways:

- Dry vacuum impregnation (DVI), in which the contact of the liquid and the impregnated material takes place in the reduced pressure phase;
- Wet vacuum impregnation (WVI), in which the contact of the liquid and the impregnated material takes place before the reduced pressure phase, under atmospheric pressure [9].

In the course of vacuum impregnation, two groups of related phenomena can be distinguished:

- Deformation relaxation phenomena, DRPs;
- Hydrodynamic mechanism, HDM [10–12].

Many different factors influence the impregnation process. Most often, they are divided into two groups: the external and internal factors [13].

In the group of external factors, the most important is the impregnation pressure (p_1). It is the driving force behind the entire process. Most often, in the impregnation of biological materials, pressures ranging from 50 to 600 hPa are used [12–19].

The t_1 and t_2 times are identified in the literature as important factors. The longer the time, the lower the pressure is maintained and the lesser the weight loss, which results in a greater degree of impregnation [20]. In the scientific literature on impregnation, the ranges of the t_1 and t_2 time variations, assumed by the researchers, are quite extensive. Some apply pressure in the range from 0 to 10 min, others conduct research in the range from 10 to 120 min [17,18,21,22]. The viscosity index, temperature and concentration of the impregnating liquid also affect the effectiveness of the impregnation [2,13,17,22].

Among the internal factors, the properties of the impregnated material and its structure should be mentioned. Porous materials with numerous voids in their tissues, such as fruit and vegetables, are particularly susceptible to impregnation. They show a higher degree of porosity compared to meat, fish and cheese [10,14,15,23].

There are many studies on the use of vacuum impregnation in food production [12–15,17–19] and there are many reports on the beneficial effect of vacuum impregnation on the structure of tissues. The use of this treatment reduces the probability of thermal damage to the organic tissue and also prevents the loss of the aroma and color of fruit and vegetables [24]. This is only possible by removing oxygen from the pores, without the need for antioxidants.

Other researchers present the possibility of introducing antioxidants or antimicrobial agents into the structure of the material, which can extend shelf life [3]. The results of studies on the application of impregnation as a pretreatment before freezing are available [25]. Impregnation is often used as a pretreatment that precedes heat treatment [26,27].

There is a scarcity of scientific data on the vacuum impregnation of cereal grains in the scientific literature [28].

Barley is subjected to steeping. In this phase, the barley reaches a moisture content of 42 to 44% [29].

Steeped barley is then directed to the malting department called the sprouting room. In modern malting houses, sprouting boxes or drums are used for sprouting barley, in which the process is completely mechanical. The barley sprouting cycle lasts 5–8 days at a temperature of 12–20 °C. The duration of the process depends on the variety of barley, the method of malting and the sort of malt [29,30].

Kilning takes place at a temperature of 30–85 °C, measured in the malt, and lasts 12–36 h for light malt, depending on the kilning method. Munich-type dark malt is kilned for 48 h, and the temperature in the final phase of this process should be 105 °C. Next, the rootlets are removed [29,30].

The use of vacuum impregnation at the steeping stage can significantly speed up this process, which can shorten the malt production cycle, because it significantly accelerates mass transfer processes in the liquid–solid system [10].

The aim of this study was to establish the effects of vacuum impregnation on the steeping process and the structural changes in barley grains in relation to some of the selected parameters of wort.

2. Materials and Methods

2.1. Materials

The research material used was malting barley grain of the Kangoo variety (Soufflet Agro Poland Ltd., Kościan, Poland). The grain used for the research was uniform in variety. The grain was cleaned and sorted, using a sieve-pneumatic separator. The grains with a fraction over 2.5 mm were tested. The moisture content of the barley was 13.7%, the protein content was 10.9% d.m. and the weight of 1000 grains was 49.8 g.

2.2. Methodology of the Vacuum Impregnation of the Grain

Water was used as an impregnating liquid in all of the cases. Wet vacuum impregnation (WVI) was applied. The process was run in a 2 dm^3 chamber coupled with a vacuum pump, facilitating the adjustment of the pressure in the chamber in the range of 5–100 kPa. To ensure complete immersion, all grains were placed in a special container. The investigations were carried out at a temperature of 20 °C, the same as that of the impregnating liquid. The grain samples were placed in a chamber filled with the liquid and, after closing the chamber, the pressure was reduced to 5 kPa, at a rate of 30 $kPa{\cdot}s^{-1}$. Immediately after the pressure reduction, the atmospheric pressure was restored in the chamber, at a rate of 30 $kPa{\cdot}s^{-1}$, and the grains were separated from the chamber. The duration of the vacuum period and the relaxation time were therefore reduced to a minimum. Concurrently, the impregnation at an atmospheric pressure (about 100 kPa), which served as a reference system, was carried out.

2.3. Methodology of the Malt Preparation

The barley was soaked in the two-phase air–water system. The water phase lasted 6 h, and the air phase lasted 18 h. The air–water soaking took place in three such cycles. In the water phase, half of the grains were soaked under atmospheric pressure. The remaining raw material, in the final stage of the water soaking, was subjected to vacuum impregnation. During the water soaking, the grain was aerated. Then, the grain was subjected to malting, which took place in a climate chamber at temperatures of 12, 14, 16 and 18 °C. The germinating grain bed was periodically mixed. The malt samples were taken once a day for the next eight days. The collected samples were subjected to convection drying, gradually increasing the temperature from 40 °C to 60 °C, at a rate of about 1 °C/h. Next, the temperature was increased, at a rate of 1 °C/min from 60 °C to 75 °C. For the final 5 h, the malt was dried at a constant temperature of 75 °C. Immediately after drying, the rootlets were separated, using a sieve with a mesh size of 1 mm. The samples prepared in this way were matured for a period of 3 months.

These experiments were carried out in triplicate.

2.4. Methodology of the Mashing

After 3 months of maturation, the malt was ground in an ML 155 laboratory mill and was subjected to mashing, to produce a congress wort.

In the congress mashing, each portion of the ground malt was transferred to a mash beaker and thoroughly mixed with a spatula. A portion of 50 g was weighted in the mash beaker. The mashing bath was set to 45 °C. A volume of 200 mL of water was stirred into each beaker and the mash was mixed using a glass rod, to avoid balling. The temperature of the mash was maintained at 45 °C for exactly 30 min. Next, the temperature of the mash was increased at 1 °C per minute for 25 min, up to 70 °C. When the temperature reached 70 °C, a further 100 mL of water with a temperature of 70 °C was added. The saccharification rate was measured from this point (after 55 min from the beginning of the mashing). A drop of the mash was transferred to a spot on the porcelain plate, and a drop of an iodine solution was added. This test was repeated at 2 min intervals until the saccharification was completed: that is, when a clear yellow spot was obtained. Following the saccharification, the mashing was terminated [31].

The mash was then filtered and cooled to 20 °C. The spent grains were washed with a small amount of water, dried outside of the beaker, and then the contents of the beaker were replenished to 450 g, with the addition of water. Then, the pH and the extract content were determined.

The determinations of the extract content in the wort were established, according to the Analytica EBC methods (2019) [31]. These parameters were expressed using a PAL1 digital refractometer. The measurements of the pH value of the congress wort were also obtained, by Analytica EBC (2019). The Elmetron CP-411 pH meter, with an accuracy of 0.01, was used to measure the pH of the wort. All testing procedures followed the Analytica EBC methods (2019).

2.5. Methodology of Measuring the Rate of the Increase in the Grain Moisture Content

Tests on the speed of the water uptake by the barley grain subjected to vacuum impregnation and soaked under atmospheric pressure were also carried out. During these tests, the grain was kept under water and aerated throughout the entire period. The grain was subjected to vacuum impregnation periodically, every 30 min. The tests were carried out at temperatures of 12, 14, 16 and 18 °C. The barley moisture content was tested with samples taken every 2 h [31]. The determinations were performed in triplicate. The moisture content was measured, according to the PN EN ISO 712:2012 standard [32].

2.6. Statistical Analysis

The statistical analysis was performed using Statistica version 13.3. A three-stage statistical analysis was performed to evaluate the statistical significance of the differences in the examined properties of the wort. In the first stage, the Shapiro–Wilk test was performed, in order to confirm that the probability distribution of the examined parameters was normal. In the second stage, multivariate analysis of variance (ANOVA) was performed, and finally, all of the results were analyzed for significance by means of Tukey's reasonable significant difference (HSD) test, at a significance level of $\alpha = 0.05$, because this test is considered to be less conservative than the Scheffé test but more conservative than the Newman–Keuls test.

3. Results

Figure 1 shows the changes in the moisture content of the Kangoo barley grain, which was vacuum-impregnated every 30 min and soaked under atmospheric pressure at the tested temperatures.

The grain subjected to cyclic vacuum impregnation did not lose its viability and absorbed more water in a shorter amount of time. The vacuum impregnation of the grain during the soaking in the tested temperatures of 12, 14, 16 and 18 °C made it possible to obtain the moisture levels required for malting after approximately 30 h, i.e., around 6 h faster than in the case of the non-impregnated grain. Table 1 presents the results of Tukey's reasonable significant difference (HSD) test used to assess the significance of the effect of the grain moistening conditions on the final moisture content.

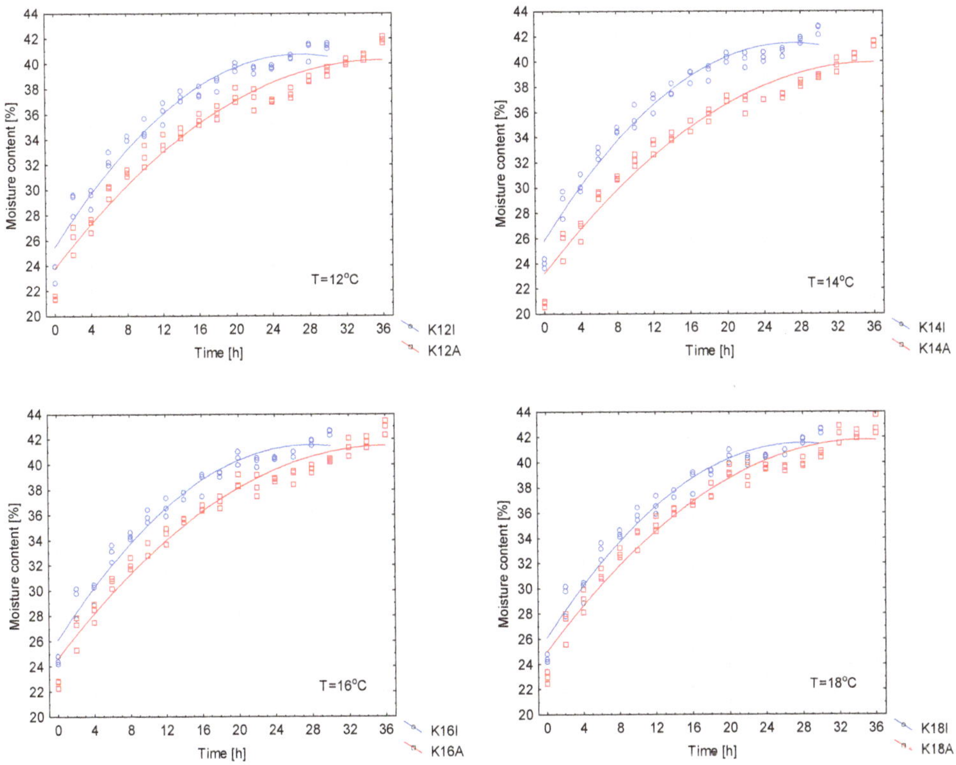

Figure 1. Changes in the moisture content during the barley grain moistening at the tested temperatures. (K12I, K14I, K16I and K18I: vacuum-impregnated grain, soaked at 12, 14, 16 and 1 °C, respectively; K12A, K14A, K16A and K18A: grain not subjected to vacuum impregnation, soaked at 12, 14, 16 and 18 °C, respectively).

Table 1. The results of the analysis of Tukey's reasonable significant difference (HSD) test of the influence of the selected factors on the moisture content of the tested barley grain cultivars.

	Factor	Value	Average Value/Homogeneous Groups *
1.	Pressure [kPa]	5	35.26 [a]
		100	30.34 [b]
2.	Temperature [°C]	12	29.53 [a]
		14	29.88 [a]
		16	30.52 [a]
		18	34.67 [b]
3.	Time [h]	0	25.22 [a]
		4	26.75 [b]
		8	28.24 [c]
		12	30.15 [d]
		16	33.26 [e]
		20	35.38 [f]
		24	35.62 [f]
		28	38.85 [g]
		32	39.18 [g]
		36	41.76 [h]

* Different letters mean statistically significant differences at the level of $\alpha = 0.05$.

The influence of the factors tested on the moisture content showed statistically significant differences. The temperature changes in the range of 12–16 °C did not cause significant changes in the final moisture content. Rather, the changes in the final moisture content observed in the case of the soaking grain at 18 °C showed statistically significant differences.

Figure 2 shows the changes in the pH value of the congress wort obtained from the Kangoo malted barley at the tested temperatures.

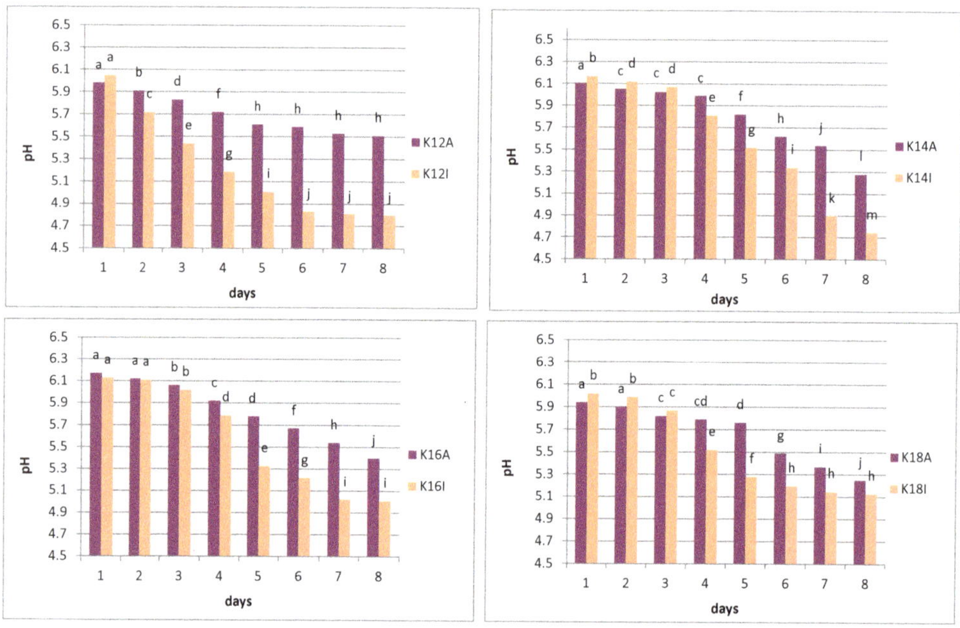

Figure 2. Changes in the pH value of the congress wort obtained from the malt created from the Kangoo variety barley. (K12I, K14I, K16I and K18I: vacuum impregnated grain, soaked at 12, 14, 16 and 18 °C, respectively; K12A, K14A, K16A and K18A: grain not subjected to vacuum impregnation, soaked at 12, 14, 16 and 18 °C, respectively). a, b, c, … - different letters mean statistically significant differences at the level of $\alpha = 0.05$.

The pH value is one of the most important parameters in the production of wort. The tested parameter changed its values, depending on the temperature and the malt production model. The highest pH value (6.2) was recorded for the wort obtained from kilned malt from barley, steeped in the atmospheric conditions, and malted for one day at a temperature of 16 °C.

The wort from the malt obtained from the grain steeped and impregnated under a pressure of 5 kPa showed a greater decrease in the value of the tested parameter. The lowest pH value was found in the wort obtained from grain steeped under atmospheric pressure and malted at 14 °C.

Figure 3 shows a diagram of the changes in the extract content in the congress wort obtained from the Kangoo variety malt that was vacuum impregnated at a temperature of 12–18 °C, under an atmospheric pressure of 5 kPa.

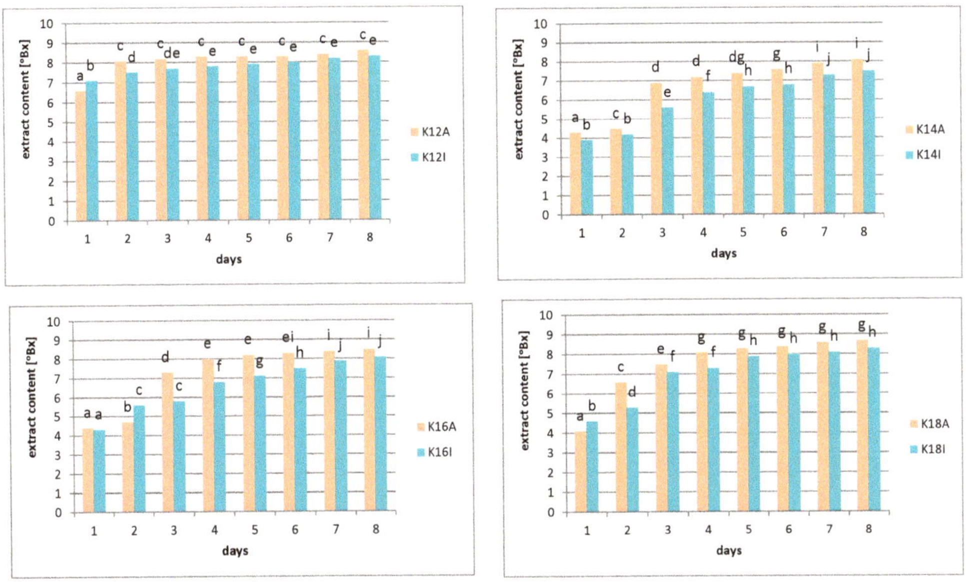

Figure 3. Changes in the extract content in the congress wort obtained from the malt created from the Kangoo variety barley. (K12I, K14I, K16I and K18I: vacuum-impregnated grain, soaked at 12, 14, 16 and 18 °C, respectively; K12A, K14A, K16A and K18A: grain not subjected to vacuum impregnation, soaked at 12, 14, 16 and 18 °C, respectively). a, b, c, . . . - different letters mean statistically significant differences at the level of α = 0.05.

Table 2 presents the results of Tukey's reasonable significant difference (HSD) test used to assess the significance of the effect of the steeping and malting conditions on the pH and extract content.

Table 2. The results of the analysis of Tukey's reasonable significant difference (HSD) test on the influence of the selected factors on the extract content and the pH.

Ordinal Number	Factor	Value	Average Extract [°Bx]/Homogeneous Groups *	Average pH/Homogeneous Groups *
1.	Pressure [kPa]	5	7.25 a	5.58 a
		100	8.15 b	5.71 b
2.	Temperature [°C]	12	7.95 a	5.67 a
		14	7.61 b	5.52 ab
		16	7.22 c	5.45 b
		18	7.31 c	5.38 c
3.	Malting time [days]	1	4.58 a	6.02 a
		2	5.25 b	5.98 ab
		3	5.78 c	5.91 b
		4	5.91 d	5.83 c
		5	6.27 de	5.71 d
		6	6.41 ef	5.57 de
		7	7.16 f	5.41 ef
		8	7.89 f	5.35 f

* Different letters mean statistically significant differences at the level of α = 0.05.

The course of changes in the extract content for the wort obtained from the malt created from the Kangoo barley variety that was steeped under atmospheric pressure and malted at 12 °C shows an increasing linear trend with values above 8°Bx, except for the kilned malt samples obtained after the first day of malting, where the extract content in the wort was 6.6°Bx. In the case of the malt obtained from grain steeped and impregnated at a pressure of 5 kPa, the lowest value of the wort extract content was 7.1°Bx, and then it increased slightly to 8.3°Bx for the kilned malt sampled on the last day of the malting. The extract content in the individual congress wort samples differed from each other by about 0.5°Bx in all of the tested samples within eight days of malting.

The malt produced at higher temperatures behaved in a similar way. Table 2 presents the results of the statistical analysis of the influence of the examined experimental factors on the extract content and pH values.

The influence of the examined factors showed statistically significant differences. The temperature changes in the range of 16–18 °C did not cause any significant changes in the extract content. To the contrary, the changes in the extract content observed at 12 °C showed statistically significant differences.

4. Discussion

The analysis performed in this study revealed no negative effects of vacuum impregnation on the brewing process or on the final product quality. In our previous studies, vacuum impregnation did not cause significant changes in the germination efficiency of the tested cereals, i.e., barley, wheat and rye. It was found that the impregnation did not affect the internal structure of the grain [28]. Other authors noticed that the integrity of plant tissue cells is unaffected despite changes in microstructure caused by vacuum impregnation [7,8].

In the current study, a decrease in the pH of the congress wort was observed during malting. Wort from malt obtained from vacuum-impregnated grain, under a pressure of 5 kPa, showed a greater decrease in pH. This has been confirmed in other studies that the optimum pH for α-amylase is 5.6 to 5.8, and for β-amylase, it is 5.4 to 5.5. Higher pH values are considered unfavorable, in contradistinction to lower pH values. At a low pH, higher amounts of fermentable sugars are obtained, which results in an increase in the degree of fermentation [30]. The pH values of the wort produced from malt, which was created from impregnated grain, were slightly lower than optimal. These low pH values do not cause enzyme inactivation and may only slightly slow down the starch saccharification process.

The literature does not indicate the optimal extract content. It should be as high as possible. In the case of malt obtained from impregnated grain, the extract content was slightly lower, but at further stages of beer production, during beer brewing, the extract content can be increased [29,30]. To produce high-quality beer, the malting industry requires malt with a high extract yield. Brewing barley is the primary raw material used in the production of beer, and its quality is critical. To meet these requirements, the barley must be able to germinate vigorously [33]. The extract is one of the most important malt quality parameters. During the mashing stage, it determines the amount of received soluble solids. The extract content slowly decreases at mash-in temperatures higher than 65 °C [34].

Obtaining the wort with the highest extract and the desired pH is the main goal of mashing. The pH value is crucial for the quality of the beverage produced. The lowest extract value was obtained in the wort produced from 100% Special X malt, whereas the highest extract values were obtained from the wort produced from 100% Vienna malt and 100% Melanoidin malt. Surprisingly, the combination of Caramel pils and Special X malt also resulted in the high extract of the wort produced. The use of a mixture of Vienna malt, melanoidin malt and Special X malt resulted in the highest extract of the wort produced from ternary mixtures. The pH values obtained by the researchers were very diverse and ranged from 6.5 to 8.4 [35].

O'Rourke [36] states that the optimum pH of the wort should be 5.6 ± 0.2. Based on the results obtained, the optimal parameters of the wort can be obtained by combining Vienna and Melanoidin malts in a 2:1 ratio.

It is an issue of utmost importance to control the pH value within the desired range in the different phases of beer production. Mash pH is an important factor affecting the isomerization and solubility of α-acids, yeast flocculation, enzymatic activity and extract yield. When it comes to beer, pH affects its flavor (acidity), flavor stability, non-biological stability and biological stability [37].

Both the raw materials at the various stages of beer production and the final product should be tested for pH. Its value has a decisive influence on the taste, prevents the development of microorganisms and is a key factor influencing the maturation of beer, its stability and its durability [38,39].

5. Conclusions

Based on the obtained research results and their analysis, the following conclusions were formulated:

1. The results of the research have shown that vacuum impregnation under a pressure of 5 kPa significantly influences the final grain moisture. In the tested temperature range of 12–18 °C, during 36 h of moistening, the grain soaked and malted under atmospheric pressure reached the final moisture level of 42%. The vacuum-impregnated Kangoo grain required 6 h less soaking time. This confirms the observation of many researchers, that vacuum impregnation accelerates the mass transfer in the solid–liquid system, despite the low susceptibility of the cereal grains to grain vacuum impregnation, in relation to, for example, fruit tissues.
2. The pH value in the obtained congress wort changed, depending on all of the tested factors (pressure, temperature and malting time), during the production of malt. The wort obtained from the malt subjected to vacuum impregnation showed a greater decrease in the pH value in the range of temperature changes of 12–18 °C. A detailed analysis of the pH changes shows that the eight-day malting caused an excessive decrease in the pH of the wort. The analysis of the pH changes also shows that the vacuum-impregnated grain should be malted for five days.
3. It was observed that, in the wort composed of non-vacuum-impregnated malt, the value of the extract content increased twofold. Such an increase was observed at temperatures of 14–18 °C.
4. The vacuum impregnation of barley grain under various conditions allows us to shorten the amount of time for moistening this grain to 42% moisture content by approximately 6 h. This is the greatest advantage of using vacuum impregnation in the malting process of grain. It allows for a significant shortening of the malt production cycle, in which soaking the grain before malting is a time-consuming process.

Author Contributions: Conceptualization, L.R.; methodology, L.R.; software, T.G.; validation, L.R. and R.R.; formal analysis, L.R. and R.R.; investigation, M.N. and T.G.; resources, K.V.; data curation, M.N.; writing—original draft preparation, L.R. and M.N.; writing—review and editing, L.R. and R.R.; visualization, M.N.; supervision, L.R.; project administration, L.R.; funding acquisition, L.R. and M.N. All authors have read and agreed to the published version of the manuscript.

Funding: This research received no external funding.

Institutional Review Board Statement: Not applicable.

Informed Consent Statement: Not applicable.

Data Availability Statement: The datasets used and/or analyzed during the current study are available from the corresponding author upon reasonable request.

Conflicts of Interest: The authors declare no conflict of interest.

References

1. D'Alfonzo, S. Vacuum impregnation. *Metal Finish.* **2013**, *99*, 507–509. [CrossRef]
2. Guillemin, A.; Degraeve, P.; Noel, C.; Saurel, R. Influence on impregnation solution viscosity and osmolarity on solute uptake during vacuum impregnation of apple cubes (var. *Granny Smith*). *J. Food Eng.* **2008**, *86*, 475–483. [CrossRef]
3. Vargas, M.; Chiralt, A.; Albors, A.; Gonzalez-Martinez, C. Effect of chitosan-based edible coatings applied by vacuum impregnation on quality preservation of fresh-cut carrot. *Postharv. Biol. Technol.* **2009**, *51*, 263–271. [CrossRef]
4. Aguilera, J.M. Why food microstructure? *J. Food Eng.* **2005**, *67*, 3–11. [CrossRef]
5. Mebatsion, H.K.; Verboven, P.; Ho, Q.T.; Verlinden, B.E.; Nicolai, B.M. Modelling fruit (micro)structures, why and how? *Trend. Food Sci. Technol.* **2008**, *19*, 59–66. [CrossRef]
6. Ramírez, C.; Troncoso, E.; Muñoz, J.; Aguilera, J.M. Microstructure analysis on pre-treated apple slices and its effect on water release during air drying. *J. Food Eng.* **2011**, *106*, 253–261. [CrossRef]
7. Derossi, T.; De Pilli, C.; Severini, C. Application of pulsed vacuum acidification for the pH reduction of mushrooms. *LWT Food Sci. Technol.* **2013**, *54*, 585–591. [CrossRef]
8. Lu, J.; Li, Y. Effects of arabinoxylan solubilization on wort viscosity and filtration when mashing with grist containing wheat and wheat malt. *Food Chem.* **2006**, *98*, 164–170. [CrossRef]
9. Shantz, T.; Versmold, R. Types of vacuum impregnation processes. Die Casting Engineer. 2013. Available online: www.diecasting.org/dce (accessed on 15 January 2022).
10. Fito, P.; Andres, A.; Chiralt, A.; Pardo, P. Coupling of hydrodynamic mechanisms and deformation-relaxation phenomena during vacuum treatments in solid porous food-liquid systems. *J. Food Eng.* **1996**, *27*, 229–241. [CrossRef]
11. Fito, P.; Chiralt, A. Food Matrix Engineering: The Use of the water-structure-functionality ensemble in dried food product development. *Food Sci. Technol. Int.* **2003**, *9*, 151–156. [CrossRef]
12. Fito, P.; Chiralt, A.; Barat, J.M.; Andres, A.; Martinez-Monzo, J.; Martinez-Navarrete, N. Vacuum impregnation for development of new dehydrated products. *J. Food Eng.* **2001**, *49*, 297–302. [CrossRef]
13. Barat, J.M.; Fito, P.; Chiralt, A. Modeling of simultaneous mass transfer and structural changes in fruit tissues. *J. Food Eng.* **2001**, *49*, 77–85. [CrossRef]
14. Derossi, A.; De Pilli, T.; Severini, C. Reduction in the pH of vegetables by vacuum impregnation: A study on pepper. *J. Food Eng.* **2010**, *99*, 9–15. [CrossRef]
15. Derossi, A.; De Pilli, T.; La Penna, M.P.; Severini, C. pH reduction and vegetable tissue structure changes of zucchini slices during pulsed vacuum acidification. *Food Sci. Technol.* **2011**, *44*, 1901–1907. [CrossRef]
16. Hofmeister, L.C.; Souza, J.A.R.; Laurindo, J.B. Use of dye solutions to visualize different aspect of vacuum impregnation of Minas Cheese. *LWT–Food Sci. Technol.* **2005**, *38*, 379–386. [CrossRef]
17. Mujica-Paz, H.; Valdez-Fragoso, A.; Lopez-Malo, A.; Palou, E.; Welti-Chanes, J. Impregnation properties of some frutis at vacuum pressure. *J. Food Eng.* **2003**, *56*, 307–314. [CrossRef]
18. Salvatori, D.; Andres, A.; Chiralt, A.; Fito, P. The response of some properties of fruits to vacuum impregnation. *J. Food Process Eng.* **1998**, *21*, 59–73. [CrossRef]
19. Zhao, Y.; Xie, J. Practical applications of vacuum impregnation in fruit and vegetable processing. *Trend. Food Sci. Technol.* **2004**, *15*, 434–451. [CrossRef]
20. Derossi, A.; De Pilli, T.; Severini, C. The application of vacuum impregnation techniques in food industry. In *Scientific, Health and Social Aspects of the Food Industry*; Valdez, B., Ed.; IntechOpen Limited: London, UK, 2012; ISBN 978-953-307-916-5.
21. Chiralt, A.; Fito, P.; Barat, J.M.; Andres, A.; Gonzales-Martinez, C.; Escriche, I.; Camacho, M.M. Use of vacuum impregnation in food salting process. *J. Food Eng.* **2001**, *49*, 141–151. [CrossRef]
22. Hironaka, K.; Kikuchi, M.; Koaze, H.; Sato, T.; Kojima, M.; Yaamamoto, K.; Yasuda, K.; Mori, M.; Tsuda, M. Ascorbic acid enrichment of whole potato tuber by vacuum impregnation. *Food Chem.* **2011**, *127*, 1114–1118. [CrossRef]
23. Gras, M.L.; Fito, P.; Vidal, D.; Albors, A.; Chiralt, A.; Andres, A. The effect of vacuum impregnation upon some properties of vegetables. *Proceed. ICEF8* **2001**, *2*, 1361–1365.
24. Perez-Cabrera, L.; Chafer, M.; Chiralt, A.; Gonzalez-Martinez, C. Effectiveness of antibrowning agents applied by vacuum impregnation on minimally processed pear. *Food Sci. Technol.* **2011**, *44*, 2273–2280. [CrossRef]
25. Sriwimon, W.; Boonsupthip, W. Utilization of partially ripe mangoes for freezing preservation by impregnation of mango juice and sugars. *Food Sci. Technol.* **2011**, *44*, 375–383. [CrossRef]
26. Bilbao-Sainz, C.; Andres, A.; Chiralt, A.; Fito, P. Microwaves phenomena during drying of apple cylinders. *J. Food Eng.* **2006**, *74*, 160–167. [CrossRef]
27. Phoon, P.Y.; Gomez Galindo, F.; Vicente, A.; Dejmek, P. Pulsed electric field in combination with vacuum impregnation with trehalose improves the freezing tolerance of spinach leaves. *J. Food Eng.* **2008**, *88*, 144–148. [CrossRef]
28. Kobus, Z.; Rydzak, L.; Guz, T.; Domin, M.; Nadulski, R.; Osmólska, E. Impact of various vacuum impregnation methods on viability of cereal grains. *Processes* **2021**, *9*, 1368. [CrossRef]
29. Briggs, D.E.; Boulton, C.A.; Brookes, P.A.; Stevens, R. *Brewing: Science and Practice*; Woodhead Publishing Ltd.: Abington, PA, USA, 2004.
30. Kunze, W. *Technology Brewing and Malting*; VLB Berlin: Berlin, Germany, 2014.

31. The European Reference Analytical Methods For Breweries By EBC. Analytica EBC 2019. Available online: https://brewup.eu/ebcanalytica (accessed on 4 January 2020).
32. *ISO 712:2009*; Cereals and Cereal Products-Determination of moisture content-Reference Method. ISO Standards: Geneva, Switzerland, 2009.
33. Frančáová, H.; Líšková, M.; Bojnanská, T.; Mareček, J. Germination index as an indicator of malting potential. *Czech J. Food Sci.* **2012**, *30*, 377–384. [CrossRef]
34. Evans, D.E.; Goldsmith, M.; Redd, K.S.; Nischwitz, R.; Lentini, A. Impact of Mashing Conditions on Extract, Its Fermentability, and the Levels of Wort Free Amino Nitrogen (FAN), β-Glucan, and Lipids. *J. Am. Soc. Brew. Chem.* **2012**, *70*, 39–49. [CrossRef]
35. Shopska, V.; Teneva, D.; Denkova-Kostova, R.; Ivanova, K.; Denev, P.; Kostov, G. Modelling of Malt Mixture for the Production of Wort with Increased Biological Value. *Beverages* **2022**, *8*, 44. [CrossRef]
36. O'Rourke, T. The role of pH in brewing. *Brew. Int.* **2002**, *2*, 21–23.
37. Kamburi, T.; Xhangolli, L. Wort production optimisation based on physico-chemical malt parameters. *Eur. J. Pure Appl. Chem.* **2016**, *3*, 29–40.
38. Bamforth, C.W. pH in brewing: An overview. *Master Brew. Assoc. Am. Tech. Q.* **2001**, *38*, 1–9.
39. Guyot-Declerck, C.; François, N.; Ritter, C.; Govaerts, B.; Collin, S. Influence of pH and ageing on beer organoleptic properties. A sensory analysis based on AEDA data. *Food Qual. Prefer.* **2005**, *16*, 157–162. [CrossRef]

applied sciences

MDPI

Article

Use of Unmalted and Malted Buckwheat in Brewing

Marius Eduard Ciocan [1], Rozália Veronika Salamon [2], Ágota Ambrus [3], Georgiana Gabriela Codină [1,*], Ancuța Chetrariu [1] and Adriana Dabija [1,*]

[1] Faculty of Food Engineering, Stefan cel Mare University of Suceava, 720229 Suceava, Romania
[2] Department of Food Science, Faculty of Economics, Socio-Human Sciences and Engineering, Sapientia Hungarian University of Transylvania, 530104 Miercurea Ciuc, Romania
[3] Doctoral School of Chemistry, Faculty of Natural Sciences, University of Pécs, 7624 Pécs, Hungary
* Correspondence: codina@fia.usv.ro (G.G.C.); adriana.dabija@fia.usv.ro (A.D.); Tel.: +40745-460-727 (G.G.C.); +40-748-845-567 (A.D.)

Abstract: With the goal of introducing specialty beers with full flavor and effective processing, craft and functional beer have revived old assortments and established new assortments in countries throughout the world. These unique selections are produced utilizing various ingredient combinations or changes to the brewing process. Given the rise in celiac disease sufferers in recent years, research investigations on the use of buckwheat in brewing, particularly gluten-free beer, have intensified. In this study, malted and unmalted buckwheat were proposed for use in brewing, and the results were contrasted with those of beer made entirely of malted barley. Commercial enzymes were added to the technical mashing process in order to optimize it and increase output yield. It was simple to incorporate buckwheat in the technological process to produce a final product that, in many respects, resembled conventional beer, making buckwheat suitable for regular beer drinkers. In terms of original extract, apparent extract, alcohol content, and energy value, the beer obtained from 100% barley malt differs from the final beer, but only minor variations were found in color, pH, and bitterness value. Ethyl alcohol, a particularly significant distinguishing characteristic, was impacted by the use of buckwheat malt and unmalted buckwheat. The beer made from buckwheat was acceptable from a sensory perspective in terms of flavor, aroma, texture, bitterness, etc. The results demonstrate that the researched raw materials can be successfully used in the production of beer, both in malted and unmalted forms. For the latter option, it is recommended to use enzyme preparations to speed up the mashing and filtration of the wort.

Keywords: gluten-free beer; buckwheat; malted buckwheat; brewing; beer properties

Citation: Ciocan, M.E.; Salamon, R.V.; Ambrus, Á.; Codină, G.G.; Chetrariu, A.; Dabija, A. Use of Unmalted and Malted Buckwheat in Brewing. *Appl. Sci.* **2023**, *13*, 2199. https://doi.org/10.3390/app13042199

Academic Editor: Antonio Valero

Received: 6 December 2022
Revised: 3 February 2023
Accepted: 7 February 2023
Published: 8 February 2023

1. Introduction

Beer production is a well-established sector with a long history. It is currently the third most consumed beverage worldwide, behind tea and water, and the most popular alcoholic beverage [1–5]. Consumers today seek out authentic, original products made from ingredients that give the completed product a distinct flavor and a touch of uniqueness [6–8]. The addition of new ingredients represents a constant challenge for the beer industry, through creativity and innovation, responds to the changing demands of consumers [4,9,10].

The traditional ingredients for beer are barley malt, water, hops, and yeast. Unmalted cereals, pseudocereals, and possibly exogenous enzymes may also be used [11–13].

The beer market is continually expanding, and producers use new raw materials in the both malted and unmalted forms to make new varieties of beer, such as other grains such as sorghum, corn, rice, and millet, as well as pseudocereals such as buckwheat, amaranth, and quinoa [8,14–16]. Every raw material is different, and its composition qualities can add a new sensory element and have both favorable and unfavorable effects on brewing. Currently, it is thought that between 85 and 90 percent of the world's beer production uses locally sourced adjuncts, which, in addition to saving money, provide authentic goods of higher quality with distinct flavors and aromas [8,12,17,18].

In studies on the topic of the production of malts and gluten-free beers, as well as functional beers, buckwheat is one of the pseudocereals that is most frequently cited as a raw material because it has consistently demonstrated outstanding results in terms of productivity, enzymatic activity, and chemical composition of the final product [13,19,20]. For the production of biscuits, snacks, pasta, and other food items, as well as bakery, pastry, and confectionery products, buckwheat is frequently utilized in the food industry as flour or grits. Without gliadin and glutenin (gluten-generating proteins) in their composition, it is used as a source of protein for products designed for patients with celiac disease [9,13,19].

The method of brewing using unmalted and malted buckwheat has been the subject of several studies: Wijngaard et al. (2005, 2006) have published results of their own research on the optimization of the process of malting buckwheat and brewing from unmalted and malted buckwheat [20–22]. Phiarais et al. (2010) obtained a beer from 100% buckwheat malt with characteristics such as wheat beer in terms of pH, nitrogen content, degree of fermentation, and alcohol content [23]. Duliński et al. (2019, 2020, 2021) studied the use of different enzymes in the technology of buckwheat beer production [24–26]. Brasil et al. (2020) showed that unmalted buckwheat could be used as a beer adjunct in association with barley malt in proportion a 45–55% (m/m) [19]. According to Cela et al. (2022), brewing with 40% unmalted buckwheat created a final product with acceptable quality attributes [18]. It was discovered that the physico-chemical composition and sensory qualities of the final product are all impacted by the use of this raw material in brewing. Because of this, buckwheat contains a higher level of proteins and polyphenols of significant importance that can generate noticeable variations in the final beer [19,27,28].

Researchers who malted buckwheat came to the conclusion that the malt's amylolytic activity was lower than that of barley malt [13]. The ability of amylolytic enzymes to hydrolyze non-fermentable starch into simple carbohydrates makes them essential for the brewing process [29]. Buckwheat malt also displayed a low percentage of fermentable carbohydrates in wort and a high friability when compared to barley malt. Exogenous enzymes may be used to promote starch breakdown and boost alcohol content to correct this issue [19,30]. On the other hand, malting requires a lot of energy. Thus brewing with a proportion of non-malting adjuncts has grown in appeal. As a result, the use of buckwheat as an unmalted adjunct in brewing has also been investigated in order to enable the creation of novel products with particularly desirable properties [31]. Since these enzyme systems are activated and generated during the malting process, the biggest disadvantage in terms of processability when including unmalted buckwheat is the decrease in enzyme activity (amylolytic, proteolytic, and cytolytic) [31].

The quality of beer made entirely from buckwheat malt has shown promise in research to date, but further research is required to optimize the malting and brewing processes. Additionally, the physicochemical and sensory characteristics of final buckwheat beer as well as buckwheat malt, have not been thoroughly studied [32]. The aim of our research was to use buckwheat grown in our country to produce a variety of gluten-free beers because making beer from unmalted and malted buckwheat is a novelty in Romania.

This study's objective was to evaluate the effects of using unmalted and malted buckwheat on the physicochemical and sensory properties of wort and beer samples. The results were compared to those of beers made entirely with barley malt.

2. Materials and Methods

2.1. Raw Materials

Buckwheat (*Fagoporium esculentum*) grown in the NE part of Romania from the 2020 harvest and buckwheat malt from the Pennsylvania Craft Malt, USA, were used in the experiments. For the comparative study, barley from the 2020 harvest and Pilsner-type barley malt from Slovakia from the Osivo Company (Zvolen, Slovakia) were used.

Also, to obtain the beer in laboratory conditions, Amarillo Yachima Chief hops (Yakima, WA, USA), 2019 harvest, Fermentis brewer's yeast type 74/30 (Marquette-lez-Lille, France), and enzyme preparation from the Novozyme Company (Bagsvaerd, Denmark), namely Termamyl Classic with thermostable α-amylase, were used.

The raw materials used in the manufacturing recipes were initially weighed, mixed, and then subjected to grinding.

2.2. Obtaining Beer in Laboratory Conditions

All the mash samples were obtained in Mash Bath R12 with PC connection (1-CUBE, Havlíckuv Brod, Czech Republic) according to EBC method 4.5.1. (Analytica EBC, European Brewery Convention, 2004) [33]. Physical and chemical analyses of wort produced by the Congress method (for samples of barley malt) and the modified Congress method (for samples of buckwheat malt) were used to determine the quality of the malt. The initial variants of the tested brewing recipes are depicted in Table 1, with variant 0 representing brewing with 100% buckwheat malt. The proportion of unmalted buckwheat used in these experiments ranged from 0 to 30%, while buckwheat malt made up the difference up to 100% (70–100%). The enzyme Termamyl Classic proportions from Table 1 were added based on recommendations from the specialized literature and industrial experience (the ideal amount of the enzyme used industrially when producing beer from unmalted and malted barley and barley is 2%). The enzyme was added at the beginning of the mashing at a temperature of 35 °C.

Table 1. Variant mashing recipes used in experiments.

Ingredient	Brewing Recipe Variant								
	0	1	2	3	4	5	6	7	8
Buckwheat malt, %	100	90	90	80	80	80	70	70	100
Unmalted buckwheat, %	0	10	10	20	20	20	30	30	0
Termamyl classic enzyme preparation, %	0	2	3	3	4	5	4	5	2

The variants shown in Table 2 were suggested for the manufacturing of beer based on the preliminary findings obtained for the mashing recipes described in Table 1.

Table 2. Experimental variants of obtaining beer in laboratory conditions.

Ingredient	Brewing Recipe Variant				
	CS	B1	B2	B3	B4
Barley malt, %	100	-	-	-	-
Buckwheat malt, %	0	70	60	50	-
Unmalted buckwheat, %	0	30	40	50	100
Termamyl classic enzyme preparation, %	2	2	2	2	2

The proposed brewing recipes were developed using knowledge from both the specialized literature and the author's own experimental brewing results. These buckwheat beers (B1, B2, B3, B4) obtained were compared with beer obtained exclusively from 100% barley malt (CS).

All laboratory experiments, both for mashing variants and for beer production, had been repeated at least three times.

2.2.1. Mashing

The malt and buckwheat samples were subjected to milling with a laboratory disc mill (the gap between the grinding discs 0.20 mm) type Perten LM 3310 (Cheltenham, UK). The Congress method was applied to the barley malt (EBC method 4.5.1) (Figure 1) [33]. The modified Congress method, shown in Figure 2, has been used with buckwheat malt and buckwheat as an unmalted ingredient [22].

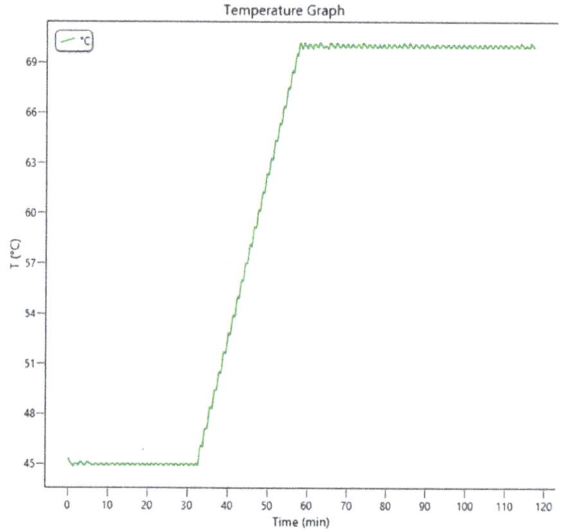

Figure 1. Congress mashing.

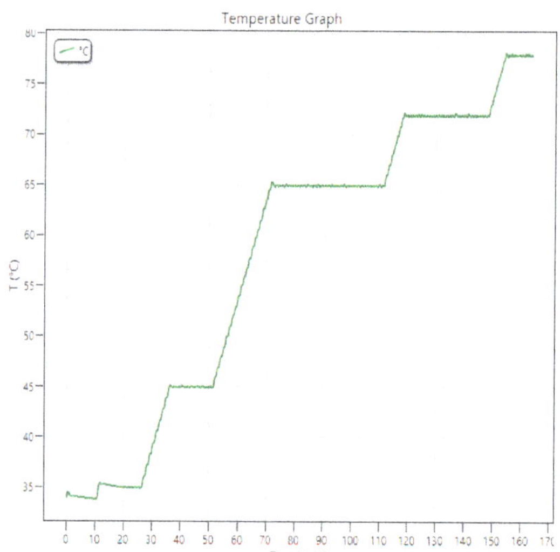

Figure 2. Modified Congress mashing.

The saccharification rate was determined with a negative iodine test—a yellow color is produced with iodine rather than a violet color. After keeping variant 0 at 72 °C for 60 min, the sample with 100% buckwheat malt did not become saccharified.

This experiment's findings suggest that using enzyme preparations to shorten the duration of saccharification of mash should be tried. Finding the right amount of enzyme for brewing needed more investigation.

2.2.2. Mash Filtration

After mashing, the contents of each mash beaker were brought to 450 g with distilled water, followed by homogenization and separate filtration through a folded filter (Sartorius with diameter 290 mm, 1602 N grade). The first 100 mL of wort was returned to the filter to give a high clarity of the obtained Congress wort. Before boiling, the quality characteristics of the wort were separately determined.

2.2.3. Wort Boiling

The obtained Congress wort from three mash beakers (2–2.2 L) was boiled with hops (1.52–1.54 g Amarillo hops with 7.8% alpha bitter acids) in Erlenmeyer flasks for 1 h. The aimed wort bitterness was 20 IBU. After boiling, the wort (1.5–1.7 L) was cooled to 20 °C and left to rest for 30 min in order to allow sedimentation of the hot tub and decanted into glass fermentation vessels with a capacity of 2 L equipped with CO_2 exhaust valves. Before inoculation with yeast, the wort was analyzed.

2.2.4. Fermentation

After cooling to 12 °C, the wort (1.5 L) was inoculated with 25 g of yeast type 74/30 Fermentis, the quality of which was previously determined using the Nucleocounter YC-100 (ChemoMetec A/S, Allerod, Denmark), the number of total cells being 24×10^8/mL biomass, of which dead cells 1.42%. The primary fermentation took place in glass fermentation vessels type Erlenmayer (Brand with 2 L volume) to 12 °C for 6 days, in an industrial refrigerator with the possibility of digital regulation, after which the young beer was bottled and subjected to secondary fermentation and maturation for 30 days at 4 °C in the same refrigerator industrial. Each beer recipe was repeated three times.

2.3. Methods of Analysis

2.3.1. Analysis of Raw Materials and Wort

The raw materials and wort properties were determined for the barley and buckwheat samples according to the standard procedures of the latest versions of Analytica EBC by the European Brewery Convention. The moisture (EBC 4.2) [34], foreign matters (EBC 4.22) [35], saccharification rate and filtration speed (EBC 4.5.1) [33], the color of wort (EBC 4.7.1) [36] and free amino nitrogen in wort (FAN, EBC 8.10.1) [37] with spectrophotometer Hach Lange DM 6000 (Hach Lange GmbH, Düsseldorf, Germany); pH of wort (EBC 8.17) WTW Inolab 720 table pH meter (Xylem Analytics, Ingolstadt, Germany) [38], original extract (EBC 8.3) [39], extract of malt, Congress mash (EBC 8.2.1) [40], fermentability, final attenuation of laboratory wort from malt (EBC 4.11.2) [41] with Anton Paar Alex (Anton Paar Austria 158 GmbH, Graz, Austria); Total nitrogen of malt (EBC 4.3.1) [42], soluble nitrogen of malt (EBC 4.9.1) [43], with Velp Scientifica UDK129 (Velp Scientifica Srl, Usmate, Italy).

Each mashing, a beaker was made for analytical determination, and the result presents the mean and standard deviation. The determinations from the same sample were performed in triplicate for analysis, and the mean values were used in this report.

2.3.2. Analysis of Beer

For the final beer, the following physico-chemical analyses were performed according to the standard procedures of the latest versions of Analytica EBC by the European Brewery Convention: the original extract and density (EBC 9.43.2) [44], the alcohol content (% *v*/*v*, % m/m; EBC 9.2.1) [45], turbidity, turbidity S25/S0, turbidity S90/S0; EBC 9.29) [46], pH (EBC 9.35) [47], CO_2 content (EBC 9.28.3) [48], O_2 content (EBC 9.37.1) [49], calories kJ/100 mL (EBC 9.45) [50] Anton Paar modular DMA meter (Anton Paar Austria 158 GmbH, Graz, Austria), color (EBC 9.6) [51] and bitterness (EBC 9.8) [52] with spectrophotometer Hach Lange DM 6000 (Hach Lange GmbH, Düsseldorf, Germany).

Three determinations from the same sample were performed in parallel for analysis, and the mean value was used in this study.

2.3.3. Sensory Analysis

The sensory analysis of the beer samples was conducted with the help of a semi-trained panel of 54 evaluators. A 9-point hedonic scale was used to evaluate the beer samples from 1 to 9, where: 1—dislikes extremely, 5—neither like nor dislike, and 9—like extremely. Nine attributes were evaluated for each beer sample as follows: appearance, color, aroma, general taste, bitterness, carbonation, body, mouthfeel, and general acceptability. The test was performed in the Sensory Analysis Laboratory of the Stefan cel Mare University of Suceava, Faculty of Food Engineering. Approximately 70 mL of beer samples coded with three-digit numbers were served to the panelists in random order at 8 °C in glass cups. Between samples, the panelists used unsalted crackers and water to cleanse their mouths to reduce taste transfer from one sample to another. The tasting area and equipment were in concordance with EBC 13.2 [53].

2.3.4. Statistical Analysis

The results were the mean of three values ± standard deviation. For the statistical evaluation of the data, we performed a one-way analysis of variance (ANOVA) and Tukey's HSD test for multiple comparisons. The differences between the means were considered statistically significant for p values ≤ 0.01. The data showing the variation of beer parameters according to the various levels of buckwheat flour and buckwheat malted flour were plotted using the design of experiment software (DOE) (Design Expert, trial version, Stat-Ease, Inc., Minneapolis, MN, USA). Using Microsoft Excel, the evaluators' mean value of the sensory data was graphically represented (Microsoft, Redmond, WA, USA). Principal component analysis (PCA) was performed using the software XLSTAT (version 2020.3.1., Addinsoft, Paris, France) to underline the correlations and differences between sensory and physicochemical characteristics of beers and their variability according to these variables.

3. Results and Discussion

3.1. Evaluation of the Quality of Buckwheat and Buckwheat Malt

The first step in assessing the potential of buckwheat as an unmalted raw material and buckwheat malt in brewing was their analysis. In Table 3, the raw materials, unmalted buckwheat, and unmalted barley, are compared in terms of their quality attributes.

Table 3. Results of the physico-chemical evaluation of the unmalted barley and buckwheat samples used in the experiments.

Characteristics	Values Obtained for:	
	Barley	Buckwheat
Moisture, %	11.00 ± 0.20	10.63 ± 0.15
Protein content, %/d.w.	10.80 ± 0.26	11.30 ± 0.17
Starch content, %/d.w.	62.00 ± 2.00	67.00 ± 1.00

Results represent mean values ± standard deviation (SD), n = 3.

It is seen that the moisture content of buckwheat grains (10.63%) is lower than the moisture content of barley grains (11%). This is a small difference and may be due to transport and storage conditions. The content of foreign matter is double in buckwheat compared to barley, which can be explained by the fact that the buckwheat came from a small producer who did not have the equipment to clean it, but this does not mean that if controlled quality buckwheat from a large producer is used, the losses will be of this magnitude.

The starch content is higher in buckwheat (67%) compared to barley (62%), which will contribute to obtaining a malt with a higher extract content. In the specialized literature, the content of starch varies between 62–91%; and the content in starch for barley varies between 52.10–69.08%. The low used buckwheat content of starch can be explained by the fact that the starch content is influenced by variety, production area, and crop year [13,14,16]. Correlated with the high content of starch, the two raw materials have a lower content of protein substances, namely barley 10.8%, and buckwheat 11.3%, which will be beneficial to the malting process. The results obtained for buckwheat are consistent with those of the studies conducted by Bobkov, S. (2016), Pirzadah et al. (2020), and Huda et al. (2021) [54–56]. Table 4 summarizes the quality characteristics of buckwheat malt and barley malt used in the study.

Table 4. Physico-chemical characteristics for barley malt and buckwheat malt.

Characteristic	Barley Malt	Buckwheat Malt
Moisture, %	5.30 ± 0.10	8.90 ± 0.09
Color, EBC-unit	3.10 ± 0.20	6.77 ± 0.10
Extract of Congress wort, °P	8.85 ± 0.50	7.30 ± 0.60
Saccharification rate, min.	8.00 ± 0.35	>90.00
Filtration speed, min.	15.00 ± 2.00	>150.00
Wort pH	6.00 ± 0.10	6.77 ± 0.21
Kolbach Index	38.00 ± 2.00	30.57 ± 0.60
Wort appearance	clear	weak opal

Results represent mean values ± standard deviation (SD), $n = 3$.

A traditional malt analysis can be subdivided into five key groups: starch conversion, carbohydrate conversion, extract content, color, and enzyme potential. Malt analysis guides the brewer in many ways by advising them on how to adjust the mashing or grinding conditions in order to prevent challenges in the technological process [57]. The obtained buckwheat malt presents comparative values with barley malt in terms of moisture and pH of the wort, with lower values being recorded for the content in the extract and the Kolbach Index. Given that the durations of saccharification and filtration were substantially longer than they were in the case of barley malt, it was suggested that enzyme preparations be used in the experiments in order to shorten these durations. Wijngaard et al. (2005) explained that due to the botanical differences between barley and buckwheat, enzyme production and, therefore, the malting process would differ, and the finished malt product would have some different physicochemical characteristics [20]. It is well known that barley is a monocotyledonous plant, and buckwheat is a dicotyledonous plant. The levels of amylolytic activity and, therefore, the yield in the extract is low compared to barley malt, but the brewing process can be optimized by adding commercial enzyme preparations [19,54]. For example, Nic Phiarais et al. (2006) showed that the addition of commercial enzyme preparations improved the filtration process and the quality of buckwheat beer by increasing extract content, yield, and total fermentable extract [58]. These effects were also obtained by Wijngaard & Arendt (2006), who developed a mashing diagram for 100% buckwheat malt by perfecting the two important parameters, temperature, and mashing time [13,22]. Myncke E. et al. (2022) investigated additionally, 40% of unmalted buckwheat mashes' lautering was enhanced by grinding the unmalted buckwheat more finely or by adding brewing enzymes, which contain α-amylase, protease, and β-glucanase. Finally, the small-scale lautering of 40% unmalted buckwheat mashes was enhanced by the addition of other enzyme solutions (β-amylase and β-glucanase). The mash filtration was unaffected by protease, in contrast [59].

3.2. Experimental Brewing with Buckwheat Malt and Buckwheat

At the start of brewing, the ground ingredients were mixed with water to form the mash. During the gradual increase in temperature (infusion mash), the enzymes in the malt and those added according to the recipe broke down the complex compounds into smaller molecules [60]. The saccharification rate indicator is a general measure of malt starch degradation, dependent on the activity of amylolytic enzymes in the mash. It is known that a particularly important factor in starch decomposition is its gelatinization temperature [60]. Giménez-Bastida et al. (2015) reported that gelatinization temperatures of buckwheat malts ranged from 65.76–67.12 °C [61].

The values of the determinations made on the resulting wort for all eight variants of brewing recipes studied are summarized in Table 5. Variant 0 is missing from Table 5 because the parameters are presented in Table 4 under the name of buckwheat malt.

Table 5. Physico-chemical properties of Congress wort for mashing variants.

Characteristic	Mashing Recipe Variant							
	1	2	3	4	5	6	7	8
Saccharification rate, minutes	5.00 ± 0.58	5.00 ± 0.58	5.00 ± 0.00	5.00 ± 0.58	5.00 ± 0.58	5.00 ± 0.00	5.00 ± 0.58	10.00 ± 0.00
Filtration speed, minutes	20.00 ± 1.0	20.00 ± 1.0	20.00 ± 1.0	20.00 ± 0.58	20.00 ± 1.0	20.00 ± 1.0	20.00 ± 1.0	30.00 ± 0.58
Extract of Congress wort, °P	7.40 ± 0.06	7.50 ± 0.12	7.50 ± 0.06	7.60 ± 0.06	7.60 ± 0.06	7.50 ± 0.12	7.60 ± 0.06	7.40 ± 0.06
Extract yield, % d.w.	64.33 ± 0.07	65.27 ± 0.04	65.27 ± 0.02	66.22 ± 0.08	66.22 ± 0.05	65.27 ± 0.07	66.22 ± 0.04	64.33 ± 0.07
Colour, EBC	7.54 ± 0.02	8.04 ± 0.03	8.22 ± 0.02	7.99 ± 0.06	8.36 ± 0.04	7.86 ± 0.02	8.19 ± 0.04	8.27 ± 0.03
pH	6.07 ± 0.05	6.10 ± 0.06	6.12 ± 0.08	6.11 ± 0.04	6.12 ± 0.02	6.15 ± 0.04	6.16 ± 0.05	6.12 ± 0.02
Wort appearance	clear	clear	clear	clear	clear	clear	clear	Faint opal

Results represent mean values ± standard deviation (SD), $n = 3$.

The duration of saccharification for each of the versions examined, independent of the amount of buckwheat malt used, was found to be 5 min during the experiment, with the exception of variant 8 (utilizing 100% buckwheat malt and 2% Termamyl enzyme), for which the duration of saccharification was 10 min. Following the mashing, the EBC method called for cooling the mash to 20 °C and adding water up to a weight of 450 g. The duration of filtration depends on the composition of the raw material, the content of β-glucans, other fibers, high molecular weight proteins, starches, and the viscosity of the mash. In studies of unconventional brewing raw materials or adjunct additives, a significant increase in the viscosity of the wort was seen, resulting in a longer filtration than when using barley malt or wheat malt [30,62]. Also, β-glucans can decrease the yield in the extract and lead to an increase in the risk of turbidity of the finished beer [30]. In this case, the specialized literature recommends the use of exogenous enzymes (β-glucanase) that allow the decomposition of β-glucans [18,63]. Recent studies showed very low levels of β-glucans for alternative cereals and pseudocereals [16,59]. The most important indicator of malt, which affects the efficiency of beer production, is the extract yield. The obtained results showed that buckwheat malt and unmalted buckwheat could be used to obtain beer. In other studies, for buckwheat malt, the mean extract yield levels were around 65% compared to 80% for the control sample (barley malt) [22]. Comparable results were obtained by Phiarais et al. (2006), who obtained an extract yield between 61.8% and 69.2% [58]. Moreover, the research conducted emphasizes the fact that the wort obtained from buckwheat and buckwheat malt presented low fermentability values and high levels of viscosity compared to the wort obtained from barley malt [22,23,64]. The color of the wort presented close values for all eight variants; the lowest value was recorded for variant 1 (7.54 EBC), and the highest value was presented by variant 5 (8.36 EBC). All variants showed higher values for color than barley malt (3.1 EBC, shown in Table 4). The pH for all variants studied showed close values in the range of 6.07–6.16, while the pH of barley Congress wort varies from 6.00–6.10, which shows that pH does not seem to have an important influence on beer quality. Lowering the pH of the wort is beneficial because many processes occur faster and better at lower pH [60]. For instance, the pH level affects the mash's viscosity, enzymatic decomposition processes, the solubilization of proteins, bitter hop compounds, and the color of the boiling wort. The pH of the wort and that of the beer produced from it are correlated. Beers with high pH values are exposed to physico-chemical disturbances due to insufficient protein coagulation during boiling. Although variant 8 showed higher values for saccharification rate and filtration speed, no considerable differences were observed for the other parameters at variants 2–7. We considered that increasing the percentage of the enzyme (3–5%) will not lead to a significant increase in extract, so from an economic point of view does not make sense. For this reason, we chose the beer recipes presented in Table 2. because we wanted to see how the mixtures of unmalted buckwheat and buckwheat malt would behave by adding only 2% Termamyl enzyme.

3.3. Study on the Production of Beer from Buckwheat and Buckwheat Malt under Laboratory Conditions

The physico-chemical parameters of beer wort in all experimental variants are summarized in Table 6.

Table 6. Physico-chemical properties of wort for the second experiment.

Characteristic	Brewing Recipe Variant				
	CS	B1	B2	B3	B4
Saccharification rate, minutes	15.00 ± 1.00 [a]	8.00 ± 1.00 [b,A]	10.00 ± 1.00 [c,A]	12.00 ± 1.00 [d,A]	10.00 ± 1.00 [e,A]
Filtration speed, minutes	15.00 ± 1.00 [a]	15.00 ± 1.00 [a,A]	15.00 ± 1.00 [a,A]	10.00 ± 1.00 [b,B]	8.00 ± 1.00 [c,D]
Extract of Congress wort, °P	9.10 ± 0.15 [a]	7.40 ± 0.10 [b,A]	7.40 ± 0.08 [c,A]	7.30 ± 0.12 [d,A]	7.10 ± 0.11 [e,B]
Extract yield, % d.w.	82.20 ± 0.30 [a]	64.33 ± 0.26 [b,A]	64.33 ± 0.34 [c,A]	63.40 ± 0.42 [d,B]	61.53 ± 0.26 [e,C]
Colour, EBC	3.30 ± 0.08 [a]	7.82 ± 0.04 [b,A]	7.76 ± 0.06 [c,A]	7.78 ± 0.08 [d,A]	8.35 ± 0.04 [e,B]
pH	6.10 ± 0.10 [a]	6.16 ± 0.12 [a,A]	6.16 ± 0.10 [a,A]	6.16 ± 0.20 [a,A]	6.21 ± 0.16 [a,A]
Total proteins, % d.w.	10.46 ± 0.08 [a]	10.75 ± 0.06 [a,A]	10.85 ± 0.08 [b,A]	10.81 ± 0.07 [a,A]	11.00 ± 0.05 [c,B]
Soluble nitrogen, mg/L	695.20 ± 1.52 [a]	630.28 ± 1.24 [a,A]	645.28 ± 1.22 [c,B]	640.60 ± 1.20 [d,C]	686.70 ± 1.16 [e,D]
Total nitrogen, % d.w.	1.68 ± 0.10 [a]	1.72 ± 0.20 [a,A]	1.74 ± 0.10 [a,A]	1.73 ± 0.40 [a,A]	1.76 ± 0.20 [a,A]
Kolbach Index	37.50 ± 0.75 [a]	30.74 ± 0.82 [b,A]	34.51 ± 0.90 [c,B]	35.02 ± 1.00 [d,B]	38.96 ± 0.82 [e,C]
FAN, mg/100 g	121.35 ± 0.84 [a]	80.29 ± 0.68 [b,A]	79.29 ± 0.50 [c,A]	80.90 ± 0.66 [d,B]	72.23 ± 0.64 [e,D]
Apparent degree of fermentation, %	83.10 ± 0.40 [a]	77.02 ± 0.50 [a,A]	77.02 ± 0.30 [a,A]	76.71 ± 0.60 [a,A]	77.40 ± 0.50 [a,B]

Results represent mean values ± standard deviation (SD), n = 3; Different letters indicate (a, b, c, d) that the result presents significant differences from the control (CS) at $p < 0.01$ level; Different letters indicate (A, B, C, D) that the result presents significant differences between the variants (B1, B2, B3, B4) at $p < 0.01$ level.

It is seen that, compared to the control sample (CS), the four experimental samples have close values in terms of wort pH, total nitrogen, and apparent degree of fermentation. Significant differences were obtained in saccharification rate, filtration speed, extract content, wort color, extract yield, soluble nitrogen, Kolbach index, and FAN. Between the variants, the saccharification rate, pH, and total nitrogen content are not had significant differences (at $p < 0.01$ level). The filtration speed is the best in B4 (1:1, unmalted buckwheat:buckwheat malt with 2% Termamyl enzyme). These data show that unmalted buckwheat has a positive effect on speed filtration. This variant shows smaller extract and extract yield, FAN content, and apparent degree of fermentation. Cela et al. (2022) showed that the yield in the extract depends on the chemical composition of the raw materials, mainly on the content of sugars and dextrin, and also on the content of nitrogenous compounds [18]. Comparing buckwheat malt to other gluten-free malts, Buiatti et al. (2018) found that buckwheat malt produced the highest extract yields for beer (40%) [65]. For this parameter, we achieved higher values between 61.53% and 64.33%. The latter is an important source of nutrition for yeast during fermentation and influence the physical and sensory properties of beer, such as colloidal stability, foam stability, and flavor profile [66]. The total nitrogen content of the wort will depend on both the raw materials used during brewing and the parameters of the brewing process [67]. The results obtained are consistent with those highlighted in other research carried out on unmalted buckwheat and buckwheat malt as raw materials in the beer industry [18,65,68]. The wort obtained was subjected to the boiling process with hops. The physico-chemical characteristics of wort after boiling with hops are summarized in Table 7.

After 60 min of boiling the samples with hops, a slight decrease in pH was observed compared to the same samples before boiling. The extracted content increased in B1, B2, B3, and B4 samples as a result of the boiling operation due to water evaporation. Compared to the wort obtained from barley malt, where the content in the extract increased after boiling to the value of 11°P, all four wort samples had the content in the extract lower, between 11 and 12°P. To improve the extracted content of wort, Cela et al. (2022) propose the gelatinization of unmalted raw materials during brewing [18]. Boiling led to an increase in color of all tested buckwheat samples from about 7–8 EBC to 11–12 EBC. After cooling to 12 °C, the wort was inoculated with yeast. The primary fermentation lasted 6 days, after which the young beer was bottled and subjected to secondary fermentation and maturation for 30 days. The final beer presented the physico-chemical characteristics displayed in Table 8.

Table 7. Physico-chemical properties of wort after boiling.

Characteristic	Brewing Recipe Variant				
	CS	B1	B2	B3	B4
Wort extract, °P	11.00 ± 0.20 ^{a}	9.10 ± 0.10 ^{b,A}	9.20 ± 0.10 ^{c,A}	9.00 ± 0.20 ^{d,A}	8.90 ± 0.16 ^{e,B}
Colour, EBC	4.40 ± 0.18 ^{a}	11.80 ± 0.14 ^{b,A}	11.00 ± 0.12 ^{c,B}	11.40 ± 0.10 ^{d,C}	12.00 ± 0.12 ^{e,D}
pH	5.98 ± 0.02 ^{a}	6.04 ± 0.04 ^{a,A}	6.05 ± 0.01 ^{a,A}	6.04 ± 0.02 ^{a,A}	6.10 ± 0.04 ^{a,A}
Bitterness value, IBU	38.90 ± 0.42 ^{a}	41.90 ± 0.38 ^{a,A}	38.60 ±0.54 ^{b,B}	39.60 ± 0.26 ^{c,C}	41.80 ± 0.32 ^{d,A}

Results represent mean values ± standard deviation (SD), $n = 3$; Different letters indicate (a, b, c, d) that the result presents significant differences from the control (CS) at $p < 0.01$ level; Different letters indicate (A, B, C, D) that the result presents significant differences between the variants (B1, B2, B3, B4) at $p < 0.01$ level.

Table 8. Physico-chemical characteristics of beer final product.

Characteristic	Brewing Recipe Variant				
	CS	B1	B2	B3	B4
Original extract, % m/m	11.05 ± 0.40	8.98 ± 0.05	8.92 ± 0.04	10.65 ± 0.05	9.52 ± 0.09
Apparent extract, % m/m	2.28 ± 0.02	2.63 ± 0.07	2.64 ± 0.06	3.13 ± 0.06	2.87 ± 0.07
Alcohol content, % *v*/*v*	4.59 ± 0.06	3.32 ± 0.09	3.29 ± 0.05	3.97 ± 0.04	3.49 ± 0.08
Alcohol content, % m/m	3.58 ± 0.05	2.60 ± 0.08	2.57 ± 0.05	3.10 ± 0.04	2.73 ± 0.08
Density, g/cm^3	1.00840 ± 0.0001	1.00841 ± 0.0003	1.00843 ± 0.0002	1.01036 ± 0.0002	1.00936 ± 0.0001
Turbidity, EBC	0.86 ± 0.02	3.19 ± 0.04	3.03 ± 0.03	1.98 ± 0.01	1.19 ± 0.01
Turbidity, S25/S0	1.05 ± 0.01	6.59 ± 0.05	4.13 ± 0.06	3.97 ± 0.04	1.99 ± 0.04
Turbidity S90/S0	1.09 ± 0.02	3.28 ± 0.02	3.05 ± 0.01	2.03 ± 0.01	1.15 ± 0.02
pH	4.53 ± 0.03	4.82 ± 0.09	4.73 ± 0.05	4.93 ± 0.05	4.84 ± 0.05
Color, EBC	5.40 ± 0.12	6.77 ± 0.18	6.11 ± 0.16	8.00 ± 0.11	6.30 ± 0.10
Bitterness value, IBU	25.80 ± 0.53	24.90 ± 0.95	18.80 ± 0.60	20.00 ± 0.69	21.50 ± 0.56
CO_2, g/L	5.12 ± 0.07	4.42 ± 0.03	4.43 ± 0.01	4.36 ± 0.01	4.49 ± 0.01
O_2, mg/L	0.01 ± 0.01	3.09 ± 0.03	3.79 ± 0.13	4.59 ± 0.13	4.47 ± 0.07
Calories, kJ/100 mL	190.00 ± 2.00	134.00 ± 2.00	134.00 ± 2.65	160.00 ± 0.58	143.00 ± 1.52

Results represent mean values ± standard deviation (SD), $n = 3$.

Insignificant differences were found in color, pH, and bitterness value. However, the beers obtained by the four variants differ from the control sample in terms of original extract, apparent extract, alcohol content, and energy value. The amount of ethyl alcohol, a crucial distinguishing factor, was influenced by utilizing buckwheat malt and unmalted buckwheat. Beer color is a consequence, among others, of products resulting from the Maillard reaction between FAN compounds and reducing carbohydrates [60]. In the study conducted by Steiner et al. (2012), a higher FAN content in 100% barley malt caused an increase in beer color due to the Maillard reaction [69]. The results obtained are in accordance with those obtained by Agu et al. (2012) and Deng et al. (2019) in their works [66,70]. For example, a study realized by Deng et al. (2019) concluded that doubling the amount of buckwheat malt as a brewing adjunct (from 20% to 40%) led to a further increase in beer color to approximately 5.1 EBC [66]. The bitter value of buckwheat beer decreased slightly compared to that of 100% barley malt beer. These differences may be due to evaporation losses when boiling the wort with hops. Because the same hopping regime was used, the isohumulone content was similar among the four varieties. The estimated global overall influence of unmalted buckwheat and buckwheat malt on the original extract, apparent extract, alcohol content, and density of the beer is shown in Figure 3.

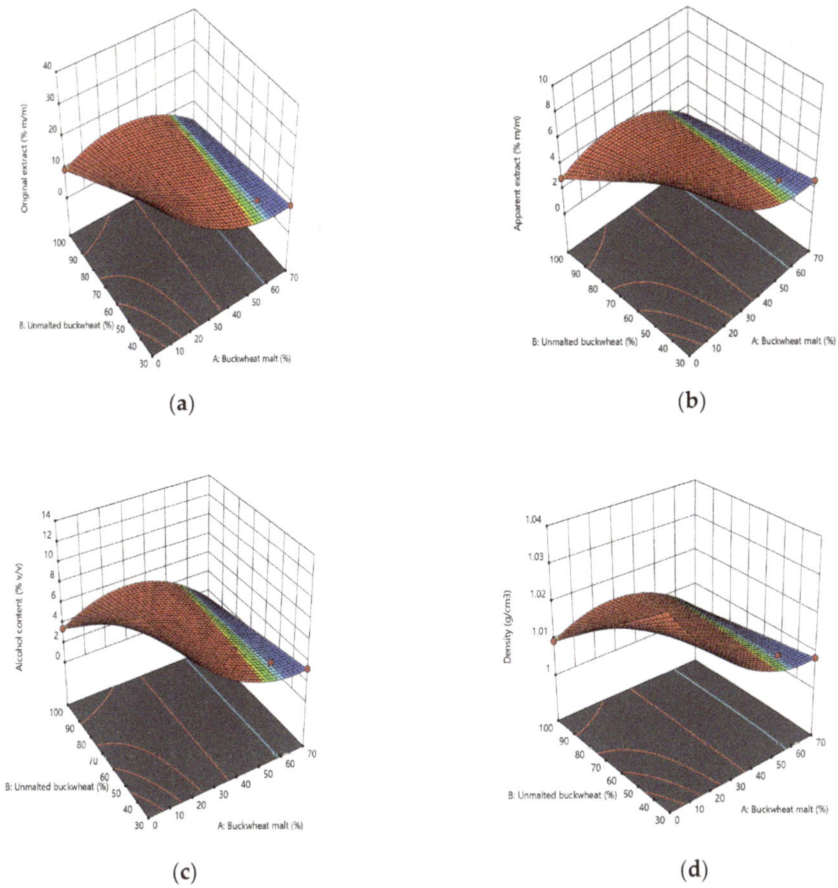

Figure 3. The graphical representations of the original extract (**a**), apparent extract (**b**), alcohol content (**c**), and density (**d**) as affected by the levels of unmalted buckwheat and buckwheat malt used in the beer recipe.

When the amount of unmalted buckwheat in the beer recipe is increased, the original and apparent extract of the beer rises; however, when buckwheat malt is added in large quantities, it falls. The beer's extract concentration is a significant qualitative indicator of it. Nitrogenous substances and carbohydrates make up the majority of their composition. The data obtained regarding the content in the extract are unexpected, and we believe that this aspect is due exclusively to the enzyme used in the mashing process (Termamyl classic). Besides these, the beer extract also contains glycerine, mineral substances, bitter substances, tannins, dyes, and organic acids. During the fermentation of wort, the alcoholic fermentation of carbohydrates takes place with the formation of ethyl alcohol and carbon dioxide. By transforming the sugars into alcohol, the density of the wort decreases, and that way, the dynamics of the fermentation can be checked by measuring the concentration of the wort extract.

A graphical representation of the beer turbidity and carbon dioxide levels is shown in Figure 4.

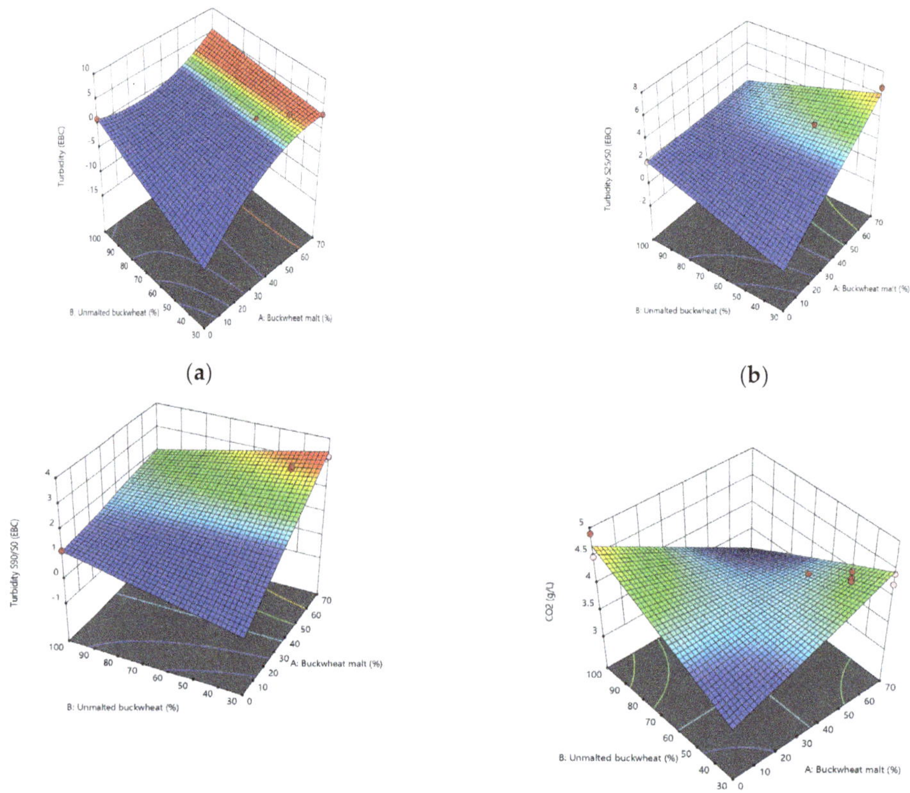

(c) (d)

Figure 4. The graphical representations of the beer turbidity (**a**), turbidity S25/S0 (**b**), turbidity S90/S0 (**c**), and carbon dioxide (**d**) as affected by the levels of unmalted buckwheat and buckwheat malt used in a beer recipe.

All turbidity readings followed a similar trend, with values falling as unmalted buckwheat malt addition levels increased. This may be explained by the fact that unmalted buckwheat contains more soluble nitrogen and polyphenolic tannins than malted buckwheat and is, therefore, more protein-rich, favoring the appearance of a beer with a higher turbidity value [71]. Proteins and other particles smaller than 1 μm, which are measured under 90° and shown in Figure 4c, scatter light predominantly. The majority of forward-scattered light is caused by particles larger than 1 μm, such as diatomaceous earth and yeast, which are measured under 25° and are shown in Figure 4b. Since our samples were analyzed in an unfiltered form, the greater turbidity S25/S0 values can be explained. Brewing with raw materials with a high Kolbach Index (which is higher for wort with high levels of unmalted buckwheat than for those with a high level of buckwheat malt), causes an increase in turbidity of the beer [18]. The absence of a malting process for buckwheat will lead to a decrease in its proteolytic activity. Therefore, when unmalted buckwheat is used in brewing as a partial replacement for buckwheat malt, there is a decrease in low-molecular nitrogen compounds and protein content, preventing haze formation. The carbon dioxide content of the beer samples varies between 4.36–4.49 g/L, while the control sample (CS) had 5.12 g/L. The absorption capacity of carbon dioxide in beer is influenced by its composition, especially by the colloid's distribution with a high contact surface [72]. For instance, some protein macromolecules may be important in stabilizing carbon dioxide, which appears to occur more frequently in malted flours [73].

Figure 5 shows the values of pH, color, bitterness, and oxygen content according to the various proportions of unmalted and malted buckwheat used in the beer recipe. There is no conclusive trend in color, pH, bitterness, and oxygen values depending on the dose of malted or unmalted buckwheat used. Bitterness is a cause of α-acids isomerization from hops during boiling. No definite trend based on the brewing recipe was seen because the identical hopping regime was followed for the manufacture of all beers. Similar findings were also reported by Phiarais et al. (2010), who came to the conclusion that the color of the beer made from buckwheat malt was likely similar to that of beers made from natural grains because no significant changes between malted and unmalted buckwheat grain were seen [23]. Since all laboratory-scale productions employed the same brewing regime, the pH and oxygen content readings were comparable across beers.

When creating products that live up to consumer expectations, sensory analysis plays a key role. As a result, studies exploring the addition of nutritious functional ingredients such as buckwheat to conventional recipes are frequently supplemented by evaluations of overall quality, appearance, scent, taste, and texture using standardized techniques involving professional judges or consumer panels. The sensory evaluation of flavor is frequently accompanied by analytical workflows aimed at isolating and determining the concentration of volatile compounds in food and determining the impact of food components on overall flavor intensity and/or final product perception. Flavor is of particular importance in influencing consumer preferences [74,75]. The sensory analysis data are shown in Figure 6.

The CS sample earned the greatest for overall acceptability, whereas the B2 sample scored the lowest, according to the data collected. All of the beer samples were enjoyed to a certain extent by the panelists. The CS, B1, and B4 samples obtained a like very much score of 8.1, 7.8, and 7.6, respectively, whereas the B3 and B2 samples obtained a mean like moderate score of 7.0 and 6.6, respectively. These results are consistent with those found by Sebestyén et al. (2013), who determined in the sensory analysis that the buckwheat beer had the aroma of roasted sunflower seeds and a chestnut-like aroma and was positively evaluated by the tasters [76]. Also, Phiarais et al. (2005) have reported that buckwheat beers were acceptable regarding taste purity, odor, mouthfeel, bitterness, and tingling [77]. The sensory characteristics that contributed to the general acceptability varied depending on the recipe for beer making. The best sensory results were obtained for the CS sample. However, for the color parameter, gluten-free beer samples obtained better sensory results. These data were in disagreement with those obtained by Brasil et al. (2020) who reported a paler beer with an increased level of buckwheat malt addition in the beer recipe [19]. From non-gluten beer samples, the highest acceptability was recorded for B1 samples which contained the highest level of buckwheat malt in its recipe. These data were in agreement with those reported by Dezelak et al. (2014), who concluded that beer from buckwheat malt has high scores from the sensory point of view [78].

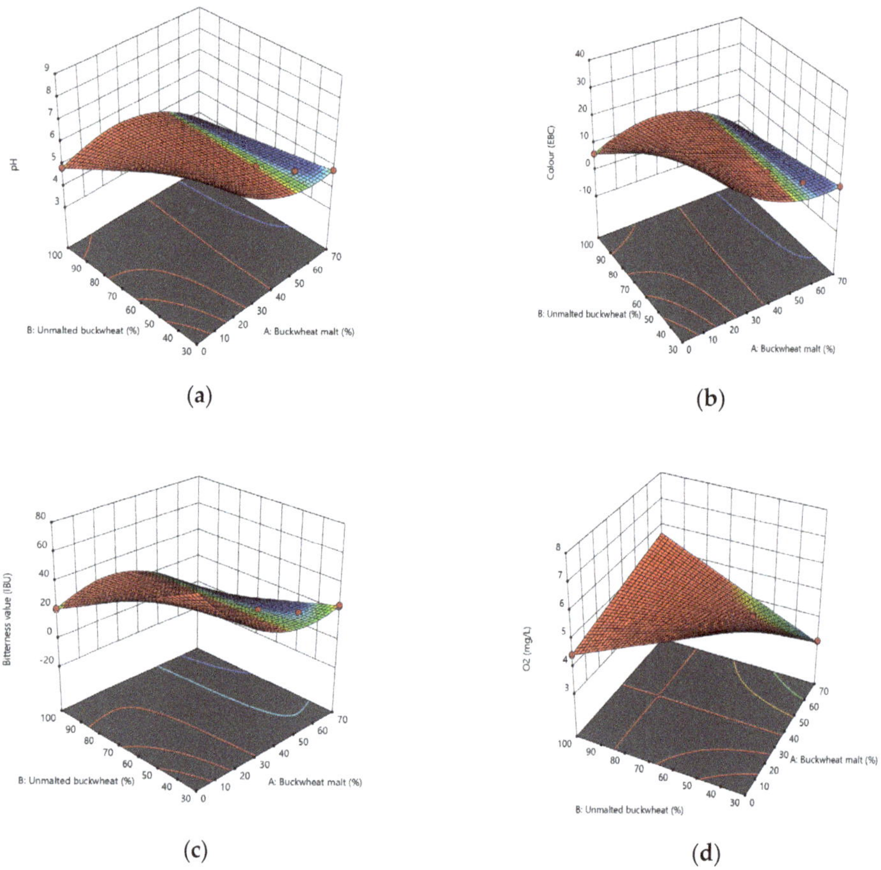

Figure 5. The graphical representations of the beer pH (**a**), color (**b**), bitterness (**c**), and oxygen (**d**) as affected by the levels of unmalted buckwheat and buckwheat malt used in the beer recipe.

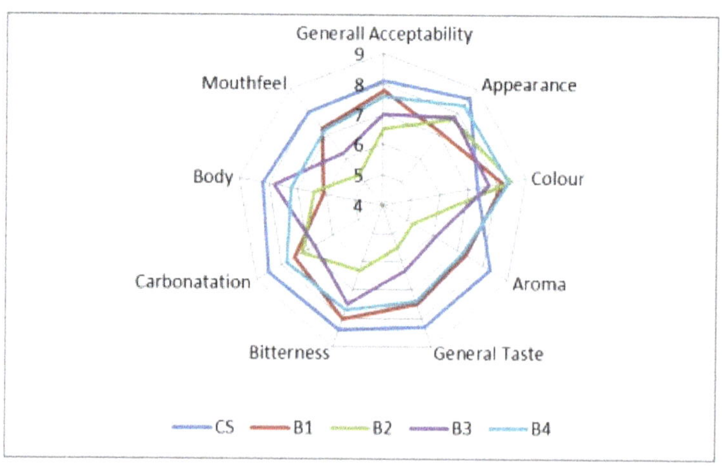

Figure 6. Sensory characteristics of the beer samples.

3.4. Principal Component Analysis of the Physico-Chemical and Sensory Characteristics of the Beer Samples

Principal Component Analysis (Figure 7) highlighted the correlation between physicochemical data and sensory characteristics of the beer samples. The first two principal components explain 69.47% of the total variance (PC1 = 37.82% and PC2 = 31.65%). The PCA graph shows a clear distinction between the beer samples in terms of their physiochemical and sensory properties. Beer made from malted barley (CS) is positioned on the graph's right side among the sensory qualities, indicating that these qualities are crucial to understanding how to appreciate it. These characteristics have a high association with this sample since, based on previously collected data, this beer was the one that consumers liked the most. These samples had a similar score for overall acceptability from consumers and were the most well-liked non-gluten beer. Due to the fact that all of the B1, B4, and CS samples are alongside the PC2 component on the PCA graph, these samples were also the closest to a malted barley beer from a sensory and physico-chemical perspective. The placement of B2 beer in the primary component area indicates that it is the most distinctive from the other beers, with this sample receiving the lowest consumer approval.

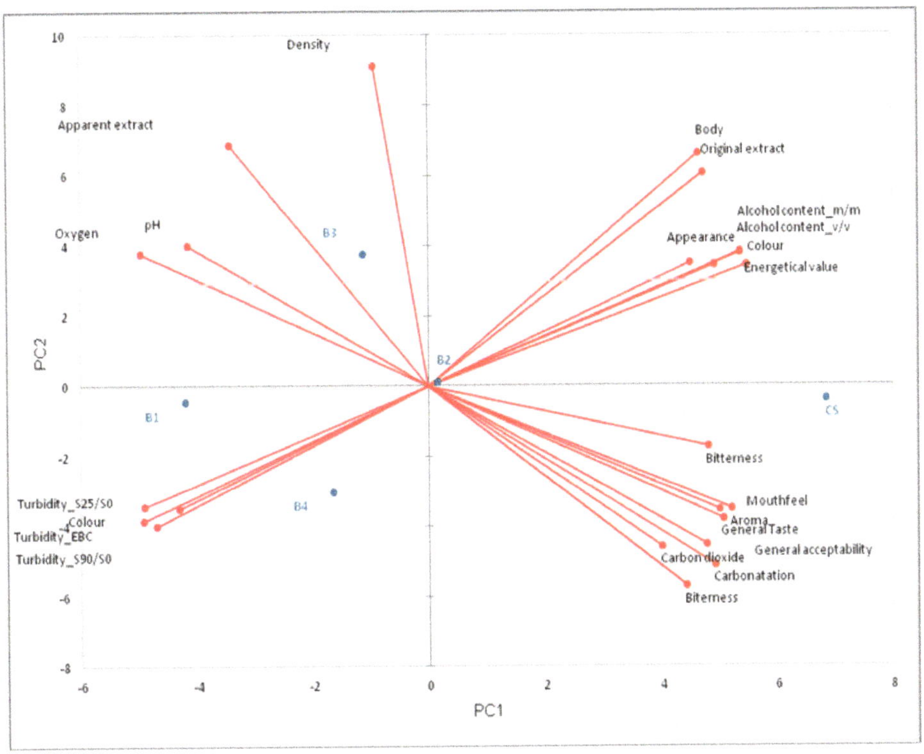

Figure 7. Principal component analysis of the physicochemical and sensory characteristics of the beer samples.

B2 and B3 are both samples that are most closely related to the PC2 component of the PCA graph, and both are located near the top of the graph. This result is consistent with the observation that both the B3 and B2 samples received similar levels of consumer appreciation based on our prior research. Relationships between physico-chemical and sensory values of beer samples show that is a good correlation between original extract and alcohol content with an r = 0.962, between body and original extract with an r = 0.968 between appearance and color (r = 0.877), between alcohol content and energy value (r = 0.999) for a level of 0.01. The original extract is one of the most important parameters from a financial point of view. This serves as an indicator of the amount of dry matter in the wort used to make the examined beer. This will have an impact on the beer's body, which consumers will feel as being heavier or denser on the tongue [18]. Also, with respect to the first principal component PC2, it may be seen direct positive correlations between body and density (r = 0.625), original extract (r = 0.968), and alcohol content (r = 0.917). During alcohol fermentation, chemical reactions take place, which transforms fermentable sugars into alcohol and carbon dioxide. Therefore, the higher the quantity of real extract is, the higher the alcohol content is. The highly significant correlation between energy value, alcohol content (r = 0.999), and original extract (r = 0.949) may be explained by the fact that the energetical value of a beer is calculated according to its alcohol content and original extract value [79]. The two plots underline a close association between the physico-chemical trait carbon dioxide and sensory carbonatation (r = 0.942) which was explainable given that beer carbonatation is experienced as a burning and stinging sensation on the tongue, being influenced by the amount of carbon dioxide and the size of the bubbles [18]. Other notable correlations were found between sensory characteristics such as general acceptability, general taste, aroma, mouthfeel, and bitterness, as well as between the measured sensory bitterness and the analytically determined bitterness characteristics (r = 0.884). In evaluating the sensory quality of beer, the highest rating was the taste and its impact on the consumers [80]. The sensory quality of beer taste is correlated to some extent with all the elements strongly associated with general acceptance (aroma, mouthfeel, bitterness).

4. Conclusions

The technological method of making beer from buckwheat and buckwheat malt is effortless, and the final product is essentially equivalent to a standard beer in many ways, making buckwheat a potential raw material for habitual beer drinkers who are looking for new sensory experiences.

The different fermentation profile for buckwheat beer with a higher first concentration of glucose is crucial for alcohol and other fermentation products, positively changing its aroma and taste. However, its high content of polyphenolic compounds can cause a slightly astringent taste and a spicy and sweet aroma appreciated by tasters.

The beer made with 50% unmalted buckwheat and 50% buckwheat malt with 2% Termamyl Classic enzyme had the best results out of all the studied recipes in terms of the key physico-chemical traits: original extract was 10.65%, apparent extract was 3.13%, alcohol content was 3.97% (v/v), and turbidity was 1.98 EBC. The sample made entirely of unmalted buckwheat with 2% Termamyl Classic enzyme was the most well-liked from a sensory perspective. We achieved our research objective to create a final beer that had good physico-chemical properties, was well-liked by consumers, and was produced at the lowest cost.

As a result of the findings, it is suggested that in subsequent experiments, beer only be produced from unmalted buckwheat with enzyme addition, which is much more advantageous from an economic and technological standpoint. This is because it is possible to produce beer from buckwheat without it being malted by simply adding enzyme preparations to the brewing process.

Author Contributions: Conceptualization, M.E.C., A.D., R.V.S. and G.G.C.; Methodology, A.C., R.V.S. and Á.A.; Formal Analysis, A.C. and A.D.; Investigation, M.E.C., R.V.S. and Á.A.; Resources, G.G.C.; Writing—Original Draft Preparation, A.D., R.V.S., G.G.C. and M.E.C.; Writing—Review and Editing, A.C. and A.D. All authors have read and agreed to the published version of the manuscript.

Funding: This work was funded by the Ministry of Research, Innovation, and Digitalization within Program 1—Development of national research and development system, Subprogram 1.2—Institutional Performance—RDI excellence funding projects, under contract no. 10PFE/2021.

Institutional Review Board Statement: Not applicable.

Informed Consent Statement: Not applicable.

Data Availability Statement: The data presented in this study are available in this article.

Conflicts of Interest: The authors declare no conflict of interest.

References

1. Betancur, M.I.; Motoki, K.; Spence, C.; Velasco, C. Factors influencing the choice of beer: A review. *Food Res. Int.* **2020**, *137*, 109367. [CrossRef] [PubMed]
2. Chetrariu, A.; Dabija, A. Pre-treatments used for the recovery of brewer's spent grain—A minireview. *J. Agroaliment. Process. Technol* **2020**, *26*, 304–312.
3. Parés Viader, R.; Yde, M.S.H.; Hartvig, J.W.; Pagenstecher, M.; Carlsen, J.B.; Christensen, T.B.; Andersen, M.L. Optimization of Beer Brewing by Monitoring α-Amylase and β-Amylase Activities during Mashing. *Beverages* **2021**, *7*, 13. [CrossRef]
4. Piva, R.C.; Verdan, M.H.; Mascarenhas Santos, M.D.S.; Batistote, M.; Cardoso, C.A.L. Manufacturing and characterization of craft beers with leaves from *Ocimum selloi* Benth. *J. Food Sci. Technol.* **2021**, *58*, 4403–4410. [CrossRef]
5. Díaz, A.B.; Durán-Guerrero, E.; Lasanta, C.; Castro, R. From the Raw Materials to the Bottled Product: Influence of the Entire Production Process on the Organoleptic Profile of Industrial Beers. *Foods* **2022**, *11*, 3215. [CrossRef]
6. Ciocan, M.; Dabija, A.; Codină, G.G. Effect of some unconventional ingredients on the production of black beer. *Ukr. Food J.* **2020**, *9*, 322–331. [CrossRef]
7. Muscolo, A.; Marra, F.; Salafia, F.; Andronaco, P.; Di Sanzo, R.; Carabetta, S.; Russo, M. Bergamot and olive extracts as beer ingredients: Their influence on nutraceutical and sensory properties. *Eur. Food Res. Technol.* **2022**, *248*, 1–11. [CrossRef]
8. Villacreces, S.; Blanco, C.A.; Caballero, I. Developments and characteristics of craft beer production processes. *Food Biosci.* **2022**, *45*, 101495. [CrossRef]
9. Orhotohwo, O.L.; Czipa, N.; Kovács, B.; Alexa, L. Impact of the use of gluten-free cereals and spices on the quality parameters of beer. *J. Microbiol. Biotechnol. Food Sci.* **2021**, *11*, e3838. [CrossRef]
10. Paiva, R.A.; Mutz, Y.S.; Conte-Junior, C.A. A Review on the Obtaining of Functional Beers by Addition of Non-Cereal Adjuncts Rich in Antioxidant Compounds. *Antioxidants* **2021**, *10*, 1332. [CrossRef]
11. Cela, N.; Condelli, N.; Caruso, M.C.; Perretti, G.; Di Cairano, M.; Tolve, R.; Galgano, F. Gluten-free brewing: Issues and perspectives. *Fermentation* **2020**, *6*, 53. [CrossRef]
12. Castro Marin, A.; Baris, F.; Romanini, E.; Lambri, M.; Montevecchi, G.; Chinnici, F. Physico-chemical and sensory characterization of a fruit beer obtained with the addition of cv. Lambrusco grapes must. *Beverages* **2021**, *7*, 34. [CrossRef]
13. Dabija, A.; Ciocan, M.E.; Chetrariu, A.; Codină, G.G. Buckwheat and Amaranth as Raw Materials for Brewing, a Review. *Plants* **2022**, *11*, 756. [CrossRef]
14. Dabija, A.; Ciocan, M.E.; Chetrariu, A.; Codină, G.G. Maize and sorghum as raw materials for brewing, a review. *Appl. Sci.* **2021**, *11*, 3139. [CrossRef]
15. Trummer, J.; Watson, H.; De Clippeleer, J.; Poreda, A. Brewing with 10% and 20% Malted Lentils—Trials on Laboratory and Pilot Scales. *Appl. Sci.* **2021**, *11*, 9817. [CrossRef]
16. Baillière, J.; Laureys, D.; Vermeir, P.; Van Opstaele, F.; De Rouck, G.; De Cooman, L.; De Clippeleer, J. 10 unmalted alternative cereals and pseudocereals: A comparative analysis of their characteristics relevant to the brewing process. *J. Cereal Sci.* **2022**, *106*, 103482. [CrossRef]
17. Habschied, K.; Živković, A.; Krstanović, V.; Mastanjević, K. Functional beer—A review on possibilities. *Beverages* **2020**, *6*, 51. [CrossRef]
18. Cela, N.; Galgano, F.; Perretti, G.; Di Cairano, M.; Tolve, R.; Condelli, N. Assessment of brewing attitude of unmalted cereals and pseudocereals for gluten free beer production. *Food Chem.* **2022**, *384*, 132621. [CrossRef]
19. Brasil, V.C.B.; Guimarães, B.P.; Evaristo, R.B.W.; Carmo, T.S.; Ghesti, G.F. Buckwheat (*Fagopyrum esculentum Moench*) characterization as adjunct in beer brewing. *Food Sci. Technol.* **2020**, *41*, 265–272. [CrossRef]
20. Wijngaard, H.H.; Ulmer, H.M.; Neumann, M.; Arendt, E.K. The effect of steeping time on the final malt quality of buckwheat. *J. Inst. Brew.* **2005**, *111*, 275–281. [CrossRef]
21. Wijngaard, H.H.; Ulmer, H.M.; Arendt, E.K. The effect of germination time on the final malt quality of buckwheat. *J. Am. Soc. Brew. Chem.* **2006**, *64*, 214–221. [CrossRef]
22. Wijngaard, H.H.; Arendt, E.K. Optimisation of a mashing program for 100% malted buckwheat. *J. Inst. Brew.* **2006**, *112*, 57–65. [CrossRef]
23. Phiarais, B.P.N.; Mauch, A.; Schehl, B.D.; Zarnkow, M.; Gastl, M.; Herrmann, M.; Arendt, E.K. Processing of a top fermented beer brewed from 100% buckwheat malt with sensory and analytical characterisation. *J. Inst. Brew.* **2010**, *116*, 265–274. [CrossRef]
24. Duliński, R.; Zdaniewicz, M.; Pater, A.; Żyła, K. Impact of Two Commercial Enzymes on the Release of Inositols, Fermentable Sugars, and Peptides in the Technology of Buckwheat Beer. *J. Am. Soc. Brew. Chem.* **2019**, *77*, 119–125. [CrossRef]
25. Duliński, R.; Zdaniewicz, M.; Pater, A.; Poniewska, D.; Żyła, K. The impact of phytases on the release of bioactive inositols, the profile of inositol phosphates, and the release of selected minerals in the technology of buckwheat beer production. *Biomolecules* **2020**, *10*, 166. [CrossRef]
26. Duliński, R.; Zdaniewicz, M.; Pater, A. Effect of Phytase Addition to Buckwheat Wort on the Selected Fermentable Sugars, Polypeptide Profile and Nitrogen Content from Free Aminoacids. *Bull. Univ. Agric. Sci. Vet. Med. Cluj-Napoca. Food Sci. Technol.* **2021**, *78*, 33–40. [CrossRef]

27. Kowalska, E.; Ziarno, M. Characterization of Buckwheat Beverages Fermented with Lactic Acid Bacterial Cultures and Bifidobacteria. *Foods* **2020**, *9*, 1771. [CrossRef]
28. Streimikyte, P.; Balciunaitiene, A.; Liapman, T.D.; Streimikyte-Mockeliune, Z.; Puzeryte, V.; Borkertas, S.; Viskelis, J. Enzymatically Hydrolysed Common Buckwheat (*Fagopyrum esculentum* M.) as a Fermentable Source of Oligosaccharides and Sugars. *Appl. Sci.* **2022**, *12*, 8210. [CrossRef]
29. Koller, H.; Perkins, L.B. Brewing and the Chemical Composition of Amine-Containing Compounds in Beer: A Review. *Foods* **2022**, *11*, 257. [CrossRef]
30. Blšáková, L.; Gregor, T.; Mešťánek, M.; Hřivna, L.; Kumbár, V. The Use of Unconventional Malts in Beer Production and Their Effect on the Wort Viscosity. *Foods* **2021**, *11*, 31. [CrossRef]
31. Yorke, J.; Cook, D.; Ford, R. Brewing with unmalted cereal adjuncts: Sensory and analytical impacts on beer quality. *Beverages* **2021**, *7*, 4. [CrossRef]
32. Dabija, A.; Ciocan, M.E.; Chetrariu, A.; Mîrzan, D. Comparative evaluation of the physico-chemical characteristics of buckwhweat malt and barley malt. *Int. Multidiscip. Sci. GeoConference SGEM* **2021**, *21*, 51–58.
33. EBC Analytica, 4.5.1—Extract of Malt: Congress Mash (IM), 2004. 22 October 2018. Available online: https://brewup.eu/ebc-analytica/malt/extract-of-malt-congress-mash/4.5.1 (accessed on 10 February 2022).
34. EBC Analytica, 4.2—Moisture content of malt, 2000. 22 October 2018. Available online: https://brewup.eu/ebc-analytica/malt/moisture-content-of-malt/4.2 (accessed on 10 February 2022).
35. EBC Analytica, 4.22—Sieving Test for Malt, 2005. 22 October 2018. Available online: https://brewup.eu/ebc-analytica/malt/sieving-test-for-malt/4.22 (accessed on 10 February 2022).
36. EBC Analytica, 4.7.1—Colour of Wort: Spectrophotometric Method (RM), 2000. 22 October 2018. Available online: https://brewup.eu/ebc-analytica/malt/colour-of-malt-sepctrophotometric-method-rm/4.7.1 (accessed on 10 February 2022).
37. EBC Analytica, 8.10.1—Free Amino Nitrogen in Wort (FAN), 2015. 23 October 2018. Available online: https://brewup.eu/ebc-analytica/wort/free-amino-nitrogen-in-wort-by-spectrophotometry-manual-method-im/8.10.1 (accessed on 10 February 2022).
38. EBC Analytica, 8.17—pH of Wort, 1999. 23 October 2018. Available online: https://brewup.eu/ebc-analytica/wort/ph-of-wort/8.17 (accessed on 10 February 2022).
39. EBC Analytica, 8.3—Extract of Wort, 2004. 23 October 2018. Available online: https://brewup.eu/ebc-analytica/wort/extract-of-wort/8.3 (accessed on 10 February 2022).
40. EBC Analytica, 8.2.1—Determination of the Specific Gravity of the Wort, 2004. 22 October 2018. Available online: https://brewup.eu/ebc-analytica/wort/specific-gravity-of-wort-using-a-pyknometer/8.2.1 (accessed on 10 February 2022).
41. EBC Analytica, 4.11.2—Fermentability, Final Attenuation of Laboratory Wort from Malt: Rapid, 1999. 22 October 2018. Available online: https://brewup.eu/ebc-analytica/malt/fermentability-final-attenuation-of-laboratory-wort-from-malt-rapid-242/4.11.2 (accessed on 10 February 2022).
42. EBC Analytica, 4.3.1.—Total Nitrogen of Malt: Kjeldahl Method (IM), 2004. 22 October 2018. Available online: https://brewup.eu/ebc-analytica/malt/total-nitrogen-of-malt-kjeldahl-method-im/4.3.1 (accessed on 10 February 2022).
43. EBC Analytica, 4.9.1.—Soluble Nitrogen of Malt: Kjeldahl Method, 1997. 22 October 2018. Available online: https://brewup.eu/ebc-analytica/malt/soluble-nitrogen-of-malt-kjeldahl-method/4.9.1 (accessed on 10 February 2022).
44. EBC Analytica, 9.43.2—Specific-Gravity-of-Beer-Using-a-Density-Meter, 2004. 23 October 2018. Available online: https://brewup.eu/ebc-analytica/beer/specific-gravity-of-beer-using-a-density-meter/9.43.2 (accessed on 10 February 2022).
45. EBC Analytica, 9.2.1—Alcohol-in-Beer-by-Distillation, 2008. 23 October 2018. Available online: https://brewup.eu/ebc-analytica/beer/alcohol-in-beer-by-distillation/9.2.1 (accessed on 10 February 2022).
46. EBC Analytica, 9.29—Haze in Beer 2015. 23 October 2018. Available online: https://brewup.eu/ebc-analytica/beer/haze-in-beer-calibration-of-haze-meters/9.29 (accessed on 10 February 2022).
47. EBC Analytica, 9.35—pH of Beer (Formerly Published as IOB Method 9.42) 2004. 23 October 2018. Available online: https://brewup.eu/ebc-analytica/beer/ph-of-beer-formerly-published-as-iob-method-9-42/9.35 (accessed on 10 February 2021).
48. EBC Analytica, 9.28.3—Carbon Dioxide in Beer: Pressure Method, 2007. 23 November 2018. Available online: https://brewup.eu/ebc-analytica/beer/carbon-dioxide-in-beer-pressure-method/9.28.3 (accessed on 10 February 2022).
49. EBC Analytica, 9.37.1—Measurement of Dissolved Oxygen by Optical Sensors, 2015. 23 November 2018. Available online: https://brewup.eu/ebc-analytica/beer/measurement-of-dissolved-oxygen-by-optical-sensors/9.37.1 (accessed on 10 February 2022).
50. EBC Analytica, 9.45—Energy Value of Beer by Calculation, 2015. 23 November 2018. Available online: https://brewup.eu/ebc-analytica/beer/energy-value-of-beer-by-calculation/9.45 (accessed on 10 February 2022).
51. EBC Analytica, 9.6—Colour of Beer: Spectrophotometric Method (IM), 2004. 23 October 2018. Available online: https://brewup.eu/ebc-analytica/beer/colour-of-beer-spectrophotometric-method-im/9.6 (accessed on 10 February 2022).
52. EBC Analytica, 9.8—Bitterness of Beer (IM), 2004. 23 November 2020. Available online: https://brewup.eu/ebc-analytica/beer/bitterness-of-beer-im/9.8 (accessed on 10 February 2022).
53. EBC Analytica, 13.2—Sensory Analysis Tasting Area Equipment Conduct of Test, 2010. 24 October 2018. Available online: https://brewup.eu/ebc-analytica/sensory/sensory-analysis-tasting-area-equipment-conduct-of-test/13.2 (accessed on 10 February 2022).
54. Bobkov, S. Biochemical and technological properties of buckwheat grains. In *Molecular Breeding and Nutritional Aspects of Buckwheat*, 1st ed.; Zhou, M., Kreft, I., Woo, S.H., Chrungoo, N., Wieslander, G., Eds.; Elsevier: Amsterdam, The Netherlands, 2016; pp. 423–440.

55. Pirzadah, T.B.; Malik, B.; Tahir, I.; Ul Rehman, R. Buckwheat journey to functional food sector. *Curr. Nutr. Food Sci.* **2020**, *16*, 134–141. [CrossRef]
56. Huda, M.N.; Lu, S.; Jahan, T.; Ding, M.; Jha, R.; Zhang, K.; Zhou, M. Treasure from garden: Bioactive compounds of buckwheat. *Food Chem.* **2021**, *335*, 127653. [CrossRef]
57. Kellershohn, J. *Brewing Materials and Processes: A Practical Approach to Beer Excellence*, 1st ed.; Elsevier: Amsterdam, The Netherlands, 2016.
58. Phiarais, B.P.N.; Schehl, B.D.; Oliveira, J.C.; Arendt, E.K. Use of response surface methodology to investigate the effectiveness of commercial enzymes on buckwheat malt for brewing purposes. *J. Inst. Brew.* **2006**, *112*, 324–332. [CrossRef]
59. Myncke, E.; Laureys, D.; Baillière, J.; Vanderputten, D.; De Clippeleer, J. Use of unmalted buckwheat for innovative beer production. In Proceedings of the Brewers Forum/38th EBC Congress, Madrid, Spain, 29 May–1 June 2022.
60. Zdaniewicz, M.; Pater, A.; Hrabia, O.; Duliński, R.; Cioch-Skoneczny, M. Tritordeum malt: An innovative raw material for beer production. *J. Cereal Sci.* **2020**, *96*, 103095. [CrossRef]
61. Giménez-Bastida, J.A.; Piskuła, M.; Zieliński, H. Recent advances in development of gluten-free buckwheat products. *Trends Food Sci. Technol.* **2015**, *44*, 58–65. [CrossRef]
62. Zhuang, S.; Shetty, R.; Hansen, M.; Fromberg, A.; Hansen, P.B.; Hobley, T.J. Brewing with 100% unmalted grains: Barley, wheat, oat and rye. *Eur. Food Res. Technol.* **2017**, *243*, 447–454. [CrossRef]
63. Byeon, Y.S.; Lim, S.T.; Kim, H.J.; Kwak, H.S.; Kim, S.S. Quality characteristics of wheat malts with different country of origin and their effect on beer brewing. *J. Food Qual.* **2021**, *2021*, 2146620. [CrossRef]
64. Arendt, E.; Dal Bello, F. *Gluten-Free Cereal Products and Beverages*, 1st ed.; Elsevier: Amsterdam, The Netherlands, 2008.
65. Buiatti, S.; Bertoli, S.; Passaghe, P. Influence of gluten-free adjuncts on beer colloidal stability. *Eur. Food Res. Technol.* **2018**, *244*, 903–912. [CrossRef]
66. Deng, Y.; Lim, J.; Lee, G.H.; Nguyen, T.T.H.; Xiao, Y.; Piao, M.; Kim, D. Brewing rutin-enriched lager beer with buckwheat malt as adjuncts. *J. Microbiol. Biotechnol.* **2019**, *29*, 877–886. [CrossRef] [PubMed]
67. Bamforth, C.W. Flavorsome components of beer. In *Brewing Materials and Processes*, 1st ed.; Elsevier: Amsterdam, The Netherlands, 2016; pp. 151–155.
68. Gallagher, E. (Ed.) *Gluten-Free Food Science and Technology*, 1st ed.; John Wiley & Sons: Oxford, UK, 2009.
69. Steiner, E.; Auer, A.; Becker, T.; Gastl, M. Comparison of beer quality attributes between beers brewed with 100% barley malt and 100% barley raw material. *J. Sci. Food Agric.* **2012**, *92*, 803–813. [CrossRef] [PubMed]
70. Agu, R.C.; Chiba, Y.; Goodfellow, V.; Mackinlay, J.; Brosnan, J.M.; Bringhurst, T.A.; Bryce, J.H. Effect of germination temperatures on proteolysis of the gluten-free grains rice and buckwheat during malting and mashing. *J. Agric. Food Chem.* **2012**, *60*, 10147–10154. [CrossRef]
71. Cadenas, R.; Caballero, I.; Nimubona, D.; Blanco, C.A. Brewing with starchy adjuncts: Its influence on the sensory and nutritional properties of beer. *Foods* **2021**, *10*, 1726. [CrossRef] [PubMed]
72. Salamon, R.V.; Dabija, A.; Ferencz, Á.; Tankó, G.; Ciocan, M.E.; Codină, G.G. The Effect of Dry Hopping Efficiency on β-Myrcene Dissolution into Beer. *Plants* **2022**, *11*, 1043. [CrossRef]
73. Gribkova, I.N.; Eliseev, M.N.; Belkin, Y.D.; Zakharov, M.A.; Kosareva, O.A. The Influence of Biomolecule Composition on Colloidal Beer Structure. *Biomolecules* **2021**, *12*, 24. [CrossRef]
74. Gorzelany, J.; Patyna, M.; Pluta, S.; Kapusta, I.; Balawejder, M.; Belcar, J. The Effect of the Addition of Ozonated and Non-Ozonated Fruits of the Saskatoon Berry (*Amelanchier alnifolia* Nutt.) on the Quality and Pro-Healthy Profile of Craft Wheat Beers. *Molecules* **2022**, *27*, 4544. [CrossRef]
75. Habschied, K.; Krstanović, V.; Mastanjević, K. Beer Quality Evaluation—A Sensory Aspect. *Beverages* **2022**, *8*, 15. [CrossRef]
76. Sebestyén, A.; Kiss, Z.; Vecseri-Hegyes, B.; Kun-Farkas, G.; Hoschke, Á. Experiences with laboratory and pilot plant preparation of millet and buckwheat beer. *Acta Aliment.* **2013**, *42*, 81–89. [CrossRef]
77. Phiarais, B.P.N.; Wijngaard, H.H.; Arendt, E.K. The impact of kilning on enzymatic activity of buckwheat malt. *J. Inst. Brew.* **2005**, *111*, 290–298. [CrossRef]
78. Deželak, M.; Zarnkow, M.; Becker, T.; Košir, I.J. Processing of bottom-fermented gluten-free beer-like beverages based on buckwheat and quinoa malt with chemical and sensory characterization. *J. Inst. Brew.* **2014**, *120*, 360–370. [CrossRef]
79. Salanță, L.C.; Coldea, T.E.; Ignat, M.V.; Pop, C.R.; Tofană, M.; Mudura, E.; Zhao, H. Functionality of special beer processes and potential health benefits. *Processes* **2020**, *8*, 1613. [CrossRef]
80. Viejo, C.G.; Fuentes, S.; Howell, K.; Torrico, D.D.; Dunshea, F.R. Integration of non-invasive biometrics with sensory analysis techniques to assess acceptability of beer by consumers. *Physiol. Behav.* **2019**, *200*, 139–147. [CrossRef]

applied sciences

MDPI

Article

Comparison of the Spreadability of Butter and Butter Substitutes

Małgorzata Ziarno [1,*], Dorota Derewiaka [2], Anna Florowska [3] and Iwona Szymańska [4]

1 Division of Milk Technology, Department of Food Technology and Assessment, Institute of Food Science, Warsaw University of Life Sciences–SGGW (WULS–SGGW), 02-787 Warsaw, Poland
2 Division of Food Quality Assessment, Department of Food Technology and Assessment, Institute of Food Science, Warsaw University of Life Sciences–SGGW (WULS–SGGW), 02-787 Warsaw, Poland
3 Division of Fats and Food Concentrates Technology, Department of Food Technology and Assessment, Institute of Food Science, Warsaw University of Life Sciences–SGGW (WULS–SGGW), 02-787 Warsaw, Poland
4 Division of Meat Technology, Department of Food Technology and Assessment, Institute of Food Science, Warsaw University of Life Sciences–SGGW (WULS–SGGW), 02-787 Warsaw, Poland
* Correspondence: malgorzata_ziarno@sggw.edu.pl; Tel.: +48-225-937-666

Abstract: There are many types of butter, soft margarine, and blends, e.g., a mixture of butter and vegetable fats, on the market as bread spreads. Among these, butter and blends of butter with vegetable fats are very popular. The consumer's choice of product is often determined by functional properties, such as texture, and the physicochemical composition of butter and butter substitutes. The aim of this study was to compare sixteen market samples of butter and butter substitutes in terms of spreadability and other selected structural (spreadability, hardness, adhesive force, and adhesiveness) and physicochemical parameters (water content, water distribution, plasma pH, color, acid value, peroxide number, saponification number, and instrumentally measured fatty acid profile) to investigate their correlation with spreadability. The parameters determined here were correlated with factors such as the type of sample, measuring temperature, and physicochemical composition. The statistical analysis revealed a very strong positive correlation between hardness and spreadability for all samples tested at 4 °C, as well as between hardness and spreadability for all samples tested 30 min after removal from the refrigerator; however, the interpretation of the results was different if the butter and butter substitute samples were subjected to a multivariate analysis separately.

Keywords: spreadable fats; texture; hardness; adhesive force; adhesiveness; fatty acid profile; multivariate analysis; functional food

Citation: Ziarno, M.; Derewiaka, D.; Florowska, A.; Szymańska, I. Comparison of the Spreadability of Butter and Butter Substitutes. *Appl. Sci.* **2023**, *13*, 2600. https://doi.org/10.3390/app13042600

Academic Editors: Georgiana Gabriela Codină, Adriana Dabija and Lăcrămioara Rusu

Received: 25 January 2023
Revised: 12 February 2023
Accepted: 15 February 2023
Published: 17 February 2023

1. Introduction

Butter is a high-fat product with at least 80%, but no more than a 90% fat content, and it is obtained from cow's milk by whipping previously obtained cream or sour cream. However, the assortment of spreads for bread on the market consists of different types of butter, soft margarine, and blends such as a mixture of butter and vegetable fats. Among the wide range of edible fats, in addition to butter and margarine, products called butter mixtures with vegetable fats are of great interest. All of these edible fats are water-in-oil emulsions [1]. The decision as to which fat to consume in spreads can therefore be somewhat difficult for some consumers, since these fats differ not only in their calorific value and therefore in their chemical composition but also in their functional properties, such as spreadability [2–5].

The study comparing the spreadability of butter and butter substitutes makes a significant scientific contribution to the field of food science and technology. Understanding the spreadability of food products is essential, as it plays a critical role in consumer acceptance and product functionality. In order to improve the spreadability of butter, several technological steps are taken, starting with the selection of raw materials with a fixed fatty acid composition, followed by technological treatments, including parameters for the physical ripening of the cream, enrichment with a low-melting-point milk fat fraction or repeatedly churning the butter [6–10]. The spreadability of butter and blends of butter and vegetable fats can be assessed based on the solid fat content (SFC) in the temperature range of 4–10 °C. In order to achieve good spreadability at a refrigeration temperature, the SFC of butter should be below 32% (by comparison, pure butterfat has an SFC of 38.8–32.4% at 5–10 °C) [6]. In addition, the textural properties of butter are determined by a number of interrelated parameters, such as the concentration, size, shape, and distribution of structural elements: fat crystals, fat globules, air bubbles, and water phase droplets [3,10–19].

Butter and butter substitutes have different physical and chemical properties, and it is important to understand how these differences affect the spreadability of these products. This study can help identify the factors that influence spreadability such as fat content, viscosity, and solid fat content. The scientific contribution and specific significance of the study comparing the spreadability of butter and butter substitutes lies in its potential to improve our understanding of the properties that affect the spreadability of food products and to inform the development of new and improved products for the food industry. A real downside to butter is its hard-to-spread texture. Many authors have attempted to study the spreadability of butter, soft margarine, mixtures of butter and vegetable fats, and shortenings using various instruments such as a penetrometer and texture analyzer (textrometer) with a penetration test, a shear test, a compression test, a texture profile analysis (TPA), DSC methods, EPR spectroscopic methods and the back extrusion test [2,9,11,16,20–31]. The results of such studies can be used to model the relationship between the structure, rheological properties, and textural properties of fat products, taking into account the morphology of the fat crystal network, the solid fat content of the product, and the properties of the fat crystal networks [10].

The aim of this study was to compare market samples of butter and butter substitutes in terms of spreadability and other selected structural, physicochemical, and chemical parameters to investigate their correlation with spreadability. The results of the study can be used to develop new and improved butter and butter substitutes that have similar spreadability characteristics without sacrificing other important properties such as flavor, texture, and nutritional value. In addition, the results of this study can have practical applications in the food industry, helping manufacturers to improve the quality and performance of their products. It can also provide valuable information to consumers to help them make informed choices when buying food.

2. Materials and Methods

2.1. Materials

The study material consisted of eight butter samples coded as follows—LMK, LaME, LoME, MEG, MEH, MMP, PME, and PrME—and eight samples of butter substitutes (spreadable fats, consisting of blends of butter with vegetable oils) coded as follows: FM, LuPM, LaM, PaEM, RMTM, SSO, ZaM, and BGP (Table 1).

These products were available in the Polish food market at the time of the analysis. All butter samples were declared by the manufacturers as products with an 82% fat content, and they were unsalted. After purchase, the product samples, packed in insulating bags, were transported to the laboratory within 0.5 h, where they were stored in a refrigerator at 4 °C until the analyses were carried out. Three independent purchases of each butter and butter substitute sample were made, representing independent replicates from three different suppliers.

Table 1. Declared composition and nutritional value of tested butter and butter substitute samples.

Sample Code	Declared Ingredients	Energy Value (in 100 g)	Fat [g]	Of Which Saturated Fatty Acids [g]	Carbohydrates [g]	Of Which Sugars [g]	Protein [g]	Salt [g]
		butter samples						
LMK	pasteurized cream, lactic acid bacteria cultures	3071 kJ/747 kcal	82	53	0.7	0.7	0.6	0.00
LaME	pasteurized cream	3095 kJ/753 kcal	83	54	0.8	0.8	0.6	0.00
LoME	pasteurized cream	3058 kJ/744 kcal	82	55	0.7	0.7	0.7	0.00
MEG	pasteurized cream, lactic acid concentrate, natural flavoring	3061 kJ/744 kcal	82	57	0.6	0.6	1.0	0.02
MEH	pasteurized cream	3068 kJ/746 kcal	82	54	1.0	1.0	1.0	0.02
MMP	pasteurized cream	3095 kJ/753 kcal	83	54	0.8	0.8	0.6	0.00
PME	pasteurized cream	3097 kJ/753 kcal	83	55	0.7	0.5	0.8	0.02
PrME	pasteurized cream	3063 kJ/745 kcal	82	57	1.0	1.0	0.7	0.03
		butter substitute samples						
FM	milk butter, vegetable oils (rapeseed, linseed), milk buttermilk, vitamins (A, D)	2807 kJ/683 kcal	75	29	1.1	1.1	0.8	0.00
LuPM	butter, rapeseed oil, water, lactic acid bacteria cultures	2905 kJ/706 kcal	78	35	0.6	0.6	<0.05	<0.01
LaM	cream, rapeseed vegetable oil, annatto bixin color, flavoring	2559 kJ/622 kcal	68	34	1.4	1.4	1.1	0.00
PaEM	vegetable fat: non-hydrogenated palm oil, sunflower oil, cream, cereal fat, emulsifiers: E471, E472c, E322, acidity regulator: lactic acid, beta-carotene, flavors	2822 kJ/686 kcal	75	33	1.6	0.9	0.6	0.04
RMTM	rapeseed and palm oils, butter, reconstituted butter, water, sea salt, lecithin, natural flavoring, lactic acid, vitamins A, D, carotenes	2994 kJ/717 kcal	80	30	0.6	0.6	<0.05	0.32
SSO	pasteurized cream, rapeseed vegetable oil, lactic acid cultures	2523 kJ/613 kcal	67	37	1.4	0.8	1.2	0.03
ZaM	cream, rapeseed oil, annatto, flavoring	2559 kJ/622 kcal	68	34	1.4	1.4	1.1	0.00
BGP	Palm, rapeseed and sunflower oils, water, anhydrous milk fat, E471, E472c, E322, salt, flavorings, E160a, E330, vitamins A, D	3034 kJ/738 kcal	82	36	0.0	0.0	0.0	0.30

2.2. Texture Characteristics of Butter and Butter Substitutes

Spreadability. The tests were performed with a TA.HD.Plus Texture Analyzer (Stable Micro Systems, Toruń, Poland). This was measured using penetration analysis with a "spreadability ring" spreadability test unit [31]. During the analysis, the upper cone was inserted into the lower container (in the form of an inverted cone) at a speed of 3 mm/s until a gap of one millimeter was obtained between the two elements of this fixture. All samples were analyzed in triplicate. Samples at 20 °C were placed in the bottom reservoir of the attachment without bubbles.

Hardness, adhesive force, and adhesiveness. The tests were performed with a TA.HD. Plus Texture Analyzer (Stable Micro Systems, Toruń, Poland). The procedure consisted of testing the penetration force at a depth of 14 mm applied to a given sample at a speed of 2 mm/s using a P/5 cylinder probe [31]. The samples were analyzed at given time intervals: straight from the fridge (at 4 °C), 30 min after removal from the refrigerator, and at 20 °C for each type of butter sample and butter substitute. All samples were analyzed in triplicate. Hardness was expressed as the maximum force necessary to obtain accurate probe deformation (N). The adhesion force was expressed as the force necessary to overcome the forces of attraction between the surface of the sample and the surface of other materials with which the food came into contact (N). Adhesiveness was expressed as the product of the force required to pull the probe from the sample and the pull time (N × s).

2.3. Physicochemical Properties of Butter and Butter Substitutes

Water content. The weight loss was calculated by determining the weight of the sample after it was dried with appropriately prepared sea sand and calculating the percentage of water content in the product [32,33]. A sample was dried in a laboratory oven at 102 ± 2 °C until a constant mass was obtained. All samples were analyzed in triplicate. Formula (1) used for the calculation was:

$$W = 100 - \frac{B \times 100\%}{A} \quad (1)$$

where W represents the water content of the sample [%]; A represents the sample weight before drying (g); and B represents the sample weight after drying (g).

Water distribution. The main principle of the method of water distribution in samples is to apply indicator paper soaked with an indicator to the freshly cut sample surface [32,34]. The indicator paper turns dark blue where it meets water droplets. All the samples were analyzed in triplicate. A point scale of 0–3 was used to determine the degree of water distribution, after which the samples were classified using the criteria given in Table 2. The analysis was carried out using commercially available indicator paper (Dysperwod, LABLACTA, Olsztyn, Poland) according to the manufacturer's instructions. This method enables the determination of whether the butter and butter substitutes had been properly kneaded and the water droplets properly dispersed in the butter and butter substitute matrix.

Table 2. Classification of butter and butter substitutes according to the degree of water distribution [32,34].

Verbal Definition of Water Distribution in Sample	The Size (Diameter) and Density of the Spots on the Indicator Paper	Class [Points]
Very bad	Diameter 3 mm 8 mm densely occurring (takes up approx. 20% of the paper surface)	0
Bad	Diameter 1 mm 3 mm moderately dense (takes up approx. 10% of the paper surface)	1
Sufficient	Diameter 0.3 mm 1 mm rare (occupies approx. 5% of the paper area)	2
Good	No spots	3

Plasma pH. The determination consisted of melting a 40.0 g weight of the butter or butter substitute sample in a water bath at 50 °C, followed by centrifugation in a laboratory centrifuge at 1100× g at 20 °C for 10 min to separate the aqueous phase (plasma) [32]. The pH of the separated aqueous phase of the butter and butter substitutes was measured with a CPO-505 pH meter (Elmetron, Zabrze, Poland) with a conventional electrode at 25 °C. All samples were analyzed in triplicate.

Color. Color components were measured at 3 or 4 locations for each sample using the reflectance method, using a Minolta CR-200 camera (Konica Minolta, Tokyo, Japan) with a D65 illuminant, 2° observer angle, and 8 mm aperture size. The parameters measured were lightness (*L**), taking values from 0 (black) to 100 (perfect white); redness (*a**), taking negative values for green color shades and positive values for red color shades; and yellowness (*b**), the proportion of blue or yellow color in the sample, taking negative values for blue and positive values for yellow. All samples were analyzed in triplicate. To determine the color differences between the butter and butter substitutes, they were compared to a predetermined standard. The standard was compared to the average results of the *a**, *b**, and *L** components obtained in the color test. The standard table (Table 3) for comparing the test samples was taken from the publication of Chudy et al. [35].

Table 3. The standard table for comparing the test samples was taken from the publication of Chudy et al. [35].

	*L**	*a**	*b**
Standard butter	91.6	5.5	24.7

To assess the changes in CIE*Lab* color, a delta E (Δ*E*) calculated according to Formula (2) was used to describe the difference between the two sample colors as follows:

$$\Delta E = \sqrt{(\Delta L)^2 + (\Delta a)^2 + (\Delta b)^2} \quad (2)$$

where Δ*L* is the difference in *L** components between the butter standard (Table 2) and the test sample; Δ*a* is the difference in *a** components between the butter standard (Table 2) and the test sample; and Δ*b* is the difference in *b** components between the butter standard (Table 2) and the test sample.

2.4. Additional Characteristics of Butter Milk Fat

Determination of the acid value. The tests were only carried out on butter fat samples according to [36]. The procedure consisted of neutralizing the free fatty acids present in the fat sample to be analyzed with a standard potassium hydroxide solution. The 10 g butter sample was weighed to the nearest 0.01 g in a 200–300 mL Erlenmeyer flask. The sample was then melted using a water bath (at 40 °C). Then, 50 mL of neutralized alcohol–ether mixture was added to the flask (it was neutralized to phenolphthalein with a KOH solution before use, to a pale pink color that did not disappear within 30 s) and mixed thoroughly. It was further titrated from the buret with a 0.1 M KOH versus phenolphthalein to give a pale pink color which persisted for 30 s. All samples were analyzed in triplicate. The acid number was calculated using Formula (3):

$$LK = \frac{(a \times 5.611)}{m} \quad (3)$$

where LK is acid value (mg KOH/g fat); a is the amount of 0.1 M KOH solution used for the titration (mL); m is the sample fat weight (g); and 5.611 is the amount of KOH contained in 1 mL of 0.1 M KOH solution (mg).

Determination of the peroxide number. The tests were only carried out on butter fat samples according to [37]. The method consisted of the quantitative determination of iodine released from potassium iodide by the action of peroxides present in the fat under study. The released iodine was titrated with a standard solution of sodium thiosulphate (VI). Approximately 1 g of butter was weighed to the nearest 0.001 g into an Erlenmeyer flask with a ground-glass stopper. Immediately, 25 mL of the chloroform–acetic acid mixture and 1 mL of saturated potassium iodide solution were added. The flask was quickly stoppered and mixed thoroughly. It was left in the dark for 5 min. After this time, 75 mL of distilled water was added using a cylinder, the stopper was rinsed thoroughly and a few drops of starch were added; after mixing, the solution was immediately titrated with 0.002 M sodium thiosulfate standard solution until the solution remained discolored for at least 30 s. At the same time, a reagent test was carried out. All samples were analyzed in triplicate. The peroxide number was calculated using Formula (4):

$$LOO = 1000 \times \frac{(a - b) \times T}{m} \tag{4}$$

where a is the volume of sodium thiosulphate solution used to titrate the sample (mL); b is the volume of sodium thiosulphate solution used to titrate the reagent sample (mL); m is the sample fat weight (g); T is the molality of the sodium thiosulphate solution used; and 1000 is the conversion factor for the volume used (mL) of sodium thiosulphate per milliequivalent of oxygen in 1 kg of fat.

Determination of the saponification number. Tests were carried out only in butter fat samples according to [38]. The method involved saponification of esterified and neutralized free fatty acids with an excess of potassium hydroxide solution, followed by quantification of unbound KOH by titration with HCl solution. A total of 2 g of butter was weighed to the nearest 0.001 g into a 100 mL ground-glass Erlenmeyer flask. The sample was then melted using a water bath. A total of 25 mL of alcoholic potassium hydroxide solution was pipetted into the sample. An aftercooler was fitted to the flask, and the flask was placed in the water bath for 30 min with occasional stirring. After this time, the aftercooler was removed and the hot solution was titrated against phenolphthalein with 0.5 M hydrochloric acid solution until the indicator color disappeared. If the volume of the solution decreased during heating, it was made up of the original volume with ethanol before titration. At the same time, a reagent test was carried out under the same conditions. All the samples were analyzed in triplicate. The saponification number was calculated using Formula (5):

$$LZ = \frac{(a - b) \times 28.055}{m} \tag{5}$$

where LZ is the saponification number (mg KOH/g fat); a is the volume of standard HCl solution used to titrate the reagent sample (mL); b is the volume of standard HCl solution used to titrate the sample (mL); m is the sample weight (g); and 28.055 is the amount of KOH present in 1 mL of 0,5 M KOH solution (mg).

Chromatographic determination of the fatty acid profile. Tests were carried out only in butter fat samples according to [39]. The process consisted of separating the components, which were divided into two phases: one was stationary (stationary phase) and the other was moving in a given direction (mobile phase). The different distribution of the mixture components between the two phases leads to different migration and separation speeds of the components. The effect of the chromatographic separation was plotted in the form of a chromatogram, showing a graph of the signal obtained at the detector as a function of time [39].

The evaporated samples were weighed to the nearest 0.001 g and then dissolved in 2 mL of hexane. A total of 0.5 mL of 2 M KOH in methanol was added to the samples and was then shaken and left for a period of 60 min, with stirring every 10 min for transesterification. Using a syringe, 1 mL of the upper hexane layer containing fatty acid methyl esters was carefully collected and transferred to a glass vial. The sample was then evaporated in a stream of nitrogen, and 0.5 mL of hexane was added.

Determination of fatty acids was carried out using a Shimadzu gas chromatograph coupled to a mass spectrometer GC–MS QP-2010S (SHIM-POL, Warsaw, Poland), using a ZB FFAP column (30 m × 0.25 mm × 0.25 µm; Phenomenex, Torrance, CA, USA). The column operating temperature was initially 40 °C for 3 min, with a programmed temperature rise at 4 °C/min to 230 °C, and finally, isotherm for 5 min. The injection chamber and ion source temperatures were 230 °C and 240 °C, respectively. The carrier gas was helium and the flow rate was 1.14 mL/min. The GC–MS coupler temperature was 225 °C. Fatty acid methyl ester analyses were carried out at an ionization energy of 70 eV. The qualitative analysis (of the obtained fatty acid methyl esters) was carried out based on a comparison of the retention times of available standards and spectra. All samples were analyzed in triplicate.

2.5. Statistical Analysis

Data were analyzed using a one-way or two-way analysis of variance (ANOVA method). Mean differences between the statistical groups were tested at a significance level of $\alpha = 0.05$. Tukey's test was used for multiple comparisons (statistical ranking) of mean responses to the sample groups (for $\alpha = 0.05$). Multivariate analysis was used to describe the relationship of multiple variables for each sample (for $\alpha = 0.05$). The statistical software Statgraphics Centurion XVII (Kraków, Poland) was used to test the data.

3. Results and Discussion

3.1. Texture Characteristics of Butter and Butter Substitutes

Research into the texture characteristics of butter and blends of butter with vegetable fats is relevant to consumers because it can provide information about the performance of these products. The texture is an important attribute for many consumers as it affects the ease of use and enjoyment of the product. By understanding the texture characteristics of butter and butter substitutes, consumers can make informed decisions about which products meet their needs and preferences. Research into the texture characteristics of butter and vegetable fat blends can play an important role in helping manufacturers to create high-quality products that meet consumer needs and preferences. By providing valuable information on the texture and performance of these products, manufacturers can ensure that their products are competitive in the marketplace and are well received by consumers.

Spreadability. The ability to spread the bread spread easily is one of its most important properties [22,40]. It is worth noting that the higher spreadability value of the butter and butter substitutes tested, as shown in Table 4, indicated poorer spreadability of the product on the bread. The highest spreadability value was obtained for the butter samples at 4 °C. If the butter was left at ambient temperature for 30 min after removal from the refrigerator, these values hardly approached the parameters obtained for butter substitute samples at 4 °C; however, for most butter substitute samples, the spreadability value was still statistically significantly better than for the butter samples. The butter samples coded as MEH and PrME were the exceptions. In their case, the spreadability value at 4 °C was the lowest of the results obtained for the butter samples, and 30 min after removal from the refrigerator, the spreadability value reached the same level as the butter substitute samples at 4 °C. By bringing the butter samples to 20 °C, the spreadability value measured reached the value originally obtained for the butter substitute samples at 4 °C.

The texture of spreadable fats, and more importantly their spreadability, is one of the most important differentiators when assessing their quality. The spreadability of butter and butter blends containing vegetable fats is determined by their chemical composition—the type of fat used in their manufacture, as well as the ratio of the aqueous phase to the fat phase and the balance between the liquid and crystalline phases [2,17]. The higher the degree of crystallization of the fat, the poorer the spreadability of the butter [17]. The spreadability of butter can be improved, among other things, by changing the fat composition (e.g., changing the diet of the animals from which the milk is obtained) [41]. Bobe et al. [30] found that butter samples from the milk of cows that had a more unsaturated composition of milk fats due to a special diet had better spreadability.

Hardness. Butter samples at 4 °C were characterized by a high hardness that was statistically significantly higher than butter substitutes at the same temperature (Table 4). Increasing the temperature of the butter samples resulted in a decrease in their hardness. Thirty minutes after removing the butter samples from the refrigerator, their hardness was already on a similar level to that of the butter substitute samples at a temperature of 4 °C. On the other hand, heating the butter samples to 20 °C led to hardness parameters comparable to those of the butter substitute samples 30 min after removal from the refrigerator.

Table 4. Texture characteristics of butter and butter substitutes.

Butter Samples	Spreadability [N × s]		
	at 4 °C	30 min after Removing it from the Refrigerator	at 20 °C
LMK	91.49 ± 3.54 [i]	34.95 ± 5.66 [f]	15.31 ± 0.79 [d]
LaME	64.76 ± 2.03 [g]	31.22 ± 2.99 [f]	4.79 ± 1.45 [c]
LoME	70.72 ± 6.04 [g,h]	30.51 ± 6.31 [f]	8.70 ± 1.22 [d,e]
MEG	94.38 ± 6.56 [i]	38.85 ± 5.24 [f]	16.40 ± 4.34 [d,e]
MEH	78.05 ± 1.11 [h]	27.30 ± 8.33 [e,f]	20.86 ± 2.46 [e]
MMP	69.69 ± 4.51 [g]	37.84 ± 3.48 [f]	5.71 ± 1.14 [b,c]
PME	94.62 ± 5.18 [i]	45.39 ± 7.50 [f,g]	17.31 ± 7.11 [d,e]
PrME	56.64 ± 4.65 [g]	28.03 ± 1.40 [e,f]	8.14 ± 1.03 [c]
Butter substitutes samples			
FM	20.38 ± 1.41 [e]	3.35 ± 1.27 [b]	n.d. [1]
LuPM	22.39 ± 0.63 [e]	8.29 ± 1.01 [c]	1.75 ± 0.40 [a]
LaM	23.18 ± 1.67 [e]	7.51 ± 1.37 [c]	0.53 ± 0.02 [a]
PaEM	30.94 ± 0.78 [f]	16.51 ± 1.10 [d,e]	14.46 ± 1.20 [d]
RMTM	13.27 ± 0.80 [d]	7.90 ± 1.86 [c]	2.12 ± 0.29 [a]
SSO	26.58 ± 0.96 [e]	16.26 ± 1.68 [d,e]	2.21 ± 0.05 [a,b]
ZaM	26.76 ± 1.10 [e]	9.35 ± 1.28 [c,d]	1.57 ± 0.55 [a]
BGP	13.21 ± 0.49 [d]	7.80 ± 1.29 [c]	2.17 ± 0.14 [a]
Butter Samples	**Hardness [N]**		
	at 4 °C	30 min after Removing it from the Refrigerator	at 20 °C
LMK	17.01 ± 0.55 [h]	6.13 ± 0.65 [d]	2.04 ± 0.17 [b,c]
LaME	12.54 ± 0.17 [f]	5.58 ± 0.48 [d]	1.02 ± 0.29 [b]
LoME	13.67 ± 0.34 [f]	5.85 ± 0.59 [d]	1.61 ± 0.22 [b]
MEG	19.28 ± 0.78 [i]	7.49 ± 0.72 [e]	3.27 ± 0.81 [c]
MEH	15.69 ± 0.07 [g]	6.14 ± 0.93 [d]	3.41 ± 0.34 [c]
MMP	14.15 ± 0.69 [f,g]	7.07 ± 0.79 [e]	1.23 ± 0.19 [b]
PME	17.56 ± 0.37 [h]	7.92 ± 0.63 [e]	3.92 ± 0.57 [c]
PrME	10.12 ± 0.82 [f]	5.21 ± 0.21 [d]	1.64 ± 0.18 [b]
Butter substitutes samples			
FM	3.85 ± 0.15 [c,d]	0.65 ± 0.20 [a]	n.d. [1]
LuPM	4.02 ± 0.06 [d]	1.46 ± 0.16 [b]	0.33 ± 0.06 [a]
LaM	4.42 ± 0.29 [d]	1.40 ± 0.21 [b]	0.12 ± 0.01 [a]
PaEM	6.91 ± 0.11 [d,e]	3.15 ± 0.13 [c]	2.58 ± 0.28 [c]
RMTM	2.61 ± 0.06 [c]	1.57 ± 0.35 [b]	0.46 ± 0.06 [a]
SSO	5.01 ± 0.15 [d]	2.85 ± 0.37 [c]	0.47 ± 0.02 [a]
ZaM	4.96 ± 0.15 [d]	1.66 ± 0.35 [b]	0.30 ± 0.09 [a]
BGP	3.79 ± 0.05 [c,d]	1.61 ± 0.33 [b]	0.38 ± 0.07 [a]

Table 4. *Cont.*

Butter Samples	Adhesive force [N]		
	at 4 °C	30 min after Removing it from the Refrigerator	at 20 °C
LMK	−5.03 ± 0.76 [a]	−2.48 ± 0.33 [c]	−0.84 ± 0.06 [d,e]
LaME	−3.82 ± 0.05 [b]	−2.16 ± 0.27 [c]	−0.50 ± 0.11 [e]
LoME	−3.79 ± 0.19 [b]	−1.92 ± 0.11 [c,d]	−1.39 ± 0.09 [d]
MEG	−3.16 ± 0.40 [b]	−2.23 ± 0.12 [c]	−1.24 ± 0.20 [d]
MEH	−3.79 ± 0.19 [b]	−1.81 ± 0.23 [d]	−1.71 ± 0.73 [d]
MMP	−3.44 ± 0.16 [b]	−2.68 ± 0.26 [c]	−0.59 ± 0.05 [f]
PME	−3.68 ± 0.45 [b]	−2.29 ± 0.07 [c]	−1.15 ± 0.45 [d]
PrME	−3.46 ± 0.14 [b]	−2.28 ± 0.07 [c]	−0.77 ± 0.09 [e]
Butter substitutes samples			
FM	−1.49 ± 0.11 [d]	−0.26 ± 0.05 [g]	n.d. [1]
LuPM	−1.42 ± 0.07 [d]	−0.59 ± 0.06 [e]	−0.16 ± 0.03 [g]
LaM	−1.46 ± 0.05 [d]	−0.58 ± 0.07 [e]	−0.08 ± 0.01 [g]
PaEM	−2.08 ± 0.13 [c]	−1.31 ± 0.05 [d]	−1.16 ± 0.10 [d]
RMTM	−1.01 ± 0.05 [d,e]	−0.66 ± 0.10 [e]	−0.24 ± 0.03 [g]
SSO	−1.47 ± 0.11 [d]	−0.95 ± 0.09 [e]	−0.24 ± 0.02 [g]
ZaM	−1.50 ± 0.06 [d]	−0.63 ± 0.08 [e]	−0.16 ± 0.03 [g]
BGP	−1.26 ± 0.04 [d]	−0.64 ± 0.08 [e]	−0.20 ± 0.03 [g]
Butter Samples	**Adhesiveness [N × s]**		
	at 4 °C	**30 min after Removing it from the Refrigerator**	**at 20 °C**
LMK	−18.46 ± 0.98 [a]	−12.16 ± 1.67 [b]	−4.12 ± 0.34 [e,f]
LaME	−17.26 ± 0.29 [a]	−10.60 ± 1.23 [b,c]	−2.53 ± 0.50 [f,g]
LoME	−9.43 ± 0.35 [c]	−8.59 ± 1.03 [c,d]	−3.55 ± 0.61 [f]
MEG	−10.82 ± 1.97 [c]	−8.81 ± 1.20 [c]	−5.64 ± 0.83 [e]
MEH	−14.10 ± 0.80 [b]	−7.93 ± 1.02 [d]	−6.93 ± 1.30 [d]
MMP	−16.59 ± 0.44 [a]	−12.68 ± 1.23 [b]	−2.93 ± 0.27 [f,g]
PME	−17.32 ± 1.57 [a]	−10.45 ± 0.70 [b,c]	−6.97 ± 1.32 [d]
PrME	−16.03 ± 0.34 [a]	−11.40 ± 0.34 [b]	−3.84 ± 0.51 [f]
Butter substitutes samples			
FM	−6.35 ± 0.32 [d]	−1.15 ± 0.17 [h]	n.d. [1]
LuPM	−6.13 ± 0.22 [d,e]	−2.61 ± 0.25 [g]	−0.86 ± 0.14 [h,i]
LaM	−5.97 ± 0.87 [d,e]	−2.65 ± 0.38 [g]	−0.52 ± 0.00 [i]
PaEM	−3.30 ± 0.83 [f]	−3.70 ± 0.76 [f]	−3.98 ± 0.45 [f]
RMTM	−3.99 ± 0.13 [f]	−3.13 ± 0.45 [f]	−1.22 ± 0.13 [h]
SSO	−3.86 ± 0.31 [f]	−3.16 ± 1.10 [f]	−1.19 ± 0.07 [h]
ZaM	−6.50 ± 0.75 [d]	−2.83 ± 0.35 [g]	−0.77 ± 0.12 [h,i]
BGP	−3.92 ± 0.12 [f]	−3.14 ± 0.70 [f]	−1.20 ± 0.05 [h]

[a,b,c,d,e,f,g,h,i]—within a given parameter that is labeled with the same letters do not differ statistically significantly at the level of $\alpha = 0.05$. [1] The analysis of samples under these conditions was not possible.

The high hardness of the butter samples at 4 °C can be explained by the higher proportion of saturated fatty acids, which contribute to the hardness and poor spreadability of butter at refrigerator temperatures, which has been confirmed by several studies [2,8,23,30,31,42–46]. Lower temperatures increase the fat solidity; however, it should be noted that both the butter samples and the butter substitute samples differed in hardness, which could indicate that the technological process parameters also determine this product quality parameter [23,44,46–48]. The results of Glibowski et al. [31] highlighted that samples with a high content of milk fat showed a stronger increase in hardness when changing the temperature conditions from room to cooling temperatures compared with samples that were predominantly vegetable fats. The authors concluded that the presence of milk fat promoted an increase in hardness. Queirs et al. [44] found that the hardness of butter depended on the crystallization of the butter at the manufacturing stage and not only on the storage temperature. Rønholt [48] found that the ratio between solid and liquid fats and the water content strongly influenced the hardness and spreadability of the product. The presence of unsaturated and liquid fats in the composition of butter and vegetable fat mixtures reduces the hardness of these products. The higher the water content, the more the ratio of solid-to-liquid fat shifts in the direction of the liquid phase, so that less fat contributes to crystal formation and thus influences product hardness and water droplet stability. It should therefore be noted that the butter substitutes included in this study were characterized by a higher water content than the butter samples, as is discussed later in this manuscript. With increasing temperature, the firmness and spreadability of the fat products analyzed decreased, i.e., the spreadability improved. The higher the temperature of the product, the more the product structure approaches that of a liquid. This can be caused, among other things, by the water content of the product [48]. As the water percentage increases, the total fat content decreases, affecting the hardness parameter [48]. As has been noted, the higher water content of the butter/vegetable fat blends allowed for a smoother and therefore more spreadable product.

Adhesive force. Adhesive force is the force between the surfaces of two different bodies to hold them together (a food product is perceived as being sticky when the adhesive force is high) [26]. In the case of butter and butter substitutes, this parameter expresses the force that allows the butter or butter substitute to spread evenly over the surface of the bread. Small statistically significant differences in the adhesive force values were found between the butter samples at 4 °C. The same observation was applied to the samples of the butter substitutes at 4 °C (Table 4). The study showed that market samples of butter had statistically significantly higher adhesion values than samples of the butter substitutes, which could be related to differences in the fatty acid composition of the butter samples and butter substitute samples [31]. As the temperature of the samples of the tested products was increased, changes in the adhesive force values towards a value close to zero were observed. These changes were statistically significant for both butter and butter substitutes as early as 30 min after removing the samples from the refrigerator.

Adhesiveness. Adhesiveness is the work required to separate a product from the surface being tested; its measurement serves to express the adhesive properties by measuring the force needed to separate them. The greater the force required to separate the two, the stickier the product. The data presented in Table 4 show that the adhesiveness of both butter and butter substitute samples was statistically significantly higher, but it was dependent on the product's temperature. The higher the temperature of the butter or butter substitute, the lower the adhesiveness, i.e., the samples were less sticky. For one of the butter substitute samples (coded as FM) raising the product temperature to 20 °C made measurement impossible as the sample had already become liquid rather than sticky. It is also worth noting that the butter and butter substitute samples differed in their adhesiveness, and this was statistically significant.

For the butter and butter substitute samples examined in this study, it can be seen that hardness, adhesive force, and adhesiveness were parameters that were partially correlated with spreadability. If their status changed, the spreadability status would also change. Some correlations between the results of measurements of the rheological properties of edible fats were also found by Glibowski et al. [31]. In their study, spreadability and cohesiveness measured at 5 °C correlated very well, but spreadability and cohesiveness at 5 °C were not very well correlated. The researchers showed a low correlation coefficient between spreadability at 20 °C and spreadability at 5 °C, and between hardness at 20 °C and hardness at 5 °C, which very clearly indicates differences in the rheological properties of edible fats at different temperatures. This was also confirmed by the low correlation coefficients between spreadability at 5 °C and apparent viscosity at 20 °C, and hardness at 5 °C and apparent viscosity at 20 °C [31].

The statistical analysis performed in this study of the spreadability, hardness, adhesive force, and adhesiveness of the butter and butter substitute samples (Figure 1a,b) revealed a completely different relationship pattern than those found by Glibowski et al. [31]. Figure 1a,b show the corrgrams (i.e., correlation plots) of the correlation matrix, with the colored cells representing the magnitude of the correlation. Correlation coefficients range from −1 to +1 and measure the strength of the linear relationship between variables (statistically significant correlations occur at the 95.0% confidence level). The colors ranged from blue for strong negative correlations to red for strong positive correlations. The interpretation of the results for butter samples and butter substitutes differed when the samples were subjected to separate multivariate analyses. Few strong correlations (whether positive or negative) were observed for the butter samples (Figure 1a) between hardness measurements at specific temperature conditions (correlation coefficients 0.96–0.86); between adhesiveness and adhesive force for samples tested 30 min after removal from the refrigerator (correlation coefficient 0.87); and between adhesiveness and spreadability or hardness for the samples tested at 20 °C (correlation coefficients −0.81 and −0.94, respectively).

Significantly stronger correlations (both positive and negative) were recorded for the butter substitute samples. This plot of correlations revealed the following strong positive correlations (Figure 1b): between spreadability and hardness at specific temperature conditions (correlation coefficients 0.89–1.00); between adhesiveness for samples tested at 20 °C and adhesive force for samples tested 30 min after removal from the refrigerator or for samples tested at 20 °C (correlation coefficients 0.90 and 0.99, respectively); and between adhesive force for samples tested at 20 °C and adhesive force for samples tested 30 min after removal from the refrigerator (correlation coefficient 0.87). Strongly negative correlations were no less important and were observed between spreadability and adhesive force for samples tested 30 min after removal from the refrigerator (correlation coefficient −0.93); between spreadability and adhesive force for samples tested at 4 °C (correlation coefficient −0.81); between spreadability for samples tested at 20 °C and adhesive force for samples tested 30 min after removal from the refrigerator or adhesive force for samples tested at 20 °C or adhesiveness for samples tested at 20 °C (correlation coefficients −0.85; −1.00 and −0.98, respectively); between hardness and adhesive force for samples tested at 4 °C (correlation coefficient −0.92); between hardness and adhesive force for samples tested 30 min after removal from the refrigerator (correlation coefficient −0.97); and between hardness for samples tested at 20 °C and adhesive force for samples tested 30 min after removal from the refrigerator or for samples tested at 20 °C or adhesiveness for samples tested at 20 °C (correlation coefficients −0.86; −1.00 and −0.98, respectively). However, it is important to remember that a high correlation coefficient does not necessarily indicate causality. It simply indicates that the two variables are related in some way. Further investigation and analysis, such as regression analysis, may be required to determine the nature of the relationship and to establish causality.

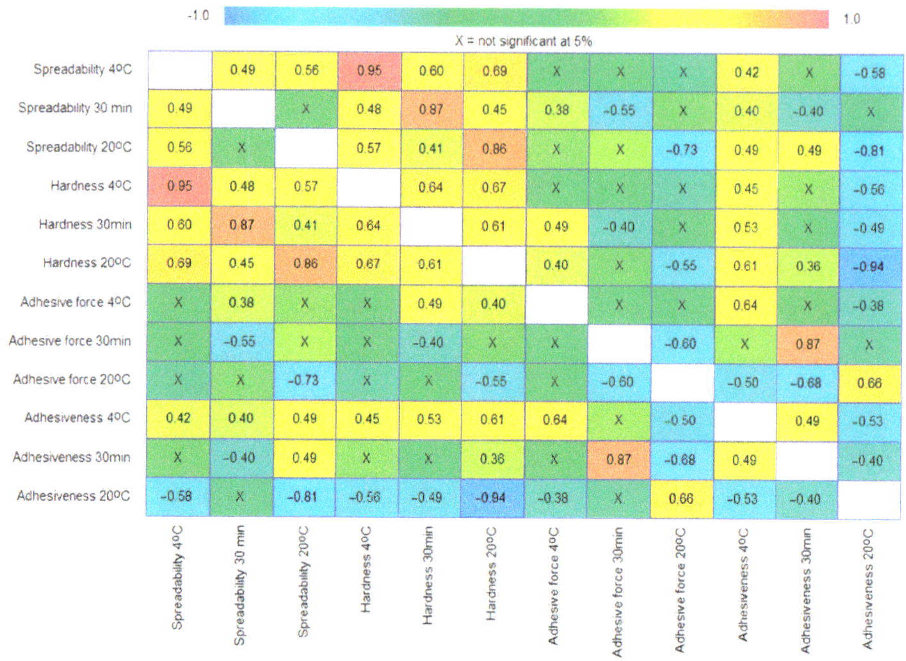

(**a**)

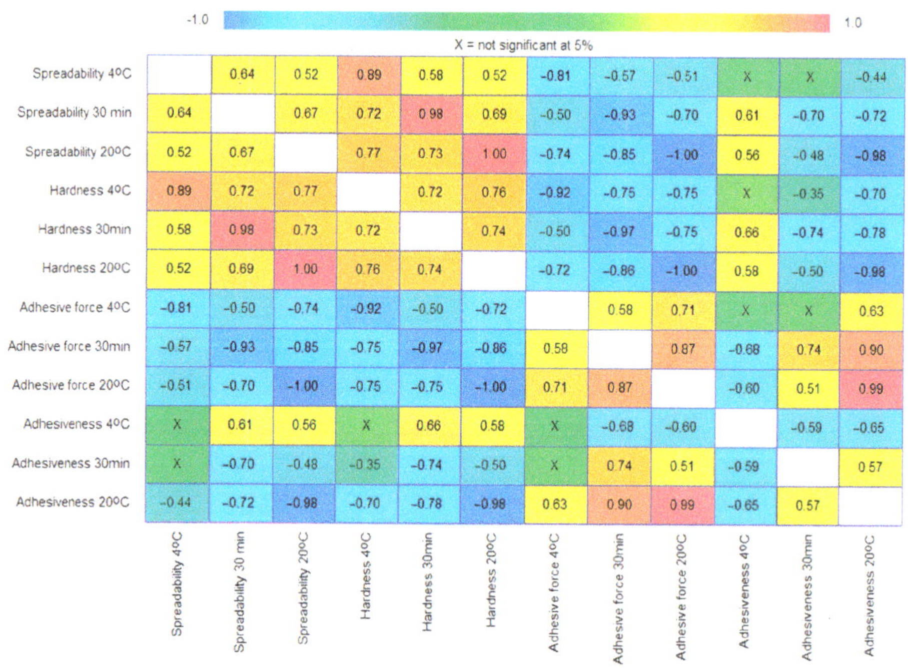

(**b**)

Figure 1. Graphs of correlation matrix showing the relationship between the spreadability, hardness, adhesive force and adhesiveness of butter and butter substitute samples (**a**) and butter substitute samples alone (**b**) measured at different temperatures (with a confidence level of 95.0%).

3.2. Physicochemical Properties of Butter and Butter Substitutes

Water content. The water contents of the tested butter (Table 5) did not exceed the set limit of 16 ± 0.5% [49,50], were in line with producers' declarations (Table 1), and, importantly, these values were not statistically significantly different from each other at the significance level $\alpha = 0.05$. Butter substitutes, on the other hand, were characterized by significantly higher water content values (in the studies presented here, the water content of the butter substitutes samples ranged from 17.93 ± 0.35% to 32.97 ± 0.33%), whereby these samples were divided into three different homogeneous groups with a significance level of $\alpha = 0.05$.

Table 5. Physicochemical properties of butter and butter substitutes.

Parameter / Butter Samples	Water Content [%]	Degree of Dispersion of Water [Points]	Plasma pH
LMK	16.12 ± 0.16^{a}	3.0 ± 0.0^{a}	$5.45 \pm 0.43^{b,c}$
LaME	15.39 ± 0.49^{a}	3.0 ± 0.0^{a}	5.94 ± 0.18^{c}
LoME	15.99 ± 0.44^{a}	3.0 ± 0.0^{a}	6.38 ± 0.36^{c}
MEG	16.08 ± 0.35^{a}	3.0 ± 0.0^{a}	5.17 ± 0.18^{b}
MEH	15.59 ± 0.36^{a}	3.0 ± 0.0^{a}	6.32 ± 0.11^{c}
MMP	15.24 ± 0.18^{a}	3.0 ± 0.0^{a}	6.38 ± 0.30^{c}
PME	15.61 ± 0.27^{a}	3.0 ± 0.0^{a}	6.27 ± 0.27^{c}
PrME	16.08 ± 0.24^{a}	3.0 ± 0.0^{a}	6.77 ± 0.16^{c}
Butter substitutes Samples			
FM	24.88 ± 0.40^{d}	3.0 ± 0.0^{a}	4.59 ± 0.28^{a}
LuPM	22.04 ± 0.40^{c}	3.0 ± 0.0^{a}	4.56 ± 0.24^{a}
LaM	32.09 ± 0.29^{e}	3.0 ± 0.0^{a}	4.54 ± 0.34^{a}
PaEM	24.95 ± 0.44^{d}	3.0 ± 0.0^{a}	4.05 ± 0.32^{a}
RMTM	19.93 ± 0.47^{c}	3.0 ± 0.0^{a}	4.50 ± 0.17^{a}
SSO	32.97 ± 0.33^{e}	3.0 ± 0.0^{a}	3.98 ± 0.08^{a}
ZaM	32.10 ± 0.42^{e}	3.0 ± 0.0^{a}	4.45 ± 0.28^{a}
BGP	17.93 ± 0.35^{b}	3.0 ± 0.0^{a}	4.43 ± 0.09^{a}

^{a,b,c,d,e}—values in the same column and marked with the same letters are not statistically significantly different at the $\alpha = 0.05$ level.

Butter and butter substitutes are physically composed of fat globules, fat crystals, air bubbles, and water droplets, all of which play a role in the physical properties of these products [51,52]. The physical and chemical properties of butter and butter substitutes (including water content and water droplet size, and textural and rheological properties, such as hardness and spreadability) are of great importance as they determine the functionality of these products [17,46,53]. Water content is closely linked to the quality of the end product, such as butter or its vegetable substitutes and blends. As studies [48,54,55] have shown, the water content of butter is influenced by the technological parameters of the creamer process and the kneading of the butter, which aims for an even distribution of water droplets that are as small as possible, in order for the butter to have the right consistency. Rønholt [48] showed that the water content is also decisive for the smear value. The water content influences the crystallization of the fat phase, and thus, also the structure of the butter [56]. The strength of the crystals formed depends on the size of the water droplets and the amount of fat crystallized. As the water content of the product increases, interactions between the water droplets can occur and the textural stability of the butter is consequently lost [57]. Similar effects are observed with butter substitutes [57].

Water distribution. Test samples of butter and butter substitutes received the maximum score in determining the degree of water dispersion (Table 5). The physical composition of butter and its vegetable substitutes varies as the different manufacturing processes result in different microstructures of these products. In addition, butter is less homogeneous and has a more complex chemical composition than its plant substitutes or blends such as margarine, which requires the use of sophisticated analytical techniques in instrumental analysis to determine water droplet size distribution [58].

The degree of water dispersion is of microbiological importance, as well as being important for the sensory properties of fat products such as butter [59]. The greater the degree of water dispersion, the more difficult it is for unwanted microflora to grow. The water content and degree of distribution can influence the course of fat crystallization, which in turn can influence the texture of the product, and thus, its spreadability [16,48,51,56,57,60].

Plasma pH. The results of the plasma pH measurements of the analyzed samples of butter and butter substitutes are summarized in Table 5. The pH of butter plasma ranged from 5.94 ± 0.18 to 6.77 ± 0.16 and was not statistically significantly different but was dependent on the butter sample. The statistical analysis allowed the butter samples to be distinguished into two groups: (a) MEG and LMK; (b) LMK, LaME, LoME, MEH, MMP, PME, and PrME (Table 5). The plasma acidity of the butter substitutes was statistically significantly different from the pH of the butter plasma and was the same for all butter substitute samples tested.

The plasma pH of butter and its vegetable substitutes (blends with other fats) is a result of the production and storage parameters of the product [61]. An important step in the production of butter, which later influences the pH value of the milk plasma, is the biological maturation of the cream, i.e., its fermentation. The lactose contained therein is converted into lactic acid, which subsequently causes the plasma of the aqueous phase to acidify and thus improves the shelf life of the product. As can be seen from the analysis of the butter samples tested in this work, the pH value indicates that the cream had not undergone biological maturation, i.e., the butter samples were made from sweet cream. The situation is different with butter substitutes, the production of which usually involves regulating the plasma pH value by adding chemical acidity regulators such as citric and lactic acids (what was claimed by some manufacturers, Table 1). It should be noted that the acidity of butter and butter substitutes is a poorly understood parameter in terms of its significant relationship to lubricity values. No available literature data were found on this topic.

Color. It was found that the butter samples tested were different from the chosen standard (Table 6). The mean values of the L^* and b^* color components for the test samples were higher than the corresponding values of the standard, while the mean value of the a^* color component was lower than that of the standard. The measured color of the butter samples according to the standard tended towards slightly greenish and lighter tones. The color component a^* did not statistically differentiate the butter and butter substitute samples.

The parameter b^* in the color analysis is often used as an indicator of the yellow-blue color bias in a sample. When discussing color results, the color tendency of parameter b^* is usually described as the amount of yellow or blue present in the sample. The b^* component divided the butter and butter substitutes into seven homogeneous groups at the 0.05 level, with the majority of butter samples ranking above the majority of butter substitutes on the CIELab scale. The magnitude of the b^* value would provide a measure of the intensity or saturation of the yellow or blue color. A positive b^* value would indicate a yellow color in the butter sample, while a negative b^* value would indicate a blue color. A high positive b^* value would indicate a strong yellow sample, while a low positive b^* value would indicate a lighter yellow color. The samples studied in this work obtained high positive values for the parameter b^*, which in most cases were statistically significantly higher for butter than for its substitutes.

Regarding the color component L^*, all butter samples and the four butter substitutes (LuPM, LaM, SSO, and ZaM) showed the same value for this parameter, which was statistically significant, while the other three butter substitutes (FM, PaEM, and RMTM) were significantly darker. It is worth noting that the butter substitutes compared to the standard for butter color components a^*, b^*, and L^* gave surprisingly similar results for each component, despite the differences in chemical composition (e.g., different fats used or water content), in textural parameters (e.g., spreadability or hardness) and due to different technological processes.

Table 6. Color components of butter and butter substitutes.

Parameter / Butter Samples	L^*	a^*	b^*	ΔE
LMK	85.72 ± 0.78 [a,b,c,d]	−6.98 ± 0.10 [a]	30.25 ± 0.40 [c,d,e]	14.89 ± 0.11 [b,c]
LaME	88.44 ± 0.39 [c,d]	−7.24 ± 0.07 [a]	31.82 ± 0.26 [e,f,g]	14.94 ± 0.10 [b,c]
LoME	88.13 ± 0.59 [c,d]	−7.48 ± 0.11 [a]	29.04 ± 0.37 [c]	14.13 ± 0.12 [b]
MEG	87.67 ± 1.41 [b,c,d]	−6.93 ± 0.07 [a]	31.57 ± 0.94 [e,f]	14.81 ± 0.17 [b,c]
MEH	87.50 ± 0.57 [b,c,d]	−7.16 ± 0.05 [a]	30.67 ± 0.37 [c,d,e]	14.59 ± 0.09 [b,c]
MMP	88.69 ± 0.19 [c,d]	−7.32 ± 0.14 [a]	32.08 ± 0.05 [e,f,g]	15.08 ± 0.16 [b,c]
PME	87.12 ± 0.49 [a,b,c,d]	−7.49 ± 0.12 [a]	33.84 ± 0.63 [f,g]	16.51 ± 0.44 [c]
PrME	89.30 ± 0.16 [d]	−7.13 ± 0.02 [a]	33.06 ± 0.14 [f,g]	15.32 ± 0.06 [b,c]
Butter Substitutes Samples				
FM	83.70 ± 1.71 [a]	−7.01 ± 0.26 [a]	25.34 ± 0.70 [b]	14.88 ± 0.68 [b,c]
LuPM	86.58 ± 2.26 [a,b,c,d]	−7.24 ± 0.16 [a]	25.55 ± 1.26 [b]	13.90 ± 0.48 [b]
LaM	87.98 ± 0.65 [c,d]	−5.97 ± 0.04 [a]	30.81 ± 0.48 [c,d,e]	13.50 ± 0.04 [b]
PaEM	85.40 ± 1.37 [a,b,c]	−6.74 ± 0.04 [a]	31.18 ± 0.78 [d,e,f]	15.23 ± 0.17 [b,c]
RMTM	84.11 ± 4.17 [a,b]	−0.63 ± 0.09 [a]	20.92 ± 2.22 [a]	15.67 ± 3.15 [c]
SSO	86.82 ± 0.87 [a,b,c,d]	−7.56 ± 0.13 [a]	26.76 ± 0.33 [b]	14.09 ± 0.14 [b]
ZaM	86.63 ± 0.20 [a,b,c,d]	−6.22 ± 0.15 [a]	29.31 ± 0.23 [c,d]	13.55 ± 0.06 [b]
BGP	86.06 ± 0.45 [a,b,c,d]	−5.04 ± 2.98 [a]	28.04 ± 0.31 [c]	12.37 ± 2.40 [a]

[a,b,c,d,e,f,g]—values in the same column and marked with the same letters are not statistically significantly different at the α = 0.05 level.

The calculated ΔE* values represented the difference between the color of the test sample and the color of the standard in CIELab space and therefore expressed the magnitude of the color change but not its direction. With regard to the expression of this parameter, the samples of butter and butter substitutes were statistically significantly different in two homogeneous groups at the α = 0.05 level. The calculated ΔE* values for the butter and butter substitute samples ranged in excess of 5, indicating large color differences to the unaided eye of an unexperienced observer between the test butter and butter substitutes and the standard color [35,62].

A multivariate analysis of the spreadability measurements, selected physicochemical properties, and the color components of the butter and butter substitute samples did not reveal any significant strong relationships between these parameters (Figure 2a,b). The only significant correlation found was between the ΔE* value and b* color compound for the butter samples (correlation coefficient 0.85). Lapčíková et al. [46] also found no general relationship between the content and composition of total fat in the samples and the values of textural parameters (i.e., springiness, cohesiveness, and stringiness). Furthermore, no correlation was to be expected for the color components, since both butter and butter substitutes can be colored (while carotenes annatto, bixin, norbixin, and curcumin are permitted in butter in the EU, as are other fat- and oil emulsions) [63].

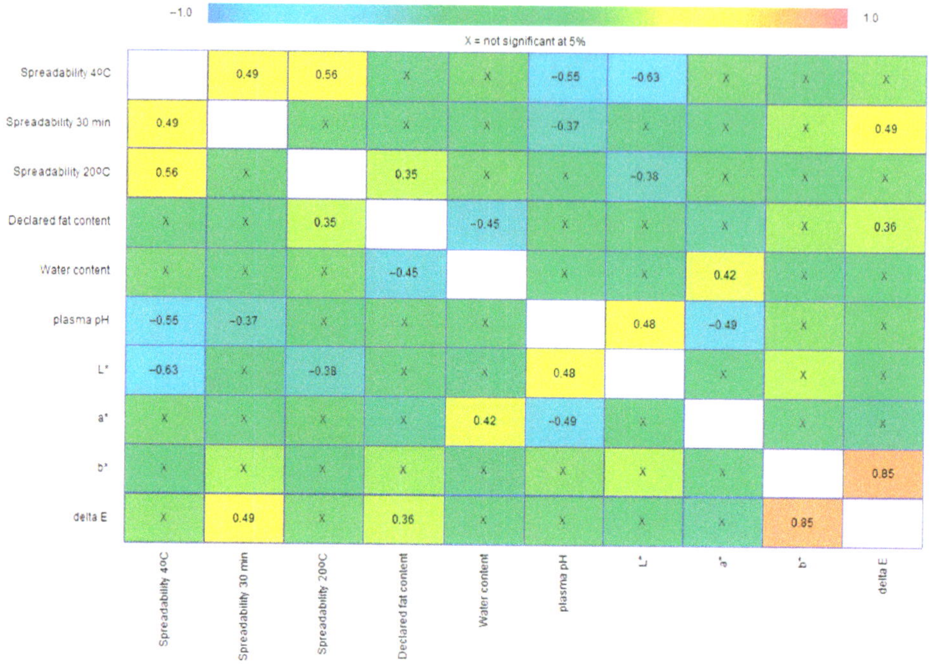

(a)

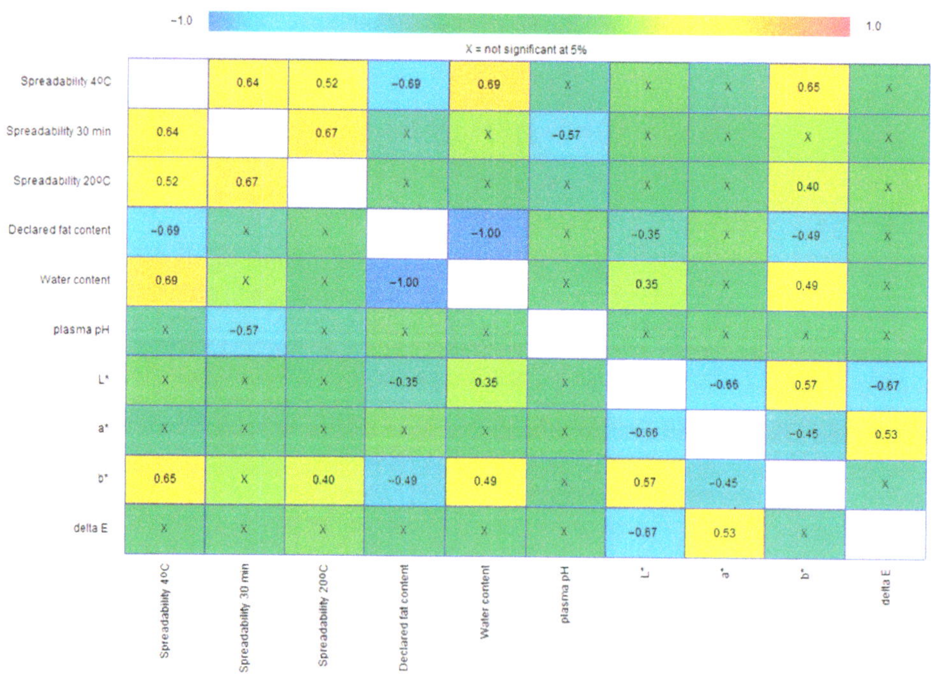

(b)

Figure 2. Graphs of correlation matrix showing the relationship between the spreadability, selected physicochemical properties, and the color components of butter and butter substitute samples (**a**) and butter substitute samples alone (**b**) measured at different temperatures (with a confidence level of 95.0%).

3.3. Additional Characteristics of Butter Milk Fat

Determination of the acid value. The acid number is defined as the number of milligrams of potassium hydroxide required to neutralize the free fatty acids in a gram of fat sample. The acid number values of the butter samples tested ranged from 1.14 to 1.54 mg KOH/g fat and did not exceed the maximum value of 2 mg KOH/g fat permitted for butter (Table 7). The acid numbers of the butter tested in this study were statistically significantly different (at a significance level of $\alpha = 0.05$). This may be due to different production dates or because the comparison involves samples from different manufacturers and technologies.

Table 7. Characteristics of butter milk fat.

Butter Samples	Acid Value [mg KOH/g fat]	Saponification Number [mg KOH/g fat]
LMK	1.27 ± 0.03 ^{b}	227.58 ± 0.43 ^{b}
LaME	1.20 ± 0.02 ^{a}	228.03 ± 0.58 ^{b,c}
LoME	1.16 ± 0.03 ^{a}	226.19 ± 0.80 ^{a}
MEG	1.31 ± 0.02 ^{b}	230.35 ± 0.32 ^{d}
MEH	1.54 ± 0.03 ^{d}	228.18 ± 0.36 ^{b,c}
MMP	1.14 ± 0.03 ^{a}	229.02 ± 0.11 ^{c}
PME	1.16 ± 0.03 ^{a}	230.30 ± 0.6 ^{d}
PrME	1.37 ± 0.02 ^{c}	231.18 ± 0.55 ^{d}

^{a,b,c,d}—values in the same column and marked with the same letters are not statistically significantly different at the $\alpha = 0.05$ level.

These data are consistent with those of other scientists [64,65]. Similar results were obtained by Bellinazo et al. [64], who examined the properties of butter during storage and obtained an acid number value of 1.08 mg KOH/1 g fat just after production. The acid value increased with the storage time and was 2.74 mg KOH/g fat after storage for 90 days.

Determination of the peroxide number. No peroxides were found at detectable levels in any of the tested samples, which was due to the good quality of the tested products. These results were consistent with the findings of other researchers [66]. This number is a measure of the peroxide content and is considered an indicator of the rancidity of the fat. The butter samples tested were products derived from sweet cream; meanwhile, Khaskheli et al. [67] showed that the peroxide number of market sweet butter (1.56 ± 0.17 mEq O2/kg fat) was significantly higher than the peroxide number of butter derived from fermented cream (1.00 ± 0.08 mEq O2/kg fat), which was determined by changes that were reported to have occurred during the storage of the tested butter samples. In comparison, the peroxide number of butter samples freshly prepared from sweet cream or fermented cream under the laboratory conditions by Khaskheli et al. [67] was 1.00 ± 0.10 mEq O2/kg fat and 1.04 ± 0.11 mEq O2/kg fat, respectively. The observed fluctuations in the peroxide number values of market butter (0.35 ± 0.24 to 1.80 ± 0.36 mEq O2/kg fat) were explained by Gonçalves and Baggio [68] by differences in the way the products were packaged, and thus, their exposure to atmospheric oxygen.

Determination of the saponification number. The saponification number values of the butter tested were in the range of 226.2–231.2 mg KOH/g fat and did not exceed the usual range specified for butter, i.e., 220–236 mg KOH/g fat (Table 7). Although the differences between the values obtained for the different butter samples were small, the values were significantly different (at a significance level of $\alpha = 0.05$). Similar results were obtained by Kahyaoğlu and Çakmakçı [69], who studied butter and obtained a saponification number of 228.1 mg KOH/g fat. Another study by Kahyaoğlu and Çakmakçı [70] showed that the saponification number increased with storage time. As the studies mentioned above have shown, the saponification number (such as the acid number) can be an indicator of the degree of freshness of the fat and, above all, of its shelf life. Determination of the saponification number in fats enables the average molecular weight of the fatty acids to be determined. Its high levels in butter are due to the presence of palmitic acid.

Chromatographic determination of the fatty acid profile. Table 8 provides a summary of the percentage of individual fatty acids found in the butter samples tested. Types of butter, which are products of animal origin, are characterized by a high percentage share of saturated fatty acids (SFA) and a low content of unsaturated fatty acids: monounsaturated fatty acids (MUFA) and polyunsaturated fatty acids (PUFA). The analyzed kinds of butter contained the following fatty acids in the highest proportion: palmitic acid (C16:0), oleic acid (C18:1 cis 9), stearic acid (C18:0), and myristic acid (C14:0).

Table 8. Percentage of fatty acids present in the fat fraction of butter samples.

Fatty Acids	Percentage of Fatty Acids Present in the Fat Fraction [%]							
Butter Samples	LMK	LaME	LoME	MEG	MEH	MMP	PME	PrME
C 4:0	0.24 ± 0.13 ^{a,b}	0.20 ± 0.04 ^{a,b}	0.17 ± 0.20 ^{a}	0.71 ± 0.04 ^{c}	0.29 ± 0.07 ^{a,b}	0.5 ± 0.11 ^{b,c}	0.69 ± 0.17 ^{c}	0.78 ± 0.08 ^{c}
C 6:0	0.52 ± 0.15 ^{a,b}	0.36 ± 0.04 ^{a}	0.43 ± 0.06 ^{a,b}	0.81 ± 0.04 ^{c}	0.54 ± 0.09 ^{a,b}	0.65 ± 0.04 ^{b,c}	0.85 ± 0.09 ^{c}	0.89 ± 0.06 ^{c}
C 8:0	0.5 ± 0.12 ^{b,c}	0.32 ± 0.06 ^{a}	0.38 ± 0.07 ^{a,b}	0.65 ± 0.04 ^{c,d}	0.48 ± 0.04 ^{a,b,c}	0.57 ± 0.01 ^{b,c,d}	0.62 ± 0.07 ^{c,d}	0.69 ± 0.04 ^{d}
C 10:0	1.89 ± 0.21 ^{a,c}	1.25 ± 0.41 ^{a}	1.46 ± 0.2 ^{a,b}	2.07 ± 0.10 ^{c}	1.71 ± 0.11 ^{a,b,c}	1.96 ± 0.08 ^{a,c}	2.15 ± 0.09 ^{c}	2.15 ± 0.04 ^{c}
C 12:0	2.67 ± 0.18 ^{a,b}	1.89 ± 0.73 ^{a}	2.15 ± 0.31 ^{a,b}	2.9 ± 0.11 ^{b}	2.55 ± 0.12 ^{a,b}	2.83 ± 0.06 ^{a,b}	2.96 ± 0.04 ^{b}	2.93 ± 0.06 ^{b}
C 14:0	9.44 ± 0.74 ^{a,b}	8.03 ± 2.79 ^{a}	9.32 ± 0.37 ^{a,b}	11.01 ± 0.14 ^{a,b}	10.44 ± 0.23 ^{a,b}	10.82 ± 0.19 ^{a,b}	11.41 ± 0.18 ^{b}	10.96 ± 0.12 ^{a,b}
C 14:1	0.13 ± 0.02 ^{a}	0.13 ± 0.05 ^{a}	0.15 ± 0.02 ^{a,b}	0.18 ± 0.01 ^{a,b,c}	0.20 ± 0.01 ^{b,c}	0.20 ± 0.01 ^{b,c}	0.22 ± 0.01 ^{c}	0.23 ± 0.01 ^{c}
C 15:0	0.90 ± 0.09 ^{a}	0.74 ± 0.29 ^{a}	0.93 ± 0.19 ^{a}	1.03 ± 0.01 ^{a}	0.98 ± 0.03 ^{a}	1.14 ± 0.05 ^{a}	1.12 ± 0.08 ^{a}	1.05 ± 0.03 ^{a}
C 16:0	32.49 ± 1.19 ^{a}	33.13 ± 2.47 ^{a}	37.15 ± 2.31 ^{b}	34.61 ± 0.22 ^{a,b}	34.56 ± 1.02 ^{a,b}	34.85 ± 0.38 ^{a,b}	35.11 ± 0.74 ^{a,b}	31.37 ± 0.39 ^{a}
C 17:0	0.54 ± 0.03 ^{a}	0.52 ± 0.17 ^{a}	0.58 ± 0.09 ^{a}	0.60 ± 0.01 ^{a}	0.65 ± 0.04 ^{a}	0.69 ± 0.07 ^{a}	0.68 ± 0.08 ^{a}	0.69 ± 0.02 ^{a}
C 18:0	12.10 ± 0.34 ^{a}	12.73 ± 1.21 ^{a}	11.39 ± 0.60 ^{a}	11.59 ± 0.11 ^{a}	11.34 ± 0.46 ^{a}	11.06 ± 0.11 ^{a}	11.17 ± 0.27 ^{a}	11.73 ± 0.06 ^{a}
C 20:0	0.13 ± 0.01 ^{a}	0.15 ± 0.01 ^{a}	0.14 ± 0.03 ^{a}	0.13 ± 0.00 ^{a}	0.17 ± 0.03 ^{a}	0.14 ± 0.02 ^{a}	0.13 ± 0.02 ^{a}	0.12 ± 0.01 ^{a}
C 14:1 cis	1.02 ± 0.10 ^{a,b}	0.90 ± 0.33 ^{a}	1.09 ± 0.05 ^{a,b,c}	1.27 ± 0.05 ^{a,b,c}	1.25 ± 0.04 ^{a,b,c}	1.37 ± 0.02 ^{b,c}	1.39 ± 0.06 ^{c}	1.39 ± 0.04 ^{c}
C 15:1	0.12 ± 0.02 ^{a}	0.14 ± 0.06 ^{a,b}	0.19 ± 0.03 ^{a,b}	0.20 ± 0.01 ^{a,b}	0.21 ± 0.02 ^{b}	0.23 ± 0.02 ^{b}	0.22 ± 0.03 ^{b}	0.22 ± 0.01 ^{b}
C 16:1 trans	0.14 ± 0.02 ^{a,b}	0.13 ± 0.01 ^{a,b}	0.13 ± 0.02 ^{a}	0.14 ± 0 ^{a,b}	0.15 ± 0.02 ^{a,b}	0.15 ± 0.01 ^{a,b}	0.14 ± 0.01 ^{a,b}	0.17 ± 0.01 ^{b}
C 16:1 cis9	1.9 ± 0.05 ^{a}	2.04 ± 0.13 ^{a}	1.98 ± 0.36 ^{a}	1.91 ± 0.01 ^{a}	1.86 ± 0.13 ^{a}	2.01 ± 0.07 ^{a}	1.91 ± 0.07 ^{a}	1.84 ± 0.03 ^{a}
∑C 16:1 cis	0.48 ± 0.05 ^{a}	0.41 ± 0.16 ^{a,b}	0.59 ± 0.11 ^{a,b}	0.58 ± 0.01 ^{a,b}	0.55 ± 0.05 ^{a,b}	0.63 ± 0.05 ^{a,b}	0.60 ± 0.03 ^{a,b}	0.64 ± 0.01 ^{b}
C 17:1 cis	0.14 ± 0.00 ^{a}	0.16 ± 0.02 ^{a}	0.19 ± 0.07 ^{a}	0.19 ± 0.03 ^{a}	0.20 ± 0.04 ^{a}	0.23 ± 0.01 ^{a}	0.20 ± 0.03 ^{a}	0.21 ± 0.02 ^{a}
C 17:1 cis izo	0.02 ± 0.01 ^{a}	0.05 ± 0.02 ^{a}	0.05 ± 0.03 ^{a}	0.04 ± 0.00 ^{a}	0.06 ± 0.01 ^{a}	0.05 ± 0.00 ^{a}	0.05 ± 0.01 ^{a}	0.04 ± 0.00 ^{a}
∑C 18:1 trans	2.13 ± 0.22 ^{b,c}	1.12 ± 0.46 ^{a}	1.32 ± 0.40 ^{a}	1.65 ± 0.02 ^{a,b}	1.30 ± 0.14 ^{a}	1.73 ± 0.05 ^{a,b}	1.49 ± 0.18 ^{a,b}	2.84 ± 0.06 ^{c}
C 18:1 cis9	25.83 ± 1.88 ^{a,b}	28.80 ± 5.12 ^{b}	25.18 ± 1.73 ^{a,b}	22.90 ± 0.21 ^{a,b}	25.34 ± 1.82 ^{a,b}	23.07 ± 0.17 ^{a,b}	22.27 ± 0.25 ^{a}	23.17 ± 0.12 ^{a,b}
C 18 1 trans9	1.58 ± 0.25 ^{a,b}	1.76 ± 0.68 ^{b}	1.13 ± 0.26 ^{a,b}	0.98 ± 0.01 ^{a,b}	1.01 ± 0.09 ^{a,b}	1.01 ± 0.0 ^{a,b}	0.89 ± 0.04 ^{a}	1.01 ± 0.01 ^{a,b}
∑C 18:1 cis	1.18 ± 0.09 ^{b}	0.81 ± 0.19 ^{a}	0.89 ± 0.20 ^{a,b}	1.10 ± 0.01 ^{a,b}	0.79 ± 0.06 ^{a}	1.11 ± 0.02 ^{a}	0.99 ± 0.06 ^{a,b}	1.19 ± 0.06 ^{b}
C 19:1 cis	0.07 ± 0.02 ^{a}	0.06 ± 0.02 ^{a}	0.08 ± 0.02 ^{a}	0.10 ± 0.01 ^{a}	0.11 ± 0.02 ^{a}	0.12 ± 0.00 ^{a}	0.10 ± 0.01 ^{a}	0.11 ± 0.01 ^{a}
C 20:1	0.10 ± 0.01 ^{a,b}	0.07 ± 0.04 ^{a}	0.08 ± 0.02 ^{a}	0.08 ± 0.01 ^{a,b}	0.13 ± 0.01 ^{b}	0.12 ± 0.02 ^{a,b}	0.10 ± 0.01 ^{a,b}	0.09 ± 0.01 ^{a,b}
C 20:1 cis	0.14 ± 0.07 ^{a}	0.20 ± 0.15 ^{a}	0.07 ± 0.07 ^{a}	0.03 ± 0.01 ^{a}	0.09 ± 0.04 ^{a}	0.04 ± 0.01 ^{a}	0.02 ± 0.01 ^{a}	0.02 ± 0.01 ^{a}
C 18:2 trans	0.12 ± 0.01 ^{b}	0.07 ± 0.02 ^{a}	0.08 ± 0.02 ^{a,b}	0.10 ± 0.00 ^{a,b}	0.08 ± 0.02 ^{a,b}	0.11 ± 0.01 ^{a,b}	0.10 ± 0.02 ^{a,b}	0.11 ± 0.01 ^{b}
C 18:2 cis9,cis12	2.46 ± 0.34 ^{a,b}	2.94 ± 1.32 ^{b}	1.89 ± 0.21 ^{a,b}	1.62 ± 0.02 ^{a,b}	1.74 ± 0.15 ^{a,b}	1.73 ± 0.09 ^{a,b}	1.40 ± 0.09 ^{a}	1.86 ± 0.02 ^{a,b}
C 18:3 cis9,cis12,cis15	0.61 ± 0.05 ^{a,b}	0.59 ± 0.13 ^{a,b}	0.52 ± 0.10 ^{a}	0.49 ± 0.02 ^{a}	0.83 ± 0.19 ^{b}	0.54 ± 0.05 ^{a,b}	0.52 ± 0.05 ^{a}	0.69 ± 0.06 ^{a,b}
C 18:2 trans9,trans11	0.42 ± 0.02 ^{c}	0.29 ± 0.08 ^{a}	0.29 ± 0.05 ^{a,b}	0.37 ± 0.01 ^{a,b,c}	0.39 ± 0.02 ^{a,b,c}	0.43 ± 0.04 ^{c}	0.41 ± 0.04 ^{b,c}	0.81 ± 0.03 ^{d}

^{a,b,c,d}—values in the same row and marked with the same letters are not statistically significantly different at the $\alpha = 0.05$ level.

Among the saturated fatty acids (SFA) found in the butter analyzed in this study, one can distinguish between short-chain fatty acids (SCFAs) and medium-chain fatty acids, characteristic of milk fat [71]. Five fatty acids classified as SCFAs were detected in all butter samples analyzed: butyric acid (C4:0), caproic acid (C6:0), caprylic acid (C8:0), capric acid (C10:0) and lauric acid (C12:0). SCFAs are also an important component of milk fat due to their biological properties and health-promoting effects [72,73]. The total saturated fat percentage share present in the butter tested differed significantly between the samples at a significance level of $\alpha = 0.05$.

In all the kinds of butter tested, among the identified MUFA were myristoleic acid (C14:1), isomers of palmitoleic acid (C16:1), and isomers of margaric acid (C17:1). However, oleic acid (C18:1 cis 9) was the predominant fatty acid. The total monounsaturated fatty acid percentage share present in the kinds of butter tested also differed significantly between the butter samples at a significance level of $\alpha = 0.05$.

The predominant polyunsaturated fatty acid (PUFA) in the butter samples of this study was linoleic acid (C18:2: cis 9, cis 12). The total polyunsaturated fatty acid percentage share in the kinds of butter studied also differed significantly between the samples, at a significance level of $\alpha = 0.05$. The fatty acid composition of butter is primarily influenced by the raw material selection, and thus, by the genetics (breed), feeding, and environmental factors (season and region) of the dairy cows that the butter comes from [72,74,75].

The rheological results obtained for the butter samples in this study did not correspond with other chemical data obtained exclusively for the butter samples and were determined by the techniques used. A multivariate analysis of spreadability measurements with acid value, saponification number, or percentage fatty acid content (percentage of saturated fatty acids, MUFA, and PUFA) of the butter samples showed no significant strong relationships between these parameters (Figure 3a). In addition, a multivariate analysis was performed to analyze the correlation between the percentage of each fatty acid identified in the butter samples and the spreadability for the butter samples measured at different temperatures (Figure 3b). In this analysis, no correlation was found between the spreadability of the butter samples and the fatty acid profile.

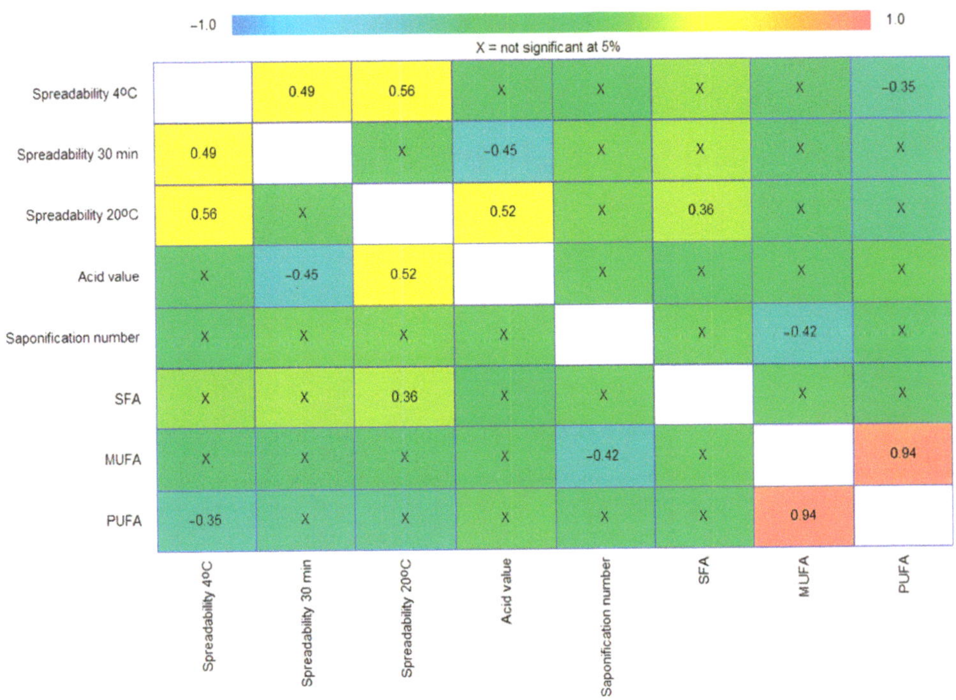

(a)

Figure 3. *Cont.*

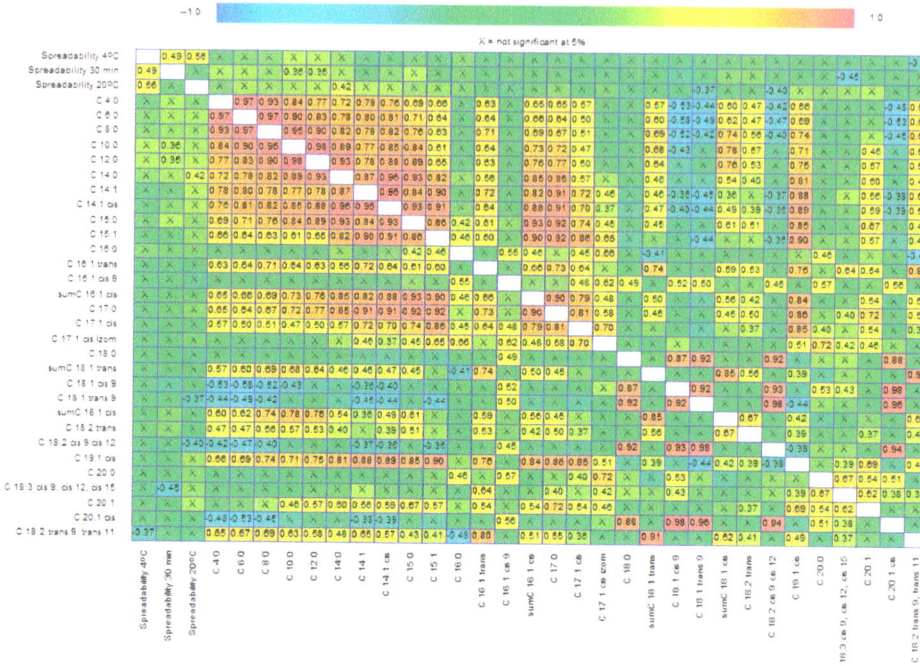

(**b**)

Figure 3. Graphs of correlation matrix showing the relationship between spreadability, acid value, saponification number and SFA, MUFA, and PUFA fatty acid profile (**a**), as well as between spreadability, and the fatty acid percentage share (the percentage for each fatty acid determined) of butter samples (**b**) measured at different temperatures (with a confidence level of 95.0%).

Meanwhile, Brunner [76] found that 80% of the differences in butter texture could be explained by differences in the composition of milk fatty acids. However, Jaeck and Pabst [77] found differences in butter texture in herds of cows fed similar diets. Meanwhile, some researchers [43,78,79] have found sufficient variability between cows fed the same feed to produce butter with different textural characteristics and a healthier fatty acid composition. This was supported by a study by Bobe et al. [30], who found that butter samples from milk from cows with a more unsaturated milk fatty acid composition were more spreadable, softer, and less sticky. Thus, the phenotypic variation in milk fatty acid composition among cows fed the same diet is sufficient to produce butter with different textural properties. Meanwhile, Lapčíková et al. [46] found no overall relationship between the composition of milk fat in the samples of butter, spreads, and shortenings available on the Czech market and the values of their textural parameters (i.e., springiness, cohesiveness, and stringiness).

4. Conclusions

The selected structural (spreadability, hardness, adhesive force, and adhesiveness) and physicochemical (water content, water distribution, plasma pH, color, acid value, peroxide number, saponification number, and fatty acid profile) parameters of the butter and butter substitutes tested in this study were correlated with factors such as the type of sample, measuring temperature and physicochemical composition.

The highest spreadability value (i.e., poorer spreadability of the product on the bread) value was obtained for butter samples at 4 °C, and they were significantly inferior to butter substitutes at the same temperature. Butter samples at 4 °C were also characterized by high hardness, which was significantly higher than butter substitutes at the same temperature.

Statistical analysis of the spreadability, hardness, adhesive strength, and adhesiveness results obtained for the butter and butter substitute samples in this study revealed correlations between the textural parameters studied. These were different for the butter and butter substitute samples tested at 4 °C, as well as between hardness and spreadability for samples tested 30 min after removal from the refrigerator. In the case of the butter samples, only very few strong correlations between the spreadability of the products and their other analyzed characteristics were found. In contrast, such correlations abounded for the butter substitute samples.

The butter substitutes had significantly higher water content values than the butter samples. No clear relationship was found between the composition of the butter and butter substitute samples and the values of the textural parameters, including spreadability. The a^*, b^*, and L^* butter color components of the butter substitutes compared with the standard gave surprisingly similar results for each component, despite the differences in chemical composition and textural parameters, and the differences due to different technological processes.

Analysis of a number of variable measurements of the spreadability and acid number, saponification number or percentage of fatty acids in the butter samples, or even the percentage of each fatty acid identified in the butter samples, did not reveal significant strong relationships between these parameters and the spreadability of the butter samples measured at different temperatures.

Author Contributions: Conceptualization. M.Z., D.D. and A.F.; methodology. M.Z., D.D., A.F. and I.S.; investigation. M.Z., D.D., A.F. and I.S.; data curation. M.Z., D.D. and A.F.; writing—original draft preparation. M.Z. and D.D.; writing—review and editing. M.Z. and D.D.; project administration. M.Z., D.D. and A.F. All authors have read and agreed to the published version of the manuscript.

Funding: This research received no external funding.

Institutional Review Board Statement: Not applicable.

Informed Consent Statement: Not applicable.

Data Availability Statement: Not applicable.

Acknowledgments: We thank the Institute of Food Sciences of the Warsaw University of Life Sciences WULSSGGW for supporting and providing the necessary infrastructure and research materials. We would especially like to thank Aleksandra Chudyga, Aleksandra Tomaszewska and Katarzyna Czarniak for their help with the analyses.

Conflicts of Interest: The authors declare no conflict of interest.

References

1. Dickinson, E. *New Physico-techniques for the Characterisation of Complex Food Systems*; Blackie Academic and Professional: London, UK, 1995; pp. 117–146.
2. Jakubczyk, E.; Gondek, E.; Samborska, K. Texture characteristics of selected fat blends. *Zesz. Probl. Post Nauk. Roln.* **2014**, *579*, 17–26. (In Polish)
3. Campos, R.; Narine, S.S.; Marangoni, A.G. Effect of Cooling Rate on the Structure and Mechanical Properties of Milk Fat and Lard. *Food Res. Int.* **2002**, *3510*, 971–981. [CrossRef]
4. Brown, W.; Langley, K.R.; Martin, A.; MacFie, H.J.H. Characterisation of Patterns of Chewing Behaviour in Human Subjects and Their Influence on Texture Perception. *J. Texture Stud.* **1994**, *25*, 455–469. [CrossRef]
5. Rohm, H.; Ulberth, F. Use of Magnitude Estimation in Sensory Texture Analysis of Butter. *J. Texture Stud.* **2007**, *20*, 409–418. [CrossRef]
6. Kim, B.H.; Akoh, C.C. A healthy cold-spreadable butter. *Nutrition* **2006**, *17*, 547–548.
7. Schäffer, B.; Szakaly, S.; Lőrniczy, D. Butter fat melting properties and butter consistency. Effect of cream ripening and modification of fatty acid composition. *Therm. Anal. Color J.* **2001**, *64*, 659–669. [CrossRef]
8. Schäffer, B.; Szakály, S.; Lőrinczy, D. Melting Properties of Butter Fat and The Consistency of Butter. Effect of modification of cream ripening and fatty acid composition. *J. Therm. Anal. Cal.* **2001**, *64*, 659–669. [CrossRef]
9. Schäffer, B.; Szakály, S. Structure of butter. 2. Influence of technological modification of the liquid/solid fat ratio on butter consistency. *Milchwissenschaft* **1988**, *43*, 561–564.

10. Vithanage, C.R.; Grimson, M.J.; Smith, B.G. The Effect of Temperature on the Rheology of Butter, A Spreadable Blend and Spreads. *J. Texture Stud.* **2009**, *40*, 346–369. [CrossRef]
11. Rousseau, D.; Marangoni, A.G. Tailoring the textural attributes of butter fat/canola oil blends via Rhizopus arrhizus lipase-catalyzed interesterification. 2. Modifications of physical properties. *J. Agricult Food Chem.* **1998**, *46*, 2375–2381. [CrossRef]
12. Tang, D.; Marangoni, A.G. Modeling the Rheological Properties and Structure of Colloidal Fat Crystal Networks. *Food Sci. Technol.* **2007**, *18*, 474–483. [CrossRef]
13. Kleyn, D.H. Textural Aspects of Butter. *Food Tech.* **1992**, *46*, 118–121.
14. Wright, A.J.; Scanlon, M.G.; Hartel, R.W.; Marangoni, A.G. Rheological Properties of Milk Fat and Butter. *J. Food Sci.* **2001**, *66*, 1056–1071. [CrossRef]
15. Fearon, A.M. Optimising Milk Fat Composition and Processing Properties. *Aust J. Dairy Tech.* **2001**, *2*, 104–108.
16. Rousseau, D.; Gosh, S.; Park, H. Comparison of the Dispersed Phase Coalescence Mechanisms in Different Table Spreads. *J. Food Sci.* **2009**, *74*, 1–7. [CrossRef]
17. Rønholt, S.; Kirkensgaard, J.J.K.; Pedersen, T.B.; Mortensen, K.; Knudsen, J.C. Polymorphism, Microstructure and Rheology of Butter. Effects of Cream Heat Treatment. *Food Chem.* **2012**, *135*, 1730–1739. [CrossRef]
18. Tondhoosh, A.; Nayebzadeh, K.; Mohamadifar, M.A.; Homayouni-Rad, A.; Hosseinoghli, H. Industrial Application of Different Heat Treatments and Cream Fat Contents for Improving the Spreadability of Butter. *Recent Pat. Food Nutr. Agric.* **2016**, *8*, 107–115. [CrossRef]
19. Lee, J.; Martini, S. Effect of Cream Aging Temperature and Agitation on Butter Properties. *J. Dairy Sci.* **2018**, *101*, 7724–7735. [CrossRef]
20. Jaworska, D.; Świderski, F.; Janicki, A. Ocena tekstury tłuszczów do smarowania pieczywa przy użyciu analizatora tekstury TA.XT2i. *Przem Spoz.* **2003**, *57*, 17–19.
21. Piska, I.; Zárubová, M.; Loužecký, T.; Karami, H.; Filip, W. Properties and Crystallization of Fat Blends. *J. Food Eng.* **2006**, *77*, 433–438. [CrossRef]
22. Litz, B.; Obert, G.; Szily, B. Examination of the Correlation of Butter Spreadability and its Fat Conformation by DSC. *J. Therm. Anal. Calor* **2006**, *84*, 425–428. [CrossRef]
23. Glibowski, P. Effect of vegetable oils on the texture of anhydrous milk fat. *Acta Agroph* **2007**, *9*, 603–612. (In Polish)
24. Siemianowski, K.; Satniewski, B.; Markiewicz, K. Comparison of the texture and microstructure of butter and fat blends available on the Polish market. *Przegl Mlecz* **2010**, *10*, 4–7. (In Polish)
25. Kaufmann, N.; Andersen, U.; Wiking, L. The Effect of Cooling Rate and Rapeseed Oil Addition on the Melting Behaviour, Texture and Microstructure of Anhydrous Milk Fat. *Int. Dairy J.* **2012**, *25*, 73–79. [CrossRef]
26. Lis, A.; Staniewski, B.; Ziajka, J. A comparison of butter texture measurements with the AP 4/2 penetrometer and TA.XT. Plus texture analyzer. *Int. J. Food Prop.* **2021**, *24*, 1744–1757. [CrossRef]
27. Narine, S.S.; Maranongi, A.G. Elastic Modulus as an Indicator of Macroscopic Hardness Fat Crystal Networks. *Lebensm.-Wiss. U.-Technol.* **2001**, *34*, 33–40. [CrossRef]
28. Bourne, M.C. Texture, Viscosity, and Food, Chapter 1. In *Food Texture and Viscosity: Concept and Measuremen*, 2nd ed.; Academic Press: New York, NY, USA, 2002; pp. 1–33.
29. Dixon, B.D. Spreadability of Butter: Determination. 1. Description and Comparison of Five Methods of Testing. *Aust J. Dairy Tech.* **1974**, *29*, 15–22.
30. Bobe, G.; Hammond, E.G.; Freeman, A.E.; Lindberg, G.L.; Beitz, D.C. Texture of butter from cows with different milk fatty acid compositions. *J. Dairy Sci.* **2003**, *86*, 3122–3127. [CrossRef]
31. Glibowski, P.; Zarzycki, P.; Krzepkowska, M. The Rheological and Instrumental Textural Properties of Selected Table Fats. *Int. J. Food Prop* **2008**, *11*, 678–686. [CrossRef]
32. Ziarno, M.; Kozłowska, M.; Ratusz, K.; Hasalliu, R. Effect of the Addition of Selected Herbal Extracts on the Quality Characteristics of Flavored Cream and Butter. *Foods* **2023**, *12*, 471. [CrossRef]
33. Evers, J.M.; Crawford, R.A.; Kissling, R.C. Determination of moisture, solids–not–fat and fat–by–difference in butter using routine methods according to ISO 8851/IDF 191—An international collaborative study and a meta–analysis. *Int. Dairy J.* **2003**, *13*, 55–65. [CrossRef]
34. Dudkiewicz, A.; Hayes, W.; Onarinde, B. Sensory quality and shelf–life of locally produced British butters compared to large–scale, industrially produced butters. *Br. Food J.* **2022**, *124*, 3220–3235. [CrossRef]
35. Chudy, S.; Bilska, A.; Kowalski, R.; Teichert, J. Colour of milk and milk products in CIE L*a*b* space. *Med. Wet* **2020**, *76*, 77–81. [CrossRef]
36. *PN-EN ISO 660:2009*; Animal and Vegetable Fats and Oils—Determination of Acid Value and Acidity. International Organization for Standardization: Geneva, Switzerland, 2009.
37. *PN-EN ISO 3960:2017*; Animal and Vegetable Fats and Oils—Determination of Peroxide Value—Iodometric (visual) endpoint determination. International Organization for Standardization: Geneva, Switzerland, 2017.
38. *PN-EN ISO 3657:2013*; Animal and Vegetable Fats and Oils Determination of Saponification Value. International Organization for Standardization: Geneva, Switzerland, 2013.
39. Derewiaka, D.; Sosińska, E.; Obiedziński, M.; Krogulec, A.; Czaplicki, S. Determination of the adulteration of butter. *Eur. J. Lip. Sci. Technol* **2011**, *113*, 1005–1011. [CrossRef]

40. Krause, A.J.; Lopetcharat, K.; Drake, M.A. Identification of the characteristics that drive consumer liking of butter. *J. Dairy Sci.* **2007**, *90*, 2091–2102. [CrossRef]
41. Lai, O.M.; Ghazali, H.M.; France, Ch.; Chong, C.L. Physical properties of lipase catalyzed transesterified blends of palm stearin and anhydrous milk fat. *Food Chem.* **2000**, *70*, 221–225. [CrossRef]
42. Ramaswamy, N.; Baer, R.J.; Schingoethe, D.J.; Hippen, A.R.; Kasperson, K.M.; Whitlock, L.A. Composition and flavor of milk and butter from cows fed fish oil, extruded soybeans, or their combination. *J. Dairy Sci.* **2001**, *84*, 2144–2151. [CrossRef]
43. Bobe, G.; Freeman, A.E.; Beitz, D.C.; Lindberg, G.L. Estimation of heritabilities for amounts of individual proteins and fatty acids in milk of Holstein cattle. *J. Anim. Sci.* **1996**, *74* (Suppl. S1), 26.
44. Queirós, M.S.; Grimaldi, R.; Gigante, M.L. Addition from milt fat positively affects the firmness of butter. *Food Res. Int.* **2016**, *84*, 69–75. [CrossRef]
45. Subroto, E.; Indiarto, R.T.; Marta, H.; Wulan, A.S. Physicochemical and sensorial properties of recombined butter produced from milk fat and fish oil blend. *Biosci. Res.* **2018**, *15*, 3720–3727.
46. Lapčíková, B.; Lapčík, L.; Valenta, T.; Kučerová, T. Functional and Quality Profile Evaluation of Butters, Spreadable Fats, and Shortenings Available from Czech Market. *Foods* **2022**, *11*, 3437. [CrossRef] [PubMed]
47. Kowalska, M.; Aljewicz, M.; Mroczek, E.; Cichosz, G. Palm oil-a cheaper alternative. *Brom. Chem. Toksykol* **2012**, *65*, 171–180. (In Polish)
48. Rønholt, S.; Kirkensgaard, J.J.K.; Mortensen, K.; Knudsen, J.C. Effect of cream cooling rate and water content on butter microstructure during four weeks of storage. *Food Hydrocol* **2014**, *34*, 169–176. [CrossRef]
49. Codex Alimentarius Commission—Procedural Manual Twenty Seventh Edition CXS 280-1973. FAO and WHO. 2019. Rome. 2019, p. 249. Available online: https://www.fao.org/3/ca2329en/ca2329en.pdf (accessed on 10 February 2023).
50. Regulation (EU) No 1308/2013 of the European Parliament and of the Council of 17 December 2013 establishing a common organisation of the markets in agricultural products and repealing Council Regulations (EEC) No 922/72, (EEC) No 234/79, (EC) No 1037/2001 and (EC) No 1234/2007. Available online: https://eur-lex.europa.eu/legal-content/EN/TXT/?uri=celex%3A32013R1308 (accessed on 10 February 2023).
51. Juriaanse, A.C.; Heertje, I. Microstructure of shortenings, margarine and butter e a review. *Food Microstr.* **1988**, *7*, 181–188.
52. Mortensen, B.K. Butter and Other Milk Fat Products. In *Encyclopedia of Dairy Sciences*, 2nd ed.; Elsevier Science and Technology: Oxford, UK, 2011; pp. 492–521.
53. Wassell, P. Bakery Fats. In *Fats in Food Technology*, 2nd ed.; Kanes, K.R., Ed.; John Wiley & Sons: Hoboken, NJ, USA, 2014; pp. 39–60.
54. Funahashia, H.; Horiuchi, J. Characteristics of the churning process in continuous butter manufacture and modelling using an artificial neural network. *Int. Dairy. J.* **2008**, *18*, 323–328. [CrossRef]
55. Staniewski, B. Selected aspects of butter quality standardisation. *Przegl Mlecz.* **2009**, *10*, 4–12. (In Polish)
56. Vereecken, J.; Foubert, I.; Meeussen, W.; Lesaffer, A.; Dewettinck, K. Fat structuring with partial acylglycerols: Effect on solid fat profiles. *Eur. J. Lip Sci. Technol.* **2009**, *111*, 259–272. [CrossRef]
57. van Dalen, G. Determination of the water droplet size distribution of fat spreads using confocal scanning laser microscopy. *J. Micr.* **2002**, *208*, 116–133. [CrossRef]
58. Van lent, K.; Vanlerberghe, B.; Van Oostveldt, P.; Thas, O.; Van der Meeren, P. Determination of water droplet size distribution in butter: Pulsed field gradient NMR in comparison with confocal scanning laser microscopy. *Int. Dairy J.* **2008**, *18*, 12–22. [CrossRef]
59. Czechowska-Liszka, M. Testing the Quality of Various Sorts of Breakfast Margarine. *Zesz Nauk. Akad Ekon Krak* **2002**, *6583*, 49–53. (In Polish)
60. Vanhoutte, B.; Dewettinck, K.; Foubert, I.; Vanlerberghe, B.; Huyghebaert, A. The effect of phospholipids and water on the isothermal crystallisation of milk fat. *Eur. J. Lip Sci. Technol* **2002**, *104*, 490–495. [CrossRef]
61. Ziarno, M.; Zaręba, D. Additives in the manufacture of butter and milkfat-based mixes. *Przem Spoz* **2008**, *10*, 24–28. (In Polish)
62. Mokrzycki, W.; Tatol, M. Color difference Delta E-A survey. *Mach. Graph. Vis.* **2011**, *20*, 383–411.
63. Regulation (EC) No 1333/2008 of the European Parliament and of the Council of 16 December 2008 on food additives. *Official Journal of the European Union L 354/16*, 16 December 2008; 16.
64. Bellinazo, P.L.; Reissig Soares Vitola, H.; Cruxen, C.E.; Kreutz Braun, C.L.; Hackbart, H.C.; Padilha da Silva, W.; Fiorentini, A.M. Probiotic butter: Viability of Lactobacillus casei strains and bixin antioxidant effect (*Bixa orellana* L.). *Food Proc. Pres.* **2019**, *43*, 1–9. [CrossRef]
65. Kashaninejad, M.; Razavi, S.; Tehrani, M.; Kashaninejad, M. Effect of extrusion conditions and storage temperature on texture, colour and acidity of butter. *Int. J. Dairy Technol.* **2017**, *70*, 102–109. [CrossRef]
66. Laikoja, K.; Teder, L.; Jõudu, I. Assessment of chemical and sensory quality of unsalted and salted sweet cream butter during storage at different temperatures and time. *Agraarteadus -J. Agric. Sci.* **2017**, *2*, 76–81. [CrossRef]
67. Khaskheli, G.B.; Khaskheli, A.A.; Magsi, A.S.; Barham, G.S.; Jamali, M.A.; Khaskheli, A.A. Study on Quality Characteristics of Sweet and Sour Cream Butter. *Proc. Pak. Acad. Sci. Life Environ. Sci.* **2020**, *57*, 71–80.
68. Gonçalves, M.F.D.; Baggio, S.R. Evaluation of quality of butter from different provenance. *Food Sci. Technol.* **2012**, *32*, 629–635. [CrossRef]
69. Kahyaoğlu, D.T.; Çakmakçı, S. Determination of the adulteration of butter with margarine by using fat constants. *J. Agric. Sci.* **2016**, *22*, 1–8. [CrossRef]

70. Kahyaoğlu, D.T.; Çakmakçı, S. A comparative study on some properties and oxidation stability during storage of butter produced from different animals' milk. *J. Food* **2018**, *43*, 283–293.
71. Rutkowska, J.; Adamska, A. Fatty Acid Composition of Butter Originated from North-Eastern Region of Poland. *Pol. J. Food Nutr. Sci.* **2011**, *61*, 187–193. [CrossRef]
72. Rutkowska, J.; Adamska, A.; Białek, M. Fatty acid profile of milk cows reared in the mountain region of Poland. *J. Dairy Res.* **2012**, *79*, 469–476. [CrossRef] [PubMed]
73. Mills, S.; Ross, R.P.; Hill, C.C.; Fitzgerald, G.F.; Stanton, C. Milk intelligence: Mining milk for bioactive substances associated with human health. *Int. Dairy J.* **2011**, *21*, 377–401. [CrossRef]
74. Blaško, J.; Kubiniec, R.; Górova, R.; Fabry, I.; Lorenz, W.; Sojak, L. Fatty acid composition of summer and winter cows' milk and butter. *J. Food Nutr. Res.* **2010**, *49*, 169–177.
75. Rutkowska, J.; Sinkiewicz, I.; Adamska, A. Fatty acid profile of milk from cows fed a TMR system. *Żywn Nauka Technol. Jakość* **2012**, *5*, 135–144. (In Polish)
76. Brunner, J.B. Physical equilibria in milk: The lipid phase. In *Fundamentals of Dairy Chemistry*, 2nd ed.; Webb, B.H., Johnson, A.H., Alford, J.A., Eds.; AVI: Westport, CT, USA, 1974; pp. 447–602.
77. Jaeck, W.; Pabst, K. Field investigations on feeding of cows and milk fat quality. *Kiel. Milch. Fors.* **1990**, *42*, 281–296.
78. Renner, E.; Kosmack, U. Genetic aspects concerning fatty acid composition of milk fat. 2. Fatty acid pattern of milk from progeny groups. *Zuechtungskunde* **1974**, *46*, 217–226.
79. Karijord, Ø.; Standal, N.; Syrstad, O. Sources of variation in composition of milk fat. *Z. Tierz Zuechtungsbiol.* **1982**, *99*, 81–93. [CrossRef]

applied sciences

MDPI

Article

Kefir Enriched with Encapsulated Volatile Oils: Investigation of Antimicrobial Activity and Chemical Composition

Mihaela Adriana Tiţa [1], Maria Adelina Constantinescu [1,*], Tiberius Ilie Opruţa [1], Cristina Bătuşaru [2], Lăcrămioara Rusu [3,*] and Ovidiu Tiţa [1,*]

1 Department of Agricultural Sciences and Food Engineering, Lucian Blaga University of Sibiu, Doctor Ion Raţiu No.7, 550012 Sibiu, Romania
2 Academy of Land Forces "Nicolae Bălcescu" of Sibiu, Revoluţiei No. 3–5, 550170 Sibiu, Romania
3 Department of Chemical Engineering and Food, Vasile Alecsandri University of Bacău, 600115 Bacău, Romania
* Correspondence: adelina.constantinescu@ulbsibiu.ro (M.A.C.); lacraistrati04@yahoo.com (L.R.); ovidiu.tita@ulbsibiu.ro (O.T.)

Abstract: The present study was designed to determine the changes in the chemical composition of kefir enriched with encapsulated volatile oils by enzymatic methods and the antimicrobial activity of volatile oils. Using encapsulated volatile oils (fennel, mint, and lavender) and cow's milk, we created three different forms of kefir. To highlight the antiseptic capacity of the volatile oils, we performed their antimicrobial analysis on three types of molds (*Geotrichum candidum*, *Penicillium expansum*, *Aspergillus niger*) and one Gram-negative bacterium (*Escherichia coli*). The technique used to determine antimicrobial activity was Kirby–Bauer. The changes in the chemical composition of kefir samples with encapsulated volatile oils were analyzed using enzymatic methods and were compared with a control sample of kefir. The main analyzed chemical compounds were lactose, D-glucose, D-galactose, acetic acid, ethanol, L-lactic acid, and L-glutamic acid. The kefir samples enriched with volatile oils obtained superior results compared to the control sample during the whole analysis period. The main advantage of using encapsulation is that the bioactive compounds of the volatile oils are gradually released in the kefir sample due to the protection provided by sodium alginate. As a result, products with high nutritional values were obtained that are beneficial to the consumer's health and have a longer shelf life thanks to the volatile oils' antimicrobial properties.

Keywords: kefir; volatile oil; antimicrobial; enzymatic analysis; bioactive compounds

Citation: Tiţa, M.A.; Constantinescu, M.A.; Opruţa, T.I.; Bătuşaru, C.; Rusu, L.; Tiţa, O. Kefir Enriched with Encapsulated Volatile Oils: Investigation of Antimicrobial Activity and Chemical Composition. *Appl. Sci.* **2023**, *13*, 2993. https://doi.org/10.3390/app13052993

Academic Editor: Antonio Valero

Received: 29 January 2023
Revised: 12 February 2023
Accepted: 24 February 2023
Published: 26 February 2023

1. Introduction

Dairy products contain a large number of proteins and are an important source of exogenous amino acids that influence metabolic processes and bring important benefits to the consumer's health [1]. There is an increased interest in producing fermented milk drinks with nutritional properties which benefit consumers' health [2]. Kefir has an acid-alcoholic taste [3] produced as a result of acid and alcoholic fermentation [4,5]. The fermentation process plays an important role because the volume of calcium, vitamins, and amino acids increases [6,7]. Following the studies carried out, it was concluded that kefir contains numerous substances beneficial to health that have antioxidant, antimicrobial, and anti-inflammatory properties [8–12]. In addition, kefir consumption lowers the risk of cardiovascular disease by attenuating the increase in protein C [13].

Current research is mainly based on functional foods that contain plant derivatives with antioxidant, antimicrobial, anticancer, and anti-inflammatory properties [14,15].

Traditional uses for lavender (*Lavandula angustifolia* Mill.) include culinary use, treatment of burns and skin wounds, and relief from headaches and digestive problems [16]. Moreover, lavender has antioxidant, antimicrobial, antispasmodic, antidepressant, analgesic, and carminative properties for various diseases [17,18]. Numerous antibacterial, antioxidant, and insecticidal activities are present in lavender volatile oil [16,19]. The most important constituents of the volatile lavender oil are 1,5-Dimethyl-1-vinyl-4-hexenyl butyrate [17], camphor, cineole, terpineol, and borneol [20]. The antimicrobial properties of lavender and lavender volatile oil have been studied by various researchers in recent years. The capacity of volatile lavender oil to prevent *Listeria innocua* (Gram-positive) was shown in the study by Marn et al. in 2016 [21]. In 2021, Benbrahim et al. showed that volatile lavender oil has antibacterial properties against antibiotic-resistant strains of *Klebsiella pneumoniae* without cytotoxicity on human lymphocytes [22].

Mint (*Mentha piperita* L.) is used in gastronomy but also in the pharmaceutical and cosmetic industry [23,24]. The volatile mint oil contains phenols and flavonoids, which have antioxidant and antibacterial effects. It is typically used to treat a variety of digestive illnesses, including diarrhea, intestinal inflammation, and nervous system issues [24,25]. Numerous studies have shown that volatile mint oil can combat a wide range of germs, including *Staphylococcus epidermidis, Escherichia coli, Staphylococcus aureus, Enterococcus faecalis, Bacillus cereus,* and *Cronobacter sakazakii* [23]. In 2013, Tsai et al. showed that volatile mint oil contains menthol, alcohol, and terpenes, these compounds giving it an antimicrobial character against *Escherichia coli, Pseudomonas aeruginosa,* and *Staphylococcus aureus* [26]. In 2019, Park et al. established the antimicrobial propriety of volatile mint oil due to its high menthol content [27]. According to Kapp et al. in 2020, the antibacterial action of mint flavoring in sweets and food supplements against *Staphylococcus aureus* and *Escherichia coli* may also help to avoid bacterial illnesses and lessen food contamination [28].

Fennel (*Foeniculum vulgare* L.) is an aromatic and therapeutic plant [29] with anti-inflammatory, analgesic, diuretic, and carminative properties [30–32]. In recent years, the real interest has begun to be given to volatile fennel oil obtained from its seeds [31]. Volatile fennel oil has a hepatoprotective, anti-inflammatory [30], antioxidant, antimicrobial [33], and antifungal character [29]. The flavonoids, phenolic compounds [32], phenylpropanoid derivatives, and monoterpenoids [30], as well as trans-anethole, fenchone, estragole, and a-phellandrene, are the most significant components of volatile fennel oil [34]. Numerous research has shown that volatile fennel oil has antibacterial properties. Anwar et al. in 2009 used the disk diffusion method to show that volatile fennel oil has antibacterial properties against *Bacillus subtilis, Aspergillus niger,* and *Escherichia coli* [31]. In 2019, using the same method, Marín et al. demonstrated the antimicrobial character of this oil on *Listeria innocua* (Gram-positive) and *Pseudomonas fluorescens* [21]. Other studies carried out in 2011, 2018, and 2021 demonstrated the antimicrobial character by associating this with the components of volatile fennel oil [33,35,36].

The purpose of our research is to create functional foods that bring nutritional benefits to the consumer by incorporating bioactive components. In 2019, we started studying dairy products enriched with different volatile oils [37]. This research has been certified by articles, and the positive results obtained have led us to analyze as many compounds in these products as possible. For the present study, we have chosen to continue the studies carried out for kefir enriched with encapsulated volatile oils. The sensory, textural acceptability, and antioxidant capacity of the products were studied, and the results were published in 2022 [38,39].

In this context, the present study was designed to determine the changes in the chemical composition of kefir enriched with encapsulated volatile oils by enzymatic methods and the antimicrobial activity of volatile oils.

Enzyme kits from the R-Biopharm company were used. Enzyme analysis is an excellent way to identify numerous compounds in dairy products because the test kits include selected, multi-tested reagents. For the current study, we decided to use these test kits because they have many advantages, the most important being the high degree of safety. These kits are recommended by the IDF (International Dairy Federation) [40]. The analyzed compounds were lactose, D-glucose, D-galactose, acetic acid, ethanol, L-lactic acid, and L-glutamic acid. The use of volatile oils with antimicrobial capacity aims to increase the finished product's shelf life. So, using encapsulated volatile oils and cow's milk, we created three different forms of kefir. Lavender, mint, and fennel were the volatile oils that were utilized.

The antimicrobial capacity of the volatile oils used was tested on several types of microorganisms that poison food products, especially dairy products. Three types of molds (*Geotrichum candidum*, *Penicillium expansum*, *Aspergillus niger*) and one Gram-negative bacterium (*Escherichia coli*) were selected.

Geotrichum candidum, the best-known species in the genus, is a yeast-like fungus tolerant of acidic environments. Its taxonomic position was revised in 2004 by De Hoog and Smith, the genus *Geotrichum* being composed of 22 species [41]. *Geotrichum candidum* thrives at temperatures between 5 and 38 °C, with an optimum temperature of 25 °C. It is tolerant to a wide range of pH values, with optimum values between 5.0 and 5.5. It has a low salt tolerance. Unlike yeasts, the growth of *G. candidum* is limited by the amount of salt. Concentrations of 1% NaCl lead to slight growth suppression. NaCl concentrations of 5–6% have an inhibitory effect [42].

Blue mold disease caused by *Penicillium expansum* is the most economically important post-harvest disease of stored fruit and vegetables. In addition to causing food spoilage, some strains of the fungus produce patulin mycotoxin [43]. Patulin is a moderately toxic mycotoxin, but according to studies, it has no carcinogenic or mutagenic effect. *Penicillium* species have a highly evolved physiology, resulting in adaptation to a very wide range of habitats. All *Penicillium* species can grow in food at a low pH of between 2 and 3. Most *Penicillium* species grow at lower temperatures, and none are thermophilic [44].

Aspergillus niger is a type of mold that causes "black mold" on the outside of certain foods, such as apricots, onions, and grapes [45]. *Aspergillus niger* is the most common species of *Aspergillus* and is more prevalent in warmer climates, both in field situations and in stored food. Black spores protect against sunlight and UV radiation, giving a competitive advantage in such habitats. *A. niger* is very commonly isolated from sun-dried products such as grapes, where it can produce ochratoxin A [46]. High temperatures and high relative humidity favor the growth of these organisms. The most favorable temperature is 25–40 °C. At temperatures below 15 °C, there is no significant growth, and under refrigerated conditions, there is no growth at all [47].

Escherichia coli is a member of the family *Enterobacteriaceae*, which are facultatively anaerobic Gram-negative rods (possessing both fermentative and respiratory metabolism) and do not produce the enzyme oxidase [48]. *Escherichia coli* are introduced into water bacteriology because it is a useful marker of fecal pollution and thus has become an important marker in food and water hygiene [49]. *Escherichia coli* is a bacterium that is carried in the gastrointestinal tract of humans and dairy animals. Milk collected for human consumption can become contaminated with *E. coli* directly through animal feces or indirectly through media, equipment, and workers [50]. The technique used to determine antimicrobial activity was Kirby–Bauer.

Due to the oils' sensitivity to environmental and technological conditions, the technique of their encapsulation was chosen. To compare the obtained results, we made a control sample of kefir without the addition of volatile oils. The analyses were completed on the first, tenth, and twenty-first days that the samples were stored. All results were statistically processed and correlated using the Pearson correlation.

2. Materials and Methods

2.1. Determination of Antimicrobial Activity for Volatile Oils

The Kirby–Bauer method was employed to ascertain the antibacterial activity. The culture medium used to activate the molds *Geotrichum candidum*, *Penicillium expansum*, and *Aspergillus niger* is DRBC Agar (Dichloran Rose Bengal Chloramphenicol Agar). A total of 31.5 g was used to prepare one liter of culture medium, and this was autoclaved for 20 min at 120 °C. After autoclaving, the medium was allowed to cool to 30 °C and poured into the test tube for activation of the molds [51]. The culture medium used for the activation of *Escherichia coli* bacteria is Blood Agar. All strains used were obtained from the culture collection of the microbiology laboratory of the Faculty of Agricultural Sciences, Food Industry and Environmental Protection, Lucian Blaga University of Sibiu. The method involves the use of specific media that are poured hot into Petri dishes. After solidification of the culture medium on the plates, the suspension of microorganisms is spread in a thin layer with a sterile buffer, which must have a concentration of 0.5 on the MacFarland scale. With sterile medical tweezers, discs impregnated with the extract under study are placed on the plate, the discs having a size of 6 mm. The amounts of extract used for impregnation were 10 µg/L. It is recommended to place the discs at a minimum distance of 30 mm between them and 15 mm from the edges of the Petri dish. Incubation was performed in a Memmert thermostat at 37 °C for 24 and 48 h. The reading was performed with a ruler measuring the resulting zone of inhibition, including the diameter of the disc. The results are expressed by giving a qualification such as sensitive, intermediate, or resistant to the studied compound or extract, according to Table 1. For each type of microorganism, two Petri dishes were made in which the three types of volatile oils were impregnated [51].

Table 1. Definition of the qualifiers sensitive, intermediate, and resistant according to the diameter of the inhibition zone [52–54].

Qualifiers	Diameter of Inhibition Zone [mm]
Sensitive	≥20
Intermediate	15–19
Resistant	≤14

The volatile oils were extracted from lavender, fennel, and mint. The plants originated from authorized crops in Sibiu county, Romania. The extraction method used was applied and described in previous research in 2019 and 2022 [37–39]. Volatile oils are very sensitive to external factors, so we wanted to protect them by encapsulation. The encapsulation method and technique were described in 2022 research by Tița et al. [39].

2.2. *Enzymatic Analyses*

The method of obtaining kefir samples with added volatile oils was identical to that presented in the 2022 research by Tița et al. [38,39].

Enzyme kits from the R-Biopharm company (Darmstadt, Germany) were used. These kits were purchased from the authorized distributor in Romania, Diamedix Impex SA. The test kits from "Boehringer Mannheim Enzymatic BioAnalysis and Food Analysis" are based on high-quality enzymes, allowing precise measurement of each compound. Enzyme analyses are recommended by the IDF, and the spectrophotometric method is used to perform them [40].

2.2.1. Determination of Lactose and D-Galactose Content

The method and sample preparation was carried out according to the method of the R-Biopharm Cat. No. 10 176 303 035 and IDF 79B (1991) and ISO 5765-2 (1999). Using a STAT FAX Model 1904 spectrophotometer, calculate the absorbance at 340 nm for each sample [55].

2.2.2. Determination of D-Glucose Content

The method and sample preparation was carried out according to the method of the R-Biopharm Cat. No. 10 986 119 035 and IDF 79B (1991) and ISO 5765-1 (1999). Using a STAT FAX Model 1904 spectrophotometer, calculate the absorbance at 340 nm for each sample [56].

2.2.3. Determination of L-Glutamic Acid Content

The method and sample preparation was carried out according to the method of the R-Biopharm Cat. No. 10 139 092 035 and ISO 4134, D, CH. Using a STAT FAX Model 1904 spectrophotometer, calculate the absorbance at 340 nm for each sample [57].

2.2.4. Determination of Acetic Acid Content

The method and sample preparation was carried out according to the method of the R-Biopharm Cat. No. 10 148 261 035 and IFU and MEBAK. Using a STAT FAX Model 1904 spectrophotometer, calculate the absorbance at 340 nm for each sample [58].

2.2.5. Determination of Ethanol Content

The method and sample preparation was carried out according to the method of the R-Biopharm Cat. No. 10 176 290 035 and D, CH, Au, F. Using a STAT FAX Model 1904 spectrophotometer, calculate the absorbance at 340 nm for each sample [59].

2.2.6. Determination of L-Lactic Acid Content

The method and sample preparation was carried out according to the method of the R-Biopharm Cat. No. 10 139 084 035 and ISO 69B (1987) and ISO 9069 (1986). Using a STAT FAX Model 1904 spectrophotometer, calculate the absorbance at 340 nm for each sample [60].

2.3. Statistical Analysis

The following statistical measures were applied to evaluate the antibacterial activity of volatile oils: mean, standard error of mean, median and standard deviation. The following statistical indicators were used to express all the data from the enzymatic analyses: mean, standard error of mean, median, standard deviation, maximum, minimum, and skewness. Correlation analysis was used to examine the statistical significance of data collected using Pearson's correlation [61]. Minitab version 14 was used to conduct all statistical analyses.

3. Results

3.1. Determination of Antimicrobial Activity

The reading was performed 24 h and 48 h after incubating. A ruler was used to take the reading while also measuring the disc's diameter and the resulting zone of inhibition. Two Petri dishes with the three different volatile oil were produced, one for each species of microorganism.

Figure 1 shows the Petri dishes with the analyzed samples. The three types of volatile oils were numbered: A, B, and C. A represents the lavender volatile oil sample, B represents the mint volatile oil sample, and C represents the fennel volatile oil sample. The Petri dishes were seeded with the four microorganisms, and their numbering was as follows: 1 represents the dish seeded with *Geotrichum candidum*, 2 represents the dish seeded with *Penicillium expansum*, 3 represents the dish seeded with *Aspergillus niger*, and 4 represents the dish seeded with *Escherichia coli*.

In terms of the inhibitory action against *Penicillium expansum* for volatile fennel oil, the mean of the zone of inhibition's diameter is 11.5 ± 0.707 mm after 24 h, and after 48 h, it is 17 ± 0.707 mm, yielding an intermediate qualification. After the first 24 h, the inhibitory zone's mean diameter in the case of volatile lavender oil is 11 ± 0 mm, yielding a high qualification. After 48 h, the mean is 14.5 ± 0.707 mm, the qualification being intermediate. In the first 24 h, the volatile mint oil has a mean diameter of the inhibition zone of 24.5 ± 0.707 mm, the qualification obtained being sensitive. After 48 h, the mean is 27 ± 0 mm; the qualification is still sensitive. The highest sensitivity of *Penicillium expansum* is to the volatile mint oil, being resistant or intermediate to the other two.

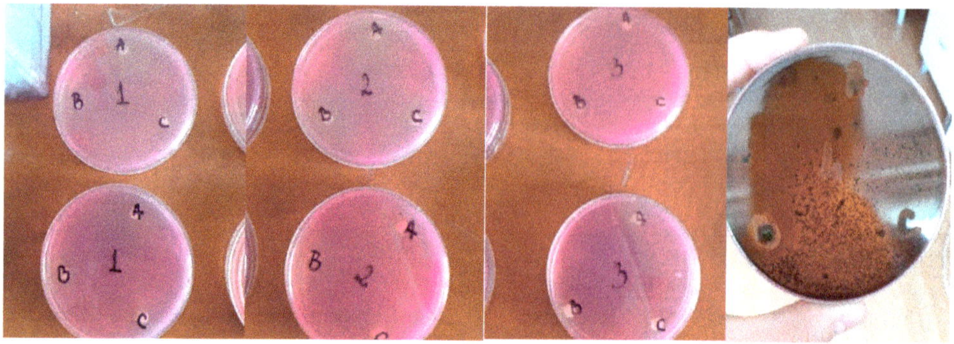

Figure 1. Zone of inhibition after 48 h: 1. *Geotrichum candidum*; 2. *Penicillium expansum*; 3. *Aspergillus niger*; 4. *Escherichia coli*; A. Lavender volatile oil; B. Mint volatile oil; C. Fennel volatile oil.

Figure 2 shows the inhibitory action of fennel, lavender, and mint volatile oils against *Penicillium expansum*, *Aspergillus niger*, *Geotrichum candidum*, and *Escherichia coli*.

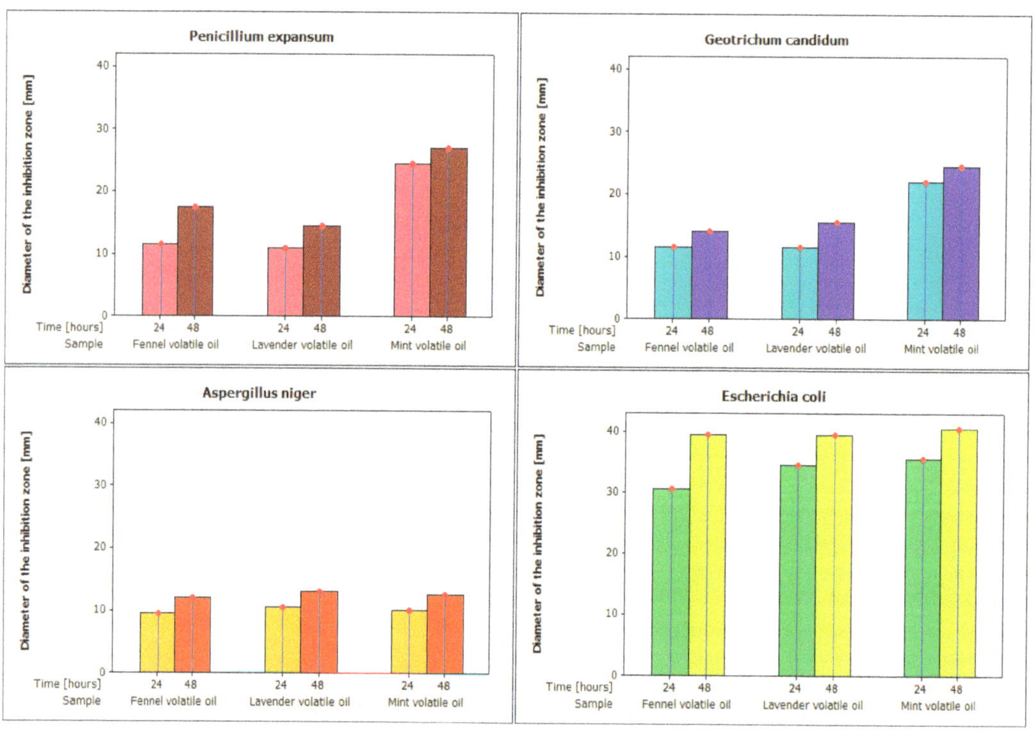

Figure 2. The inhibitory action of fennel, lavender, and mint volatile oils against *Penicillium expansum*, *Aspergillus niger*, *Geotrichum candidum*, and *Escherichia coli*.

Regarding the fennel volatile oil's ability to inhibit *Aspergillus niger*, the inhibition zone's average diameter over the first 24 h was 9.5 ± 0.707 mm, resulting in the qualification of resistance. After 48 h, the inhibitory zone's mean diameter is 12 ± 0 mm; resistance is therefore qualified. After 24 h, the mean for the volatile oil of lavender is 10.5 ± 0.707 mm, indicating that it is resistant. After 48 h, the mean is 13 ± 0 mm, the qualification obtained being resistant. The volatile mint oil has a mean diameter of the inhibition zone of 10 ± 0 mm in the first 24 h, with the qualifier being resistant. After 48 h, the mean is 12.5 ± 0.707 mm, the qualification obtained being resistant. The inhibitory action increases between 24 and 48 h for all three volatile oils. All three oils indicate *Aspergillus niger* resistance, with volatile lavender oil producing the greatest results and fennel volatile oil producing the worst.

After 24 h, the fennel volatile oil's inhibition activity against *Geotrichum candidum* has a mean diameter of the inhibition zone of 11.5 ± 0.707 mm, falling into the category of being resistant. After 48 h, the mean is 14 ± 0 mm; the qualification is still resistant. The average diameter of the zone of inhibition of volatile lavender oil after 24 h is 11.5 ± 0.707 mm, with the descriptor "resistant." After 48 h, the mean is 15.5 ± 0.707 mm, the qualification to which it falls being intermediate. For volatile mint oil, the mean after 24 h is 22 ± 0 mm, the qualifier being sensitive. After 48 h, the mean is 24.5 ± 0.707 mm, the qualification is still sensitive. The inhibitory action increases between 24 and 48 h, the highest increase with volatile lavender oil. Among all three types of oils, *Geotrichum candidum* is only sensitive to mint volatile oil after both 24 and 48 h. Fennel volatile oil has the strongest resistance, and volatile lavender oil becomes resistant after 24 h and intermediate after 48 h.

After 24 h, the mean width of the inhibition zone for the fennel volatile oil's inhibition action against *Escherichia coli* is 30.5 ± 0.707 mm, with the qualifier being sensitive. The mean is 39.5 ± 0.707 mm after 48 h, with the qualifying being sensitive. After 24 h, the inhibition zone's mean diameter for volatile lavender oil is 34.5 ± 0.707 mm, with the descriptor "sensitive." The mean is 39.5 ± 0.707 mm after 48 h, with the qualifying being sensitive. After 24 h, the mean for the volatile oil of mint is 35.5 ± 0.707 mm, with the descriptor "sensitive". The mean after 48 h is 40.5 ± 0.707 mm, with sensitive being the qualifier. The inhibitory action of all three varieties of volatile oils increases between 24 and 48 h. *Escherichia coli* show susceptibility to all three types of volatile oils after 24 and 48 h. The volatile oils of mint and fennel caused the greatest and the least amount of sensitivity, respectively.

3.2. Determination of Lactose Content

Table A1 (Appendix A) presents the results of determining the lactose content for each kefir sample, as well as the statistical processing of the obtained data. For each kefir sample, three replicates were performed on each day of analysis.

Figure 3 compares the variations in lactose concentration between the control sample and the cow's milk kefir samples that have been enhanced with encapsulated volatile oils.

Regarding the control sample, the first day of storage produced the highest lactose content value, with a mean of 4.579 ± 0.0006 g/100 g. On day 20, the lowest value of the lactose content was recorded, the mean being 3.682 ± 0.0015 g/100 g. The lactose content in the kefir sample with volatile lavender oil increased on the first day of storage, with a mean of 4.683 ± 0.0017 g/100 g. The 20th day of storage saw the lowest lactose content value, with a mean of 4.581 ± 0.0015 g/100 g. The kefir with volatile mint oil had a mean content of 4.664 ± 0.0012 g/100 g, with the highest value being recorded on the first day of storage. The mean lactose content was 4.567 ± 0.0012 g/100 g on the 20th day of storage, the lowest value recorded. The kefir sample with volatile fennel oil had the greatest lactose concentration on the first day of storage, with a mean of 4.607 ± 0.001 g/100 g. The mean lactose content was 4.506 ± 0.0012 g/100 g on the 20th day of storage, the lowest number recorded.

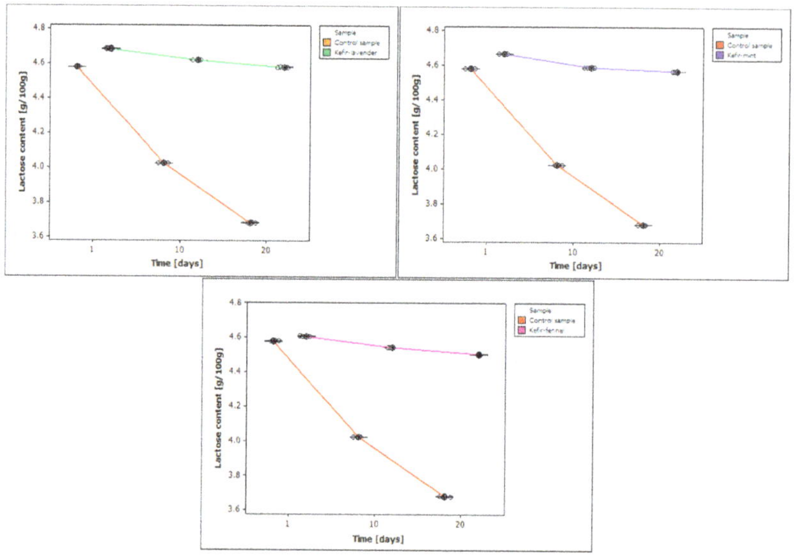

Figure 3. Comparative variation of lactose content determined by the enzymatic method in kefir samples with encapsulated volatile oils and the control sample.

3.3. Determination of D-Galactose Content

Table A2 (Appendix A) presents the results of determining the D-galactose content for each kefir sample, as well as the statistical processing of the obtained data. For each kefir sample, three replicates were performed on each day of analysis.

Figure 4 compares the variance in D-galactose content between the control sample and the samples of cow's milk kefir that has been enhanced with encapsulated volatile oils.

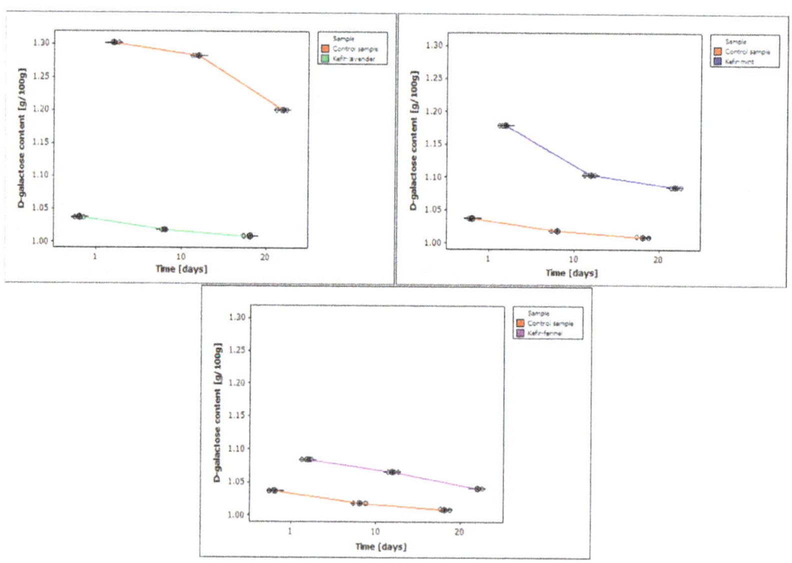

Figure 4. Comparative variation of D-galactose content determined by the enzymatic method in kefir samples with encapsulated volatile oils and the control sample.

On the first day, the D-galactose level in the control sample was greatest, with a mean of 1.0378 ± 0.0001 g/100 g. The 20th day had the lowest D-galactose level, with a mean of 1.0098 ± 0.0002 g/100 g. With a mean of 1.3021 ± 0.0001 g/100 g, the kefir sample with volatile lavender oil had the greatest D-galactose level on the first day. On the twentieth day, the mean amount of D-galactose was 1.2014 ± 0.0003 g/100 g. The D-galactose content in the kefir sample with volatile mint oil was highest on the first day of storage, and the mean value was 1.1793 ± 0.0001 g/100 g. On day 20 of storage, 1.0856 ± 0.0002 g/100 g was recorded as the lowest D-galactose concentration value. The mean of the D-galactose content from the first day is 1.0848 ± 0.0001 g/100 g for the kefir with volatile fennel oil. The mean D-galactose content is 1.0662 ± 0.0002 g/100 g on the 10th day of storage and 1.0415 ± 0.0001 g/100 g on the 20th day.

3.4. Determination of D-Glucose Content

Table A3 (Appendix A) presents the results of determining the D-glucose content for each kefir sample, as well as the statistical processing of the obtained data. For each kefir sample, three replicates were performed on each day of analysis.

Figure 5 compares the variations in D-glucose content between the control sample and the kefir samples that have been enhanced with encapsulated volatile oils.

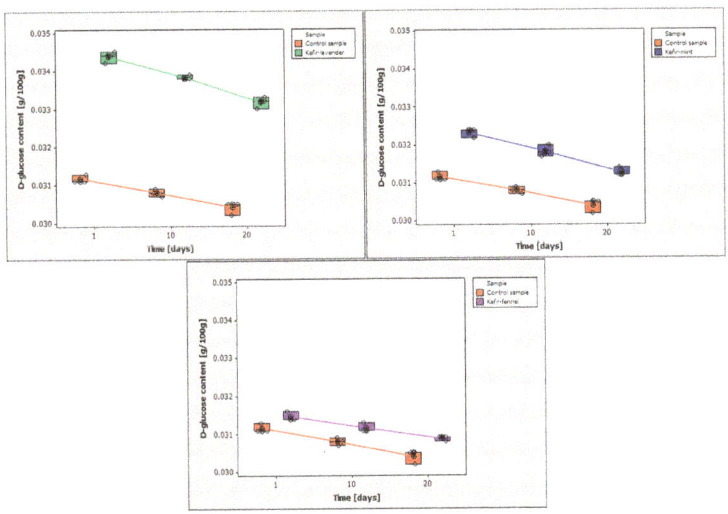

Figure 5. Comparative variation of D-glucose content determined by the enzymatic method in kefir samples with encapsulated volatile oils and the control sample.

All of the kefir samples see a drop in D-glucose content during the 20-day storage period. On the first day of storage, the control sample has a D-glucose level of 0.0312 ± 0.0001 g/100 g. The mean is 0.0308 ± 0.0001 g/100 g on day 10 and 0.0304 ± 0.0002 g/100 g on day 20 of storage. The mean D-glucose concentration for the kefir with volatile lavender oil on the first day of storage is 0.0344 ± 0.0002 g/100 g. The mean D-glucose content is 0.0338 ± 0.0001 g/100 g on day 10 of storage and 0.0332 ± 0.0002 g/100 g on day 20. The mean D-glucose concentration for the kefir with volatile oil of mint is 0.0323 ± 0.0001 g/100 g on the first day of storage. The mean is 0.0318 ± 0.0002 g/100 g on day 10 and 0.0313 ± 0.0001 g/100 g on day 20 of storage. The average amount of D-glucose in the kefir sample containing volatile fennel oil on the first day of storage is 0.0315 ± 0.0001 g/100 g. The mean D-glucose content is 0.0312 ± 0.0001 g/100 g on day 10 and 0.0309 ± 0.0001 g/100 g on day 20 of storage.

3.5. Determination of L-Glutamic Acid Content

Table A4 (Appendix A) presents the results of determining the L-glutamic acid content for each kefir sample, as well as the statistical processing of the obtained data. For each kefir sample, three replicates were performed on each day of analysis.

Figure 6 compares the variations in L-glutamic acid content between the control sample and the samples of cow's milk kefir that have been enhanced with encapsulated volatile oils.

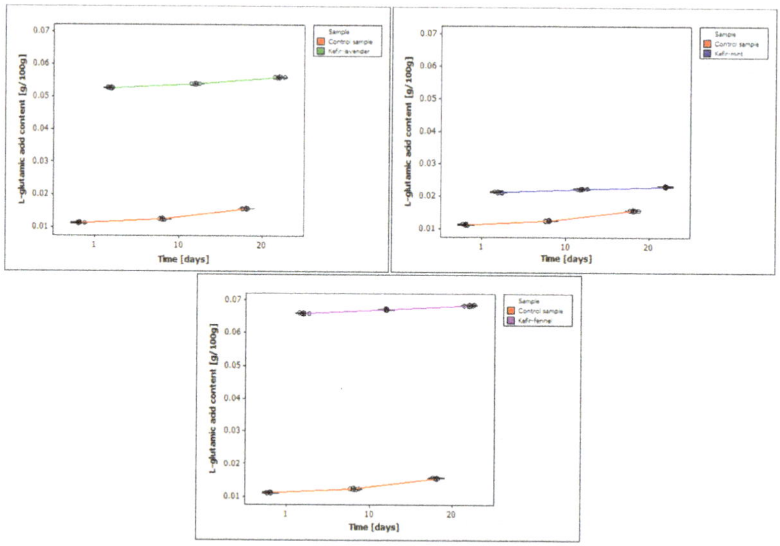

Figure 6. Comparative variation of L-glutamic acid content determined by the enzymatic method in kefir samples with encapsulated volatile oils and the control sample.

All of the kefir samples see an increase in L-glutamic acid content during the entire storage period. On the first day of storage, the control sample's mean level of L-glutamic acid was 0.0113 ± 0.0002 g/100 g. The mean L-glutamic acid concentration is 0.0125 ± 0.0001 g/100 g on day 10 of storage and 0.0157 ± 0.0001 g/100 g on day 20. The mean L-glutamic acid level of the kefir with volatile lavender oil on the first day of storage is 0.0526 ± 0.0001 g/100 g. The mean is 0.0538 ± 0.0001 g/100 g on day 10 and 0.0559 ± 0.0001 g/100 g on day 20 of storage. The mean L-glutamic acid level in the kefir sample containing the volatile oil of mint is 0.0213 ± 0.0002 g/100 g on the first day of storage. The mean is 0.0224 ± 0.0002 g/100 g on day 10 and 0.0231 ± 0.0002 g/100 g on day 20 of storage. The mean L-glutamic acid level on the first day of storage is 0.0661 ± 0.0001 g/100 g for the kefir with volatile fennel oil. The mean is 0.0672 ± 0.0002 g/100 g on the 10th day of storage, and it is 0.0686 ± 0.0001 g/100 g on the 20th day.

3.6. Determination of Acetic Acid Content

Table A5 (Appendix A) presents the results of determining the acetic acid content for each kefir sample, as well as the statistical processing of the obtained data. For each kefir sample, three replicates were performed on each day of analysis.

Figure 7 compares the variations in acetic acid levels between the control sample and the kefir samples that have been enhanced with encapsulated volatile oils.

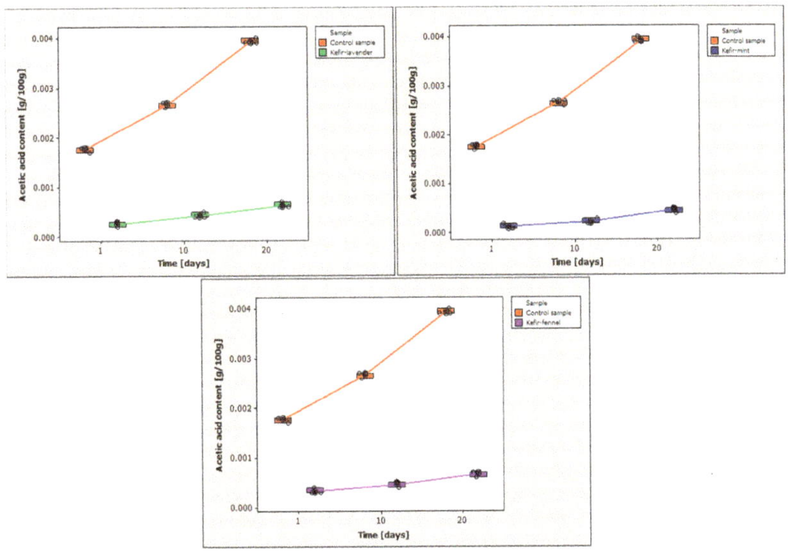

Figure 7. Comparative variation of acetic acid content determined by the enzymatic method in kefir samples with encapsulated volatile oils and the control sample.

The acetic acid content of the kefir samples rises after 20 days of storage. The average amount of acetic acid in the control sample on the first day of storage was 0.0018 ± 0.0001 g/100 g. The mean acetic acid content is 0.0027 ± 0.0001 g/100 g on the 10th day of deposition and 0.0039 ± 0.0001 g/100 g on the 20th day.

The mean of the acetic acid content from the first day of storage for the kefir sample with volatile lavender oil is 0.0003 ± 0.0001 g/100 g. The mean acetic acid content is 0.0006 ± 0.0001 g/100 g on day 20 and 0.0004 ± 0.0001 g/100 g on day 10. The mean acetic acid content of the kefir sample containing the volatile oil of mint is 0.0001 ± 0.0001 g/100 g on the first day of storage. The average acetic acid concentration on day 20 of storage is 0.0005 ± 0.0001 g/100 g. The average amount of acetic acid in the kefir sample containing volatile fennel oil on the first day of storage is 0.0003 ± 0.0001 g/100 g. The mean acetic acid content is 0.0005 ± 0.0001 g/100 g on the 10th day of storage and 0.0007 ± 0.0001 g/100 g on the 20th day.

3.7. Determination of Ethanol Content

Table A6 (Appendix A) presents the results of determining the ethanol content for each kefir sample, as well as the statistical processing of the obtained data. For each kefir sample, three replicates were performed on each day of analysis.

Figure 8 compares the variations in ethanol content between the control sample and the cow's milk kefir samples that have been enhanced with encapsulated volatile oils.

All of the examined kefir samples have an increase in ethanol content over the period. The average amount of ethanol in the control sample on the first day of storage was 0.0005 ± 0.0001 g/100 g. The mean ethanol content is 0.0007 ± 0.0001 g/100 g on the 10th day of storage and 0.0010 ± 0.0001 g/100 g on the 20th day.

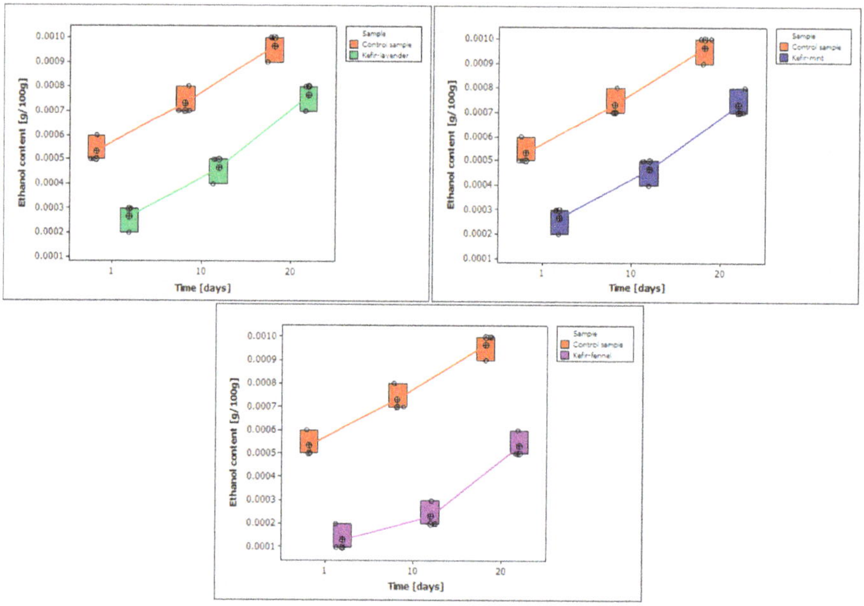

Figure 8. Comparative variation of ethanol content determined by the enzymatic method in kefir samples with encapsulated volatile oils and the control sample.

The mean ethanol concentration from the first day of storage for the sample of kefir with volatile lavender oil is 0.0003 ± 0.0001 g/100 g. The average ethanol content is 0.0005 ± 0.0001 g/100 g on the 10th day of storage and 0.0008 ± 0.0001 g/100 g on the 20th day. The mean of the ethanol concentration from the first day of storage is 0.0003 ± 0.0001 g/100 g in the case of the kefir with volatile mint oil. The ethanol content is 0.0005 ± 0.0001 g/100 g on day 10 and 0.0007 ± 0.0001 g/100 g on day 20 of storage. The mean ethanol concentration of the kefir sample with volatile fennel oil on the first day of storage is 0.0001 ± 0.0001 g/100 g. The mean ethanol content is 0.0002 ± 0.0001 g/100 g on the 10th day of storage and 0.0005 ± 0.0001 g/100 g on the 20th day.

3.8. Determination of L-Lactic Acid Content

Table A7 (Appendix A) presents the results of determining the ethanol content for each kefir sample, as well as the statistical processing of the obtained data. For each kefir sample, three replicates were performed on each day of analysis.

Figure 9 compares the variations in L-lactic acid content between the control sample and the samples of cow's milk kefir that have been enhanced with encapsulated volatile oils.

The amount of L-lactic acid increases during the 20-day storage period. The mean L-lactic acid content of the control sample on the first day of storage is 0.7014 ± 0.0001 g/100 g. L-lactic acid content is 0.7028 ± 0.0001 g/100 g on day 10 of storage and 0.7039 ± 0.0001 g/100 g on day 20. The mean of the L-lactic acid content from the first day of storage is 0.7021 ± 0.0001 g/100 g for the kefir sample containing volatile lavender oil. The mean L-lactic acid content is 0.7026 ± 0.0001 g/100 g on day 10 of storage and 0.7029 ± 0.0001 g/100 g on day 20.The mean L-lactic acid level from the first day of storage is 0.7007 ± 0.0001 g/100 g for the kefir sample containing volatile mint oil. The mean L-lactic acid content is 0.7009 ± 0.0001 g/100 g on the 10th day of storage and 0.7011 ± 0.0001 g/100 g on the 20th day. The mean L-lactic acid level of the sample containing volatile fennel oil on the first day of storage is 0.7019 ± 0.0002 g/100 g. The mean is 0.7023 ± 0.0002 g/100 g on day 10 and 0.7025 ± 0.0001 g/100 g on day 20.

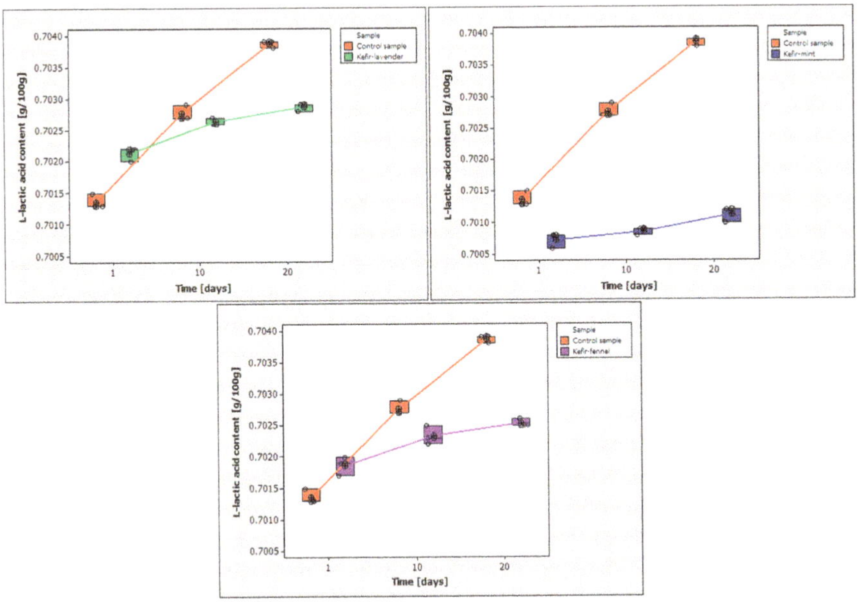

Figure 9. Comparative variation of L-lactic acid content determined by the enzymatic method in kefir samples with encapsulated volatile oils and the control sample.

4. Discussion

The many components in the plants used in this study contribute to their antibacterial properties. The most important compounds of fennel are trans-anethole and α-pinene [62]. Trans-anethole has strong antifungal activity by inhibiting the mycelial growth of a wide range of fungi and could be used as a preservative in food preparation and processing [63,64]. The tested antimicrobial potential of α-pinene is moderate [65]. The antimicrobial capacity of volatile fennel oil is also given by compounds such as β-myrcene, eucalyptol [66], γ-Terpinene [67,68], camphor [69], terpinen-4-ol [70], and linalool [71]. γ-Terpinene is a very potent antimicrobial agent against Gram-positive and Gram-negative bacteria [67], especially *Escherichia coli* and *Staphylococcus aureus* bacteria [68]. Lavender volatile oil compounds with antimicrobial character are α-pinene [65], eucalyptol [66], nerolidol [72], p-Cymene [73], β-caryophyllene, camphor [69], terpinen-4-ol [70], linalool [71], and linalyl acetate [74]. Studies have shown that nerolidol has potent antibacterial activity against *Staphylococcus aureus* [72]. The antimicrobial character of volatile mint oil is provided by compounds such as α-pinene [65], β-myrcene, eucalyptol [66], eugenol [75,76], p-Cymene [73], γ-Terpinene [67,68], β-caryophyllene [69], pulegone [77], menthone [78], sabinene [79], and linalool [71]. Eugenol has been reported to have high antibacterial activity [75], inhibiting the growth of several test microorganisms such as *Escherichia coli*, *Aeromonas hydrophila*, and *Salmonella typhi* [76].

The study aimed to produce a functional dairy product that brings nutritional benefits to the consumer's health. The use of volatile oils with antimicrobial capacity aims to increase the finished product's shelf life. The antimicrobial capacity of the volatile oils used was initially tested on several types of microorganisms affecting food products, especially dairy products.

Following the antimicrobial determination, volatile mint oil has the highest antibacterial activity against the chosen microorganisms, with *Aspergillus niger* being the only microorganism that showed resistance to it. The most sensitive bacteria to this kind of oil include *Escherichia coli*, *Penicillium expansum*, and *Geotrichum candidum*. *Escherichia coli* is quite susceptible to the antibacterial effects of lavender and fennel volatile oils. The antibacterial activity of volatile lavender oil is low to moderate against *Aspergillus niger* and intermediate against *Geotrichum candidum* and *Penicillium expansum*. The remaining chosen microorganisms are only weakly inhibited by the fennel volatile oil's antimicrobial properties.

Using encapsulated volatile oils and cow's milk, we created three different forms of kefir. We analyzed different chemical compounds in kefir using enzymatic methods. Enzyme kits from the R-Biopharm company were used. The analyzed compounds were lactose, D-glucose, D-galactose, acetic acid, ethanol, L-lactic acid, and L-glutamic acid.

The kefir with lavender oil has the highest lactose concentration, while the control sample has the lowest amount. The lactose level, as assessed by the enzymatic method, gradually declines during storage in the case of kefir samples enhanced with volatile oils, especially between days 10 and 20. The lactose concentration of the control sample is lower than that of the kefir samples with oils, and it decreases more quickly throughout storage.

The D-galactose content of the four kefir samples reduces throughout storage. The kefir sample with volatile lavender oil has the highest concentration of D-galactose, while the control sample has the lowest concentration. The loss of D-galactose happens more gradually in kefir samples that contain volatile oils, particularly between the first and tenth days of storage. Over the remaining storage term, the fall gradually quickens. In the case of the control sample, the decline speeds up during the 20-day storage period.

The kefir sample with volatile lavender oil has the highest D-glucose concentration on day 1 of storage, while the control sample has the lowest. On days 10 and 20 of storage, similar outcomes were attained. The D-glucose level of kefir samples with volatile oils falls gradually over the first 10 days of storage in contrast to the subsequent days when this drop significantly accelerates. For the control sample, the decline would happen more quickly during the storage period.

During the 20 days of storage, the L-glutamic acid concentration of the kefir samples rises. The fennel volatile oil kefir sample had higher glutamic acid content than the control sample, which had a lower level. About kefir samples containing volatile oils, the rise in L-glutamic acid concentration is gradual for the first 10 days of storage before increasing rapidly over the next 10. When compared to the control sample, the level of L-glutamic acid rises more quickly throughout storage.

All four samples of kefir had an increase in acetic acid levels due to storage. The control sample had the greatest acetic acid content throughout all three days of examination, whereas the kefir sample with volatile mint oil had the lowest. About kefir samples containing volatile oils, the increase in acetic acid level is slower during the first 10 days of storage than it is for the control sample, which has an accelerated increase throughout the storage period.

The four varieties of kefir all contain more ethanol after 20 days of storage. The kefir sample with volatile fennel oil has the lowest level of ethanol on day 1 of storage, and the control sample has the highest level. The control sample has the greatest ethanol content on days 10 and 20, while the kefir sample with volatile fennel oil has the lowest. Kefir samples containing volatile oils show a slower increase in ethanol level during the first 10 days of storage compared to the last 10 days of storage. The greatest levels of ethanol concentration were found in the control sample, and during the storage period, the growth accelerated at a faster rate.

The kefir sample with volatile lavender oil has the highest value of the L-lactic acid content on the first day of storage, whereas the kefir sample with volatile mint oil has the lowest value. The control sample has the highest L-lactic acid level on days 10 and 20, whereas the kefir sample with volatile mint oil has the lowest. When it comes to kefir samples that contain volatile oils, the increase in L-lactic acid content happens more slowly during the first 10 days of storage than it does during the final days. For the control sample, it happens more quickly, all during storage time.

5. Conclusions

This research aims to develop functional foods enriched with bioactive compounds. The food product that we decided to enrich with bioactive components is cow's milk kefir. Volatile oils extracted from lavender, mint, and fennel are excellent sources of nutritive compounds, as they have antimicrobial and antioxidant properties. Due to the fact that volatile oils are very sensitive to the action of external factors, they were encapsulated in a natural polymer matrix (sodium alginate). Finally, three types of kefir with volatile oil capsules were obtained, as well as a kefir control sample.

To highlight, on the one hand, the nutritional intake and the influence of bioactive compounds in the finished product and, on the other hand, the advantages of using encapsulation, the enzymatic analysis of several chemical compounds, were used.

The kefir samples enriched with volatile oils obtained superior results compared to the control sample during the whole analysis period.

Encapsulating the volatile oils in the sodium alginate and adding the capsules to kefir resulted in a value-added finished product.

The main advantage of using encapsulation is that the bioactive compounds of the volatile oils are gradually released in the kefir sample due to the protection provided by sodium alginate.

All these aspects show that the obtained products have a high nutritional value, bringing benefits to the health of the consumer. The completed product has a longer shelf life since volatile oils have antibacterial properties.

Author Contributions: Conceptualization, M.A.T., M.A.C. and O.T.; methodology, M.A.T., T.I.O., C.B. and L.R.; software, M.A.C.; validation, M.A.T., M.A.C. and L.R.; formal analysis, C.B. and L.R.; investigation, M.A.T. and M.A.C.; resources, O.T.; data curation, M.A.T., M.A.C. and O.T.; writing—original draft preparation, M.A.C. and O.T.; writing—review and editing, M.A.T., T.I.O. and C.B.; visualization, M.A.T., O.T. and L.R.; supervision, M.A.T.; project administration, M.A.T.; funding acquisition, M.A.T. All authors have read and agreed to the published version of the manuscript.

Funding: This research was funded by Lucian Blaga University of Sibiu and Hasso Plattner Foundation, grant number LBUS-IRG-2021-07.

Institutional Review Board Statement: Not applicable.

Informed Consent Statement: Not applicable.

Data Availability Statement: Not applicable.

Acknowledgments: We would like to express our sincere gratitude to the Research Center in Biotechnology and Food Engineering (C.C.B.I.A.), the Lucian Blaga University of Sibiu, and the Hasso Plattner Foundation for their support provided throughout the research period.

Conflicts of Interest: The authors declare no conflict of interest.

Appendix A

Table A1. Determination of lactose content.

Sample	Day 1 Mean	Day 1 Standard error of mean	Day 1 Standard deviation	Day 1 Minimum	Day 1 Median	Day 1 Maximum	Day 1 Pearson's correlation	Day 1 Skewness	Day 10 Mean	Day 10 Standard error of mean	Day 10 Standard deviation	Day 10 Minimum	Day 10 Median	Day 10 Maximum	Day 10 Pearson's correlation	Day 10 Skewness	Day 20 Mean	Day 20 Standard error of mean	Day 20 Standard deviation	Day 20 Minimum	Day 20 Median	Day 20 Maximum	Day 20 Pearson's correlation	Day 20 Skewness
Kefir-lavender	4.683	0.0010	0.0017	4.681	4.684	4.684	1.000	−1.730	4.621	0.000	0.0006	4.621	4.621	4.622	−0.500	1.730	4.581	0.001	0.0015	4.580	4.581	4.583	−0.500	0.940
Kefir-mint	4.664	0.0007	0.0012	4.663	4.665	4.665	1.000	−1.730	4.588	0.000	0.0006	4.588	4.588	4.589	1.000	1.730	4.567	0.001	0.0012	4.566	4.568	4.568	0.189	−1.730
Kefir-fennel	4.607	0.0006	0.0010	4.606	4.607	4.608	−0.189	0.000	4.545	0.000	0.0006	4.544	4.545	4.545	0.500	−1.730	4.506	0.001	0.0012	4.505	4.505	4.507	−0.189	1.730
Control sample	4.579	0.0003	0.0006	4.578	4.579	4.579		−1.730	4.025	0.000	0.0006	4.025	4.025	4.026		1.730	3.682	0.001	0.0015	3.681	3.682	3.684		0.940

Table A2. Determination of D-galactose content.

Sample	Day 1 Mean	Day 1 Standard error of mean	Day 1 Standard deviation	Day 1 Minimum	Day 1 Median	Day 1 Maximum	Day 1 Pearson's correlation	Day 1 Skewness	Day 10 Mean	Day 10 Standard error of mean	Day 10 Standard deviation	Day 10 Minimum	Day 10 Median	Day 10 Maximum	Day 10 Pearson's correlation	Day 10 Skewness	Day 20 Mean	Day 20 Standard error of mean	Day 20 Standard deviation	Day 20 Minimum	Day 20 Median	Day 20 Maximum	Day 20 Pearson's correlation	Day 20 Skewness
Kefir-lavender	1.3021	0.0001	0.0001	1.3020	1.3022	1.3022	0.5000	−1.7300	1.2832	0.0001	0.0002	1.2830	1.2832	1.2833	−0.1890	−0.9400	1.2014	0.0002	0.0003	1.2011	1.2015	1.2016	0.9900	−1.4600
Kefir-mint	1.1793	0.0001	0.0001	1.1792	1.1794	1.1794	0.5000	−1.7300	1.1038	0.0000	0.0001	1.1037	1.1038	1.1038	1.0000	−1.7300	1.0856	0.0001	0.0002	1.0855	1.0855	1.0858	−0.9450	1.7300
Kefir-fennel	1.0848	0.0001	0.0001	1.0847	1.0849	1.0849	0.5000	−1.7300	1.0662	0.0001	0.0002	1.0660	1.0662	1.0663	0.9450	−0.9400	1.0415	0.0001	0.0001	1.0414	1.0416	1.0416	0.9450	−1.7300
Control sample	1.0378	0.0001	0.0001	1.0377	1.0377	1.0379		1.7300	1.0189	0.0001	0.0001	1.0188	1.0190	1.0190		−1.7300	1.0098	0.0001	0.0002	1.0096	1.0098	1.0099		−0.9400

Table A3. Determination of D-glucose content.

Sample	Day 1 Mean	Day 1 Standard error of mean	Day 1 Standard deviation	Day 1 Minimum	Day 1 Median	Day 1 Maximum	Day 1 Pearson's correlation	Day 1 Skewness	Day 10 Mean	Day 10 Standard error of mean	Day 10 Standard deviation	Day 10 Minimum	Day 10 Median	Day 10 Maximum	Day 10 Pearson's correlation	Day 10 Skewness	Day 20 Mean	Day 20 Standard error of mean	Day 20 Standard deviation	Day 20 Minimum	Day 20 Median	Day 20 Maximum	Day 20 Pearson's correlation	Day 20 Skewness
Kefir-lavender	0.0344	0.0001	0.0002	0.0342	0.0344	0.0345	−0.9450	−0.9400	0.0338	0.0000	0.0001	0.0338	0.0338	0.0339	−0.8660	1.7300	0.0332	0.0001	0.0002	0.0330	0.0332	0.0333	0.9450	−0.9400
Kefir-mint	0.0323	0.0001	0.0001	0.0322	0.0324	0.0324	−1.0000	−1.7300	0.0318	0.0001	0.0002	0.0317	0.0318	0.0320	−0.3270	0.9400	0.0313	0.0001	0.0001	0.0312	0.0312	0.0314	−1.0000	1.7300
Kefir-fennel	0.0315	0.0001	0.0001	0.0314	0.0314	0.0316	1.0000	1.7300	0.0312	0.0001	0.0001	0.0311	0.0311	0.0313	0.8660	1.7300	0.0309	0.0000	0.0001	0.0308	0.0309	0.0309	1.0000	−1.7300
Control sample	0.0312	0.0001	0.0001	0.0311	0.0311	0.0313		1.7300	0.0308	0.0001	0.0001	0.0307	0.0308	0.0309		0.0000	0.0304	0.0001	0.0002	0.0302	0.0305	0.0305		−1.7300

Table A4. Determination of L-glutamic acid content.

Sample	Time																							
	Day 1								Day 10								Day 20							
	Mean	Standard error of mean	Standard deviation	Minimum	Median	Maximum	Pearson's correlation	Skewness	Mean	Standard error of mean	Standard deviation	Minimum	Median	Maximum	Pearson's correlation	Skewness	Mean	Standard error of mean	Standard deviation	Minimum	Median	Maximum	Pearson's correlation	Skewness
Kefir-lavender	0.0526	0.0000	0.0001	0.0526	0.0525	0.0527	−0.6930	1.7300	0.0538	0.0000	0.0001	0.0538	0.0538	0.0539	0.5000	1.7300	0.0559	0.0001	0.0001	0.0558	0.0559	0.0560	−0.1860	0.0000
Kefir-mint	0.0213	0.0001	0.0002	0.0211	0.0213	0.0214	0.5770	−0.9400	0.0224	0.0001	0.0002	0.0222	0.0223	0.0226	0.2770	1.2900	0.0231	0.0001	0.0002	0.0230	0.0231	0.0233	−0.1890	0.9400
Kefir-fennel	0.0661	0.0001	0.0001	0.0660	0.0660	0.0662	−0.6930	1.7300	0.0672	0.0001	0.0002	0.0671	0.0672	0.0674	0.1890	0.9400	0.0686	0.0001	0.0001	0.0685	0.0685	0.0687	−0.5000	1.7300
Control sample	0.0113	0.0001	0.0002	0.0111	0.0112	0.0115		1.2900	0.0125	0.0001	0.0001	0.0124	0.0126	0.0126		−1.7300	0.0157	0.0000	0.0001	0.0157	0.0157	0.0158		1.7300

Table A5. Determination of acetic acid content.

Sample	Time																							
	Day 1								Day 10								Day 20							
	Mean	Standard error of mean	Standard deviation	Minimum	Median	Maximum	Pearson's correlation	Skewness	Mean	Standard error of mean	Standard deviation	Minimum	Median	Maximum	Pearson's correlation	Skewness	Mean	Standard error of mean	Standard deviation	Minimum	Median	Maximum	Pearson's correlation	Skewness
Kefir-lavender	0.0003	0.00003	0.0001	0.0002	0.0003	0.0003	1.0000	−1.7300	0.0004	0.00003	0.0001	0.0004	0.0004	0.0005	0.5000	1.7300	0.0006	0.00003	0.0001	0.0006	0.0006	0.0007	−0.5000	1.7300
Kefir-mint	0.0001	0.00003	0.0001	0.0001	0.0001	0.0002	0.5000	1.7300	0.0002	0.00003	0.0001	0.0002	0.0002	0.0003	0.5000	1.7300	0.0005	0.00003	0.0001	0.0004	0.0005	0.0005	−1.0000	−1.7300
Kefir-fennel	0.0003	0.00003	0.0001	0.0003	0.0003	0.0004	0.5000	1.7300	0.0005	0.00003	0.0001	0.0004	0.0005	0.0005	1.0000	−1.7300	0.0007	0.00003	0.0001	0.0006	0.0007	0.0007	0.5000	−1.7300
Control sample	0.0018	0.00003	0.0001	0.0017	0.0018	0.0018		−1.7300	0.0027	0.00003	0.0001	0.0026	0.0027	0.0027		−1.7300	0.0039	0.00003	0.0001	0.0039	0.0039	0.0040		1.7300

Table A6. Determination of ethanol content.

Sample	Time																							
	Day 1								Day 10								Day 20							
	Mean	Standard error of mean	Standard deviation	Minimum	Median	Maximum	Pearson's correlation	Skewness	Mean	Standard error of mean	Standard deviation	Minimum	Median	Maximum	Pearson's correlation	Skewness	Mean	Standard error of mean	Standard deviation	Minimum	Median	Maximum	Pearson's correlation	Skewness
Kefir-lavender	0.0003	0.00003	0.0001	0.0002	0.0003	0.0003	0.5000	−1.7300	0.0005	0.00003	0.0001	0.0004	0.0005	0.0005	0.5000	−1.7300	0.0008	0.00003	0.0001	0.0007	0.0008	0.0008	−0.5000	−1.7300
Kefir-mint	0.0003	0.00003	0.0001	0.0002	0.0003	0.0003	0.5000	−1.7300	0.0005	0.00003	0.0001	0.0004	0.0005	0.0005	0.5000	−1.7300	0.0007	0.00003	0.0001	0.0007	0.0007	0.0008	−1.0000	1.7300
Kefir-fennel	0.0001	0.00003	0.0001	0.0001	0.0001	0.0002	−0.5000	1.7300	0.0002	0.00003	0.0001	0.0002	0.0002	0.0003	−0.5000	1.7300	0.0005	0.00003	0.0001	0.0005	0.0005	0.0006	0.5000	1.7300
Control sample	0.0005	0.00003	0.0001	0.0005	0.0005	0.0006		1.7300	0.0007	0.00003	0.0001	0.0007	0.0007	0.0008		1.7300	0.0010	0.00003	0.0001	0.0009	0.0010	0.0010		−1.7300

Table A7. Determination of L-lactic acid content.

Sample	Time																							
	Day 1								Day 10								Day 20							
	Mean	Standard error of mean	Standard deviation	Minimum	Median	Maximum	Pearson's correlation	Skewness	Mean	Standard error of mean	Standard deviation	Minimum	Median	Maximum	Pearson's correlation	Skewness	Mean	Standard error of mean	Standard deviation	Minimum	Median	Maximum	Pearson's correlation	Skewness
Kefir-lavender	0.7021	0.00007	0.0001	0.7020	0.7022	0.7022	0.5000	−1.7300	0.7026	0.00003	0.0001	0.7026	0.7026	0.7027	1.0000	1.7300	0.7029	0.00003	0.0001	0.7028	0.7029	0.7029	1.0000	−1.7300
Kefir-mint	0.7007	0.00007	0.0001	0.7006	0.7008	0.7008	0.5000	−1.7300	0.7009	0.00003	0.0001	0.7008	0.7009	0.7009	−1.0000	−1.7300	0.7011	0.00007	0.0001	0.7010	0.7012	0.7012	1.0000	−1.7300
Kefir-fennel	0.7019	0.00009	0.0002	0.7017	0.7019	0.7020	0.1890	−0.9400	0.7023	0.00009	0.0002	0.7022	0.7023	0.7025	0.9450	0.9400	0.7025	0.00003	0.0001	0.7025	0.7025	0.7026	−1.0000	1.7300
Control sample	0.7014	0.00007	0.0001	0.7013	0.7013	0.7015		1.7300	0.7028	0.00007	0.0001	0.7027	0.7027	0.7029		1.7300	0.7039	0.00003	0.0001	0.7038	0.7039	0.7039		−1.7300

References

1. SeKiełczewska, K.; Dąbrowska, A.; Bielecka, M.M.; Dec, B.; Baranowska, M.; Ziajka, J.; Zhennai, Y.; Żulewska, J. Protein Preparations as Ingredients for the Enrichment of Non-Fermented Milks. *Foods* **2022**, *11*, 1817. [CrossRef] [PubMed]
2. Sabokbar, N.; Khodaiyan, F. Characterization of pomegranate juice and whey based novel beverage fermented by kefir grains. *J. Food Sci. Technol.* **2015**, *52*, 3711–3718. [CrossRef] [PubMed]
3. Carullo, G.; Governa, P.; Spizzirri, U.G.; Biagi, M.; Sciubba, F.; Giorgi, G.; Loizzo, M.R.; Di Cocco, M.E.; Aiello, F.; Restuccia, D. Sangiovese cv pomace seeds extract-fortified kefir exerts anti-inflammatory activity in an in vitro model of intestinal epithelium using caco-2 cells. *Antioxidant* **2020**, *9*, 54. [CrossRef] [PubMed]
4. Kesenkaş, H.; Dinkçi, N.; Seçkin, K.; Kinik, Ö.; Gönç, S. Antioxidant properties of kefir produced from different cow and soy milk mixtures. *J. Agric. Sci.* **2011**, *17*, 253–259.
5. Liu, J.R.; Chen, M.J.; Lin, C.W. Antimutagenic and antioxidant properties of milk-kefir and soymilk-kefir. *J. Agric. Food Chem.* **2005**, *53*, 2467–2474. [CrossRef]
6. Aiello, F.; Restuccia, D.; Spizzirri, U.G.; Carullo, G.; Leporini, M.; Loizzo, M.R. Improving kefir bioactive properties by functional enrichment with plant and agro-food waste extracts. *Fermentation* **2020**, *6*, 83. [CrossRef]
7. Farag, M.A.; Jomaa, S.A.; El-wahed, A.A.; El-seedi, H.R. The many faces of kefir fermented dairy products: Quality characteristics, flavour chemistry, nutritional value, health benefits, and safety. *Nutrients* **2020**, *12*, 346. [CrossRef]
8. Darvishzadeh, P.; Orsat, V.; Martinez, J.L. Process Optimization for Development of a Novel Water Kefir Drink with High Antioxidant Activity and Potential Probiotic Properties from Russian Olive Fruit (*Elaeagnus angustifolia*). *Food Bioprocess Technol.* **2021**, *14*, 248–260. [CrossRef]
9. Paredes, J.L.; Escudero-Gilete, M.L.; Vicario, I.M. A new functional kefir fermented beverage obtained from fruit and vegetable juice: Development and characterization. *LWT* **2022**, *154*, 112728. [CrossRef]
10. Liu, J.R.; Lin, Y.Y.; Chen, M.J.; Chen, L.J.; Lin, C.W. Antioxidative activities of kefir, Asian-australas. *J. Anim. Sci.* **2005**, *18*, 567–573.
11. Can, E.; Kurtoğlu, İ.Z.; Benzer, F.; Erişir, M.; Kocabaş, M.; Kızak, V.; Kayim, M.; Çelik, H.T. The Effects of Different Dosage of Kefir with Different Durations on Growth Performances and Antioxidant System in the Blood and Liver Tissues of Çoruh Trout (Salmo coruhensis). *Turk. J. Fish. Aquat. Sci.* **2012**, *12*, 277–283.
12. Łopusiewicz, Ł.; Drozłowska, E.; Siedlecka, P.; Mężyńska, M.; Bartkowiak, A.; Sienkiewicz, M.; Zielińska-Bliźniewska, H.; Kwiatkowski, P. Development, characterization, and bioactivity of non-dairy kefir-like fermented beverage based on flaxseed oil cake. *Foods* **2019**, *8*, 544. [CrossRef] [PubMed]
13. Saadi, L.O.; Zaidi, F.; Oomah, B.D.; Haros, M.; Yebra, M.J.; Hosseinian, F. Pulse ingredients supplementation affects kefir quality and antioxidant capacity during storage. *LWT* **2017**, *86*, 619–626. [CrossRef]
14. Bătușaru, C.M. Sustainability of the small business environment in Romania in the context of increasing economic competitiveness. *Manag. Sustain. Dev.* **2019**, *11*, 37–41. [CrossRef]
15. Sabahi, S.; Mortazavi, S.A.; Nassiri, M.; Ghazvini, K.; Shahidi, F.; Abbasi, A. Production of Functional Ice Cream Fortified by Immunoglobulin Y against Escherichia coli O157:H7 and Helicobacter Pylori. *Biointerface Res. Appl. Chem.* **2023**, *13*, 188.
16. Radu, D.; Râpeanu, G.; Bahrim, G.E.; Stănciuc, N. Investigations on thermal degradation of phytochemicals from lavender extract. *Ann. Univ. Dunarea Jos Galati. Fascicle VI-Food Technol.* **2019**, *43*, 33–47.
17. Aboutaleb, N.; Jamali, H.; Abolhasani, M.; Pazoki Toroudi, H. Lavender oil (*Lavandula angustifolia*) attenuates renal ischemia/reperfusion injury in rats through suppression of inflammation, oxidative stress and apoptosis. *Biomed. Pharmacother.* **2019**, *110*, 9–19. [CrossRef]
18. Wang, D.; Yuan, X.; Liu, T.; Liu, L.; Hu, Y.; Wang, Z.; Zheng, Q. Neuroprotective activity of lavender oil on transient focal cerebral ischemia in mice. *Molecules* **2012**, *17*, 9803–9817. [CrossRef]
19. Radu, D.; Alexe, P.; Stănciuc, N. Overview on the potential role of phytochemicals from lavender as functional ingredients. *Ann. Univ. Dunarea Jos Galati Fascicle VI Food Technol.* **2020**, *44*, 173–188. [CrossRef]
20. Taşkaya, L.; Hasanhocaoğlu Yapici, H.; Metin, C.; Alparslan, Y. The effect of lavender (*Lavandula stoechas*) on the shelf life of a traditional food: Hamsi kaygana. *Food Sci. Technol.* **2018**, *38*, 711–718. [CrossRef]
21. Marín, I.; Sayas-Barberá, E.; Viuda-Martos, M.; Navarro, C.; Sendra, E. Chemical Composition, Antioxidant and Antimicrobial Activity of Essential Oils from Organic Fennel, Parsley, and Lavender from Spain. *Foods* **2016**, *5*, 18. [CrossRef]
22. Benbrahim, C.; Barka, M.S.; Basile, A.; Maresca, V.; Flamini, G.; Sorbo, S.; Carraturo, F.; Notariale, R.; Piscopo, M.; Khadir, A.; et al. Chemical composition and biological activities of oregano and lavender essential oils. *Appl. Sci.* **2021**, *11*, 5688. [CrossRef]
23. Parham, S.; Kharazi, A.Z.; Bakhsheshi-Rad, H.R.; Nur, H.; Ismail, A.F.; Sharif, S.; RamaKrishna, S.; Berto, F. Antioxidant, antimicrobial and antiviral properties of herbal materials. *Antioxidants* **2020**, *9*, 1309. [CrossRef] [PubMed]
24. Wu, Z.; Tan, B.; Liu, Y.; Dunn, J.; Martorell Guerola, P.; Tortajada, M.; Cao, Z.; Ji, P. Chemical Composition and Antioxidant Properties of Essential Oils from Peppermint, Native Spearmint and Scotch Spearmint. *Molecules* **2019**, *24*, 2825. [CrossRef] [PubMed]
25. Hejna, M.; Kovanda, L.; Rossi, L.; Liu, Y. Mint oils: In vitro ability to perform anti-inflammatory, antioxidant, and antimicrobial activities and to enhance intestinal barrier integrity. *Antioxidants* **2021**, *10*, 1004. [CrossRef]
26. Tsai, M.L.; Wu, C.T.; Lin, T.F.; Lin, W.C.; Huang, Y.C.; Yang, C.H. Chemical composition and biological properties of essential oils of two mint species. *Trop. J. Pharm. Res.* **2013**, *12*, 577–582. [CrossRef]
27. Park, C.H.; Yeo, H.J.; Baskar, T.B.; Park, Y.E.; Park, J.S.; Lee, S.Y.; Park, S.U. In vitro antioxidant and antimicrobial properties of flower, leaf, and stem extracts of Korean mint. *Antioxidants* **2019**, *8*, 75. [CrossRef]

28. Kapp, K.; Orav, A.; Roasto, M.; Raal, A.; Püssa, T.; Vuorela, H.; Tammela, P.; Vuorela, P. Composition and antibacterial effect of mint flavorings in candies and food supplements. *Planta Med.* **2020**, *86*, 1089–1096. [CrossRef]
29. Mazandrani, H.A.; Javadian, S.R.; Bahram, S. The effect of encapsulated fennel extracts on the quality of silver carp fillets during refrigerated storage. *Food Sci. Nutr.* **2016**, *4*, 298–304. [CrossRef]
30. Sakr, S.A.; Shalaby, S.Y.; Beder, R.H. Ameliorative Effect of Fennel Oil on Cyclophosphamide Induced Hepatotoxicity in Albino Rats. *Br. J. Pharm. Res.* **2017**, *17*, 1–12. [CrossRef]
31. Anwar, F.; Ali, M.; Hussain, A.I.; Shahid, M. Antioxidant and antimicrobial activities of essential oil and extracts of fennel (*Foeniculum vulgare* Mill.) seeds from Pakistan. *Flavour. Fragr. J.* **2009**, *24*, 170–176. [CrossRef]
32. Farid, A.; Kamel, D.; Abdelwahab Montaser, S.; Mohamed Ahmed, M.; El Amir, M.; El Amir, A. Synergetic role of senna and fennel extracts as antioxidant, anti-inflammatory and anti-mutagenic agents in irradiated human blood lymphocyte cultures. *J. Radiat. Res. Appl. Sci.* **2020**, *13*, 191–199. [CrossRef]
33. Ahmad, B.S.; Talou, T.; Saad, Z.; Hijazi, A.; Cerny, M.; Kanaan, H.; Chokr, A.; Merah, O. Fennel oil and by-products seed characterization and their potential applications. *Ind. Crops Prod.* **2018**, *111*, 92–98. [CrossRef]
34. Ahmed, A.F.; Shi, M.; Liu, C.; Kang, W. Comparative analysis of antioxidant activities of essential oils and extracts of fennel (*Foeniculum vulgare* Mill.) seeds from Egypt and China. *Food Sci. Hum. Wellness* **2019**, *8*, 67–72. [CrossRef]
35. Shahat, A.A.; Ibrahim, A.Y.; Hendawy, S.F.; Omer, E.A.; Hammouda, F.M.; Abdel-Rahman, F.H.; Saleh, M.A. Chemical composition, antimicrobial and antioxidant activities of essential oils from organically cultivated fennel cultivars. *Molecules* **2011**, *16*, 1366–1377. [CrossRef]
36. Korinek, M.; Handoussa, H.; Tsai, Y.H.; Chen, Y.Y.; Chen, M.H.; Chiou, Z.W.; Fang, Y.; Chang, F.R.; Yen, C.H.; Hsieh, C.F.; et al. Anti-Inflammatory and Antimicrobial Volatile Oils: Fennel and Cumin Inhibit Neutrophilic Inflammation via Regulating Calcium and MAPKs. *Front. Pharmacol.* **2021**, *12*, 674095. [CrossRef]
37. Tița, O.; Constantinescu, M.A.; Tița, M.A.; Georgescu, C. Use of yoghurt enhanced with volatile plant oils encapsulated in sodium alginate to increase the human body's immunity in the present fight against stress. *Int. J. Environ. Res. Public Health* **2020**, *17*, 7588. [CrossRef]
38. Tița, M.A.; Constantinescu, M.A.; Tița, O.; Mathe, E.; Tamošaitienė, L.; Bradauskienė, V. Food Products with High Antioxidant and Antimicrobial Activities and Their Sensory Appreciation. *Appl. Sci.* **2022**, *12*, 790. [CrossRef]
39. Tița, O.; Constantinescu, M.A.; Tița, M.A.; Opruța, T.I.; Dabija, A.; Georgescu, C. Valorization on the Antioxidant Potential of Volatile Oils of *Lavandula angustifolia* Mill., *Mentha piperita* L. and *Foeniculum vulgare* L. in the Production of Kefir. *Appl. Sci.* **2022**, *12*, 10287. [CrossRef]
40. R-BIOPHARM AG–Constituents. 2021. Available online: https://food.r-biopharm.com/analytes/constituents/ (accessed on 21 February 2021).
41. Eliskases-Lechner, F.; Guéguen, M.; Panoff, J.M. Yeasts and Molds | Geotrichum candidum. In *Encyclopedia of Dairy Sciences*, 2nd ed.; Fuquay, J.W., Fox, P.F., McSweeney, P.L.H., Eds.; Academic Press: Cambridge, MA, USA, 2011; pp. 765–771.
42. Eliskases-Lechner, F. Geotrichum candidum. In *Encyclopedia of Dairy Sciences*, 1st ed.; Fuquay, J.W., Fox, P.F., Roginski, H., Eds.; Academic Press: Cambridge, MA, USA, 2002; pp. 1229–1234.
43. Errampalli, D. Penicillium expansum (Blue Mold). In *Postharvest Decay*, 1st ed.; Bautista-Banos, S., Ed.; Academic Press: Cambridge, MA, USA, 2014; pp. 189–231.
44. Pitt, J.I. Penicillium | Penicillium and Talaromyces: Introduction. In *Encyclopedia of Food Microbiology*, 2nd ed.; Batt, C.A., Tortorello, M.L., Eds.; Academic Press: Cambridge, MA, USA, 2014; pp. 6–13.
45. What Is Aspergillus Niger? Available online: https://www.news-medical.net/life-sciences/What-is-Aspergillus-niger.aspx (accessed on 21 March 2021).
46. Hocking, A.D. Aspergillus and related teleomorphs. In *Food Spoilage Microorganisms*; Blackburn, C.d.W., Ed.; Woodhead Publishing Ltd.: Cambridge, UK, 2006; pp. 451–487.
47. Ladaniya, M.S. Postharvest diseases and their management. In *Citrus Fruit. Biology, Technology and Evaluation*, 1st ed.; Academic Press: Cambridge, MA, USA, 2008; pp. 417–449.
48. Desmarchelier, P.; Fegan, N. Escherichia coli. In *Encyclopedia of Dairy Sciences*, 1st ed.; Fuquay, J.W., Fox, P.F., Roginski, H., Eds.; Academic Press: Cambridge, MA, USA, 2002; pp. 948–954.
49. Percival, S.L.; Williams, D.W. Escherichia coli. In *Microbiology of Waterborne Diseases*, 2nd ed.; Percival, S.L., Yates, M.V., Williams, D.W., Chalmers, R.M., Gray, N.F., Eds.; Academic Press: Cambridge, MA, USA, 2014; pp. 89–117.
50. Desmarchelier, P.; Fegan, N. Pathogens in Milk | Escherichia coli. In *Encyclopedia of Dairy Sciences*, 2nd ed.; Fuquay, J.W., Fox, P.F., McSweeney, P.L.H., Eds.; Academic Press: Cambridge, MA, USA, 2011; pp. 60–66.
51. Lengyel, E.; Panaitescu, M. Chemical Compounds from Thymus Vulgaris and their Antimicrobial Activity. *Manag. Sustain. Dev.* **2019**, *11*, 25–28.
52. Bauer, A.W.; Kirby, W.M.M.; Sherris, J.C.; Turck, M. Antibiotic Susceptibility Testing by a Standardized Single Disk Method. *Am. J. Clin. Pathol.* **1966**, *45*, 493–496. [CrossRef] [PubMed]
53. Clinical and Laboratory Standards Institute. Performance Standards for Antimicrobial Disk Susceptibility Tests: Approved Standard-Eleventh Edition. Available online: File:///D:/Downloads/01-CLSI-M02-A11-2012%20.pdf (accessed on 10 February 2021).
54. Hudzicki, J. Kirby-Bauer Disk Diffusion Susceptibility Test Protocol. Available online: https://asm.org/getattachment/2594ce26-bd44-47f6-8287-0657aa9185ad/Kirby-Bauer-Disk-Diffusion-Susceptibility-Test-Protocol-pdf.pdf (accessed on 25 April 2021).

55. Lactose/D-Galactose. Available online: https://food.r-biopharm.com/wp-content/uploads/2012/06/roche_ifu_lactose-galactose_en_10176303035_2017-08.pdf (accessed on 16 May 2021).
56. Lactose/D-Glucose. Available online: https://food.r-biopharm.com/wp-content/uploads/2012/06/roche_ifu_lactose-glucose_en_10986119035_2017-09.pdf (accessed on 16 May 2021).
57. L-Glutamic Acid. Available online: https://food.r-biopharm.com/wp-content/uploads/2012/06/roche_ifu_glutamic-acid_10139092035_en-v7_2019-06.pdf (accessed on 16 May 2021).
58. Acetic Acid. Available online: https://food.r-biopharm.com/wp-content/uploads/2012/06/roche_ifu_acetic-acid_en_10148261035_2017-08.pdf (accessed on 16 May 2021).
59. Ethanol. Available online: https://food.r-biopharm.com/wp-content/uploads/2012/06/roche_ifu_ethanol_10176290035_en-v6_2019-11.pdf (accessed on 16 May 2021).
60. L-Lactic Acid. Available online: https://food.r-biopharm.com/wp-content/uploads/2012/06/roche_ifu_lactic-acid-l_en_10139084035_2017-08.pdf (accessed on 16 May 2021).
61. Pearson's Correlation Using Minitab. Available online: https://statistics.laerd.com/minitab-tutorials/pearsons-correlation-using-minitab.php (accessed on 26 March 2021).
62. Zeng, H.; Chen, X.; Liang, J. In vitro antifungal activity and mechanism of essential oil from fennel on dermatophyte species. *J. Med. Microbiol.* **2015**, *64*, 93–103. [CrossRef] [PubMed]
63. Ibáñez, M.D.; Blázquez, M.A. Phytotoxicity of essential oils on selected weeds: Potential hazard on food crops. *Plants* **2018**, *7*, 79. [CrossRef] [PubMed]
64. Zhang, S.; Chen, X.; Devshilt, I.; Yun, Q.; Huang, C.; An, L.; Dorjbat, S.; He, X. Fennel main constituent, trans-anethole treatment against LPS-induced acute lung injury by regulation of Th17/Treg function. *Mol. Med. Rep.* **2018**, *18*, 1369–1376. [CrossRef]
65. Soković, M.; Glamočlija, J.; Marin, P.D.; Brkić, D.; van Griensven, L.J.L.D. Antibacterial effects of the essential oils of commonly consumed medicinal herbs using an in vitro model. *Molecules* **2010**, *15*, 7532–7546. [CrossRef]
66. Falowo, A.B.; Mukumbo, F.E.; Idamokoro, E.M.; Afolayan, A.J.; Muchenje, V. Phytochemical Constituents and Antioxidant Activity of Sweet Basil (*Ocimum basilicum L.*) Essential Oil on Ground Beef from Boran and Nguni Cattle. *Int. J. Food Sci.* **2019**, *2019*, 2628747. [CrossRef]
67. Qaralleh, H.N. Chemical composition and antibacterial activity of origanum ramonense essential oil on the β-lactamase and extended-spectrum β-lactamase urinary tract isolates. *Bangladesh J. Pharmacol.* **2018**, *13*, 280–286. [CrossRef]
68. Li, Z.H.; Cai, M.; Liu, Y.S.; Sun, P.L.; Luo, S.L. Antibacterial Activity and Mechanisms of Essential Oil from *Citrus medica L. var. sarcodactylis*. *Molecules* **2019**, *24*, 1577. [CrossRef]
69. De Oliveira, J.R.; Camargo, S.E.A.; De Oliveira, L.D. *Rosmarinus officinalis* L. (rosemary) as therapeutic and prophylactic agent. *J. Biomed. Sci.* **2019**, *26*, 5. [CrossRef]
70. Dore, S.; Ferrini, A.M.; Appicciafuoco, B.; Massaro, M.R.; Sotgiu, G.; Liciardi, M.; Cannas, E.A. Efficacy of a terpinen-4-ol based dipping for post-milking teat disinfection in the prevention of mastitis in dairy sheep. *J. Essent. Oil Res.* **2019**, *31*, 19–26. [CrossRef]
71. Tylkowski, B.; Tsibranskaa, I.; Kochanova, R.; Peeva, G.; Giamberini, M. Concentration of biologically active compounds extracted from *Sideritis* ssp. L. by nanofiltration. *Food Bioprod. Process.* **2011**, *89*, 307–314. [CrossRef]
72. Cazella, L.N.; Glamoclija, J.; Soković, M.; Gonçalves, J.E.; Linde, G.A.; Colauto, N.B.; Gazim, Z.C. Antimicrobial activity of essential oil of *Baccharis dracunculifolia* DC (Asteraceae) aerial parts at flowering period. *Front. Plant Sci.* **2019**, *10*, 27. [CrossRef] [PubMed]
73. Hassanzadazar, H.; Taami, B.; Aminzare, M.; Daneshamooz, S. Bunium persicum (Boiss.) B. Fedtsch: An overview on phytochemistry, therapeutic uses and its application in the food industry. *J. Appl. Pharm. Sci.* **2018**, *8*, 150–158.
74. Moon, P.D.; Han, N.R.; Lee, J.S.; Kim, H.M.; Jeong, H.J. Effects of Linalyl Acetate on Thymic Stromal Lymphopoietin Production in Mast Cells. *Molecules* **2018**, *23*, 1711. [CrossRef]
75. Golkar, P.; Moattar, F. Essential Oil Composition, Bioactive Compounds, and Antioxidant Activities in *Iberis amara* L. *Nat. Prod. Commun.* **2019**, *14*, 1711. [CrossRef]
76. Prasetya, N.B.A.; Ngadiwiyana; Ismiyarto; Sarjono, P.R. Synthesis of copolymer eugenol crosslinked with divinyl benzene and preliminary study on its antibacterial activity. *IOP Conf. Ser. Mater. Sci. Eng.* **2019**, *509*, 012102. [CrossRef]
77. Slama, H.B.; Cherif-Silini, H.; Bouket, A.C.; Qader, M.; Silini, A.; Yahiaoui, B.; Alenezi, F.N.; Luptakova, L.; Triki, M.A.; Vallat, A.; et al. Screening for fusarium antagonistic bacteria from contrasting niches designated the endophyte bacillus halotolerans as plant warden against fusarium. *Front. Microbiol.* **2019**, *10*, 3236. [CrossRef]
78. Tahghighi, A.; Maleki-Ravasan, N.; Dinparast Djadid, N.; Alipour, H.; Ahmadvand, R.; Karimian, F.; Yousefinejad, S. GC-MS analysis and anti-mosquito activities of *Juniperus virginiana* essential oil against *Anopheles stephensi* (Diptera: Culicidae). *Asian Pac. J. Trop. Biomed.* **2019**, *9*, 168–175. [CrossRef]
79. Matulyte, I.; Marksa, M.; Ivanauskas, L.; Kalvenien, Z.; Lazauskas, R.; Bernatoniene, J. GC-MS analysis of the composition of the extracts and essential Oil from *Myristica fragrans* Seeds Using Magnesium Aluminometasilicate as Excipient. *Molecules* **2019**, *24*, 1062. [CrossRef]

applied sciences

MDPI

Article

Physical and Antioxidant Properties of Innovative Gluten-Free Bread with the Addition of Hemp Inflorescence

Anna Pecyna [1], Agnieszka Buczaj [1], Renata Różyło [2,*] and Zbigniew Kobus [1]

[1] Department of Technology Fundamentals, University of Life Sciences in Lublin, 28 Głęboka Str., 20-612 Lublin, Poland; anna.pecyna@up.lublin.pl (A.P.); agnieszka.buczaj@up.lublin.pl (A.B.); zbigniew.kobus@up.lublin.pl (Z.K.)
[2] Department of Food Engineering and Machines, University of Life Sciences in Lublin, 28 Głęboka Str., 20-612 Lublin, Poland
* Correspondence: renata.rozylo@up.lublin.pl

Abstract: Hemp inflorescences from byproducts have been proposed as an addition to gluten-free rice bread. The scope of the research was to bake a control loaf of bread as well as bread loaves containing 1%, 2%, 3%, 4%, and 5% dried and crushed hemp inflorescence (HI). The loaves of bread were evaluated in terms of their physical and sensory properties, polyphenol and flavonoid contents, and DPPH and FRAP antioxidant activities. The study's findings revealed that the addition of HI influenced changes in the physical properties of the bread loaves, such as increased specific volume, decreased bread hardness, increased elasticity, and chewiness of the breadcrumb, especially when the additive concentrations were greater than 3%. The addition of HI significantly increased the total amount of polyphenols, flavonoids, and antioxidant activity in the bread. The sensory evaluation revealed that gluten-free bread can be produced with a maximum of 2% HI without affecting its taste and aroma.

Keywords: hemp; *Cannabis sativa* L.; inflorescence; bread physical properties; gluten-free bread

Citation: Pecyna, A.; Buczaj, A.; Różyło, R.; Kobus, Z. Physical and Antioxidant Properties of Innovative Gluten-Free Bread with the Addition of Hemp Inflorescence. *Appl. Sci.* **2023**, *13*, 4889. https://doi.org/10.3390/app13084889

Academic Editors: Georgiana Gabriela Codină, Adriana Dabija and Lăcrămioara Rusu

Received: 11 March 2023
Revised: 11 April 2023
Accepted: 12 April 2023
Published: 13 April 2023

1. Introduction

Hemp (*Cannabis sativa* L.), belonging to the *Cannabaceae* family, is one of the world's oldest cultivated plants. It is used for textiles, medicinal purposes, food [1–6], and for fodder purposes [7,8]. As a byproduct of hemp harvesting, hemp inflorescences are used to make niche products for the cosmetics, pharmaceutical, nutraceutical, and environmentally friendly insecticide industries [9,10]. Hemp metabolites include flavonoids, fatty acids, terpenes, amino acids, terpenophenols, and sugars, which are abundant in inflorescences, and they are gaining popularity [11]. As a result, hemp inflorescences have been identified as a potentially novel food that requires further investigation [12].

Cannabis sativa L., also known as *Cannabis sativa* var. *spontanea*, contains few psychoactive cannabinoids and is a good source of terpenoids and polyphenols, including flavonoids, as well as phenolic acids, phenolic amides, and lignoamides with significant health benefits [13–15]. The inflorescences of industrial hemp varieties, such as Kompolti (grown, among other things, in the Italian Alps), are particularly high in cannabidiolic acid (CBDA), which decarboxylates spontaneously to cannabidiol (CBD) under favorable environmental conditions [16].

Numerous aromatic compounds are found in cannabis inflorescences, including pinene, pinene, myrcene, terpinolene, caryophyllene, humulene, and caryophyllene. Hemp has a pleasant aroma due to the presence of volatile compounds such as terpenes, aliphatic compounds, and phenylpropanoids/benzenoids, which justifies the use of these hemp-derived aromatic compounds in the food and cosmetics industries, such as the production of essential oils as natural flavor and fragrance additives [17,18].

There are many compounds in cannabis that include terpenoids, sugars, alkaloids, stilbenoids, quinones, and cannabinoids. Due to the presence of bioactive compounds such as terpenes and cannabidiol, hemp (*Cannabis sativa* L.) has become widely used in a variety of industries [19]. The term "cannabinoid" refers to hemp-derived terpenophenols.

Hemp has the nutritional potential to be used in the food industry as well. It is used in the baking industry, for example, in the production of bread. Bread has previously included hemp flour, hemp protein concentrates, or hemp pomace. The addition of these raw hemp materials to bread improves its antioxidant properties (e.g., by increasing the content of polyphenols), protein content [20–23], as well as its increase its mineral content.

According to Bădărău et al. [24], the most suitable raw hemp materials for use in bread processing are 15% hemp flour, 4% hemp seed, and 8% hemp oil, all of which can be added to the dough. Aside from improving bread's nutritional value, the benefits of using hemp products (particularly variants with 15% and 25% hemp flour) as the second auxiliary ingredient in bread include rinsing out the volume of the bread, improving the structure of the crumb and increasing the bread's durability, providing an interesting crumb color due to the content of natural sugars, and increasing the bread's organic acids. Furthermore, the addition of 15% hemp products has the greatest effect on the characteristics of the dough by improving the crumb quality (porosity and elasticity). As a result, the addition of hemp products can be used in baking to produce unique and highly demanded bread products.

Other research has shown that up to 10% hemp flour can be used in wheat bread [21], although too much hemp flour contributes to a decrease in the organoleptic evaluation of the bread. Furthermore, the addition of hemp flour changes the color of the crumb by raising its browning index [20]. Several studies have shown that bread with the addition of 5%, 10%, 15%, and 20% hemp flour has better nutritional properties compared to wheat bread, but the best texture was characterized by the addition of 5% hemp flour [25]. Hemp and hemp seed cakes were also tested as an addition to wheat bread. The results suggested that the addition of these hemp by-products in a proportion greater than 1% caused unfavorable changes in the quality of the bread [26].

Gluten-free bread, as a healthy alternative to wheat bread, may also include hemp flour and hemp protein concentrate, which are natural nutritional and structure-forming gluten-free bread ingredients. Hemp flour was used to make sourdough, and it was tested as an addition to gluten-free bread. The bread's staining was reduced by using hemp sourdough. Hemp flour has improved the acceptability of gluten-free corn/rice bread [27]. In a past study, gluten-free bread with 5% hemp flour did not differ significantly from the control bread and was accepted by consumers [28].

Korus et al. [29] found that replacing some of the starch with hemp flour made bread dough weaker and more likely to deform, while a 20% share of hemp protein concentrate made bread dough stronger. The level of fiber increased from 15.2 to 61.0 $g \cdot kg^{-1}$, and the level of dietary fiber increased from 29.3 to 90.0 $g \cdot kg^{-1}$. Hemp protein supplementation improved the color of the crumb, reducing its brightness from 62.3 to 40.8 and increasing the bread volume from 633 to 878 mL. Hemp flour and hemp protein concentrate are the most commonly used ingredients in gluten-free bread production. Although various hemp products have been added to gluten-free bread, hemp inflorescence has not been added to this type of bread.

Due to the growing number of people suffering from the consumption of cereals containing gluten, there is a demand for gluten-free bread on the market. Most often, rice and corn flour are used for their production. However, consumers perceive gluten-free bread to be less appealing than traditional wheat bread due to its unsatisfactory texture and taste. Gluten-free bread also has a lower nutritional value. Various additives are used to improve the quality of gluten-free bread and they affect the physical properties, taste, and content of bioactive substances [30–35].

As previously stated, hemp inflorescence is a valuable raw material in terms of health and it has not previously been used in bread production, which is why it was proposed for inclusion in gluten-free bread. The scope of the work included determining the changes in the physical and sensory properties as well as the antioxidant potential of gluten-free bread with added hemp inflorescence. Furthermore, we aim to recommend the most nutritionally valuable and acceptable recipe for customers.

2. Materials and Methods

2.1. Materials

The ingredients for the preparation of the control bread were gluten-free rice flour (Melvit, Warsaw, Poland). The rice flour, according to the manufacturer's declaration, contained 6 g/100 g of protein, 1.4 g/100 g of fiber, 1 g/100 g of fat, 77 g/100 g of carbohydrates, and 14.6 g/100 g moisture content. The hemp inflorescence that was used as an addition came from crops grown on a farm in the Lublin Province of Poland. The Futura hemp variety with a CBD (cannabidiol) content of 0.09% and a THC (tetrahydrocannabinol) content of 0.01% was used for the study. Salt, guar gum (NatVita, Mirków, Poland), and dry instant yeast (Instaferm, Lallemand Iberia, Setúbal Municipality, Portugal) were also used in this study to make the breads (Table A1).

2.2. Process of Making the Gluten-Free Bread

A single-phase method was used to produce the gluten-free bread [36]. The gluten-free rice flour used to make the control loaf was used exclusively, and the recipe also included guar gum, dry instant yeast, and salt. Hemp inflorescences, both dried and shredded, were added in various amounts to the control bread recipe (0%, 1%, 2%, 3%, 4%, and 5%). Together with the addition of the hemp inflorescence, the amount of water was increased from 120% to 136.7% [37].

A laboratory spiral mixer (Clatronic KM 3630, Clatronic International GmBh, Kempen, Germany) was used to mix the dough for 5 min at 200 rpm. The dough was then divided into 300 g pieces and placed into molds where it fermented and proved for 40 min at 30 °C with 75% humidity. A laboratory oven (Sadkiewicz Instruments, Bydgoszcz, Poland) was used to bake the bread for 40 min at 240 degrees. The baking test was repeated three times.

2.3. Determination of the Basic Physical Parameters of the Bread

After measuring the volume of the bread using the millet seed displacement method, the specific volume (bread volume divided by weight) was calculated. The pH level of the breadcrumb was determined using a pH meter (TESTO 206-ph2, Pruszków, Poland) [38]. Color measurements were taken on an L*a*b* scale (4Wave CR30-16) (Planeta, Tychy, Poland), and ΔE was calculated. The following parameters were measured: L*—lightness, a*—changes in the color from green to red, and b*—changes in the color from blue to yellow (higher values of a* and b* indicated a greater intensity of red and yellow, respectively). The total color change, ΔE, was calculated as follows:

$$\Delta E = \sqrt{(\Delta L)^2 + (\Delta a)^2 + (\Delta b)^2} \quad (1)$$

where ΔL, Δa, and Δb are the indices of the differences in the color of the surfaces of the samples compared with the control bread. In total, three replications of the measurements of the physical characteristics of the bread were carried out [39].

2.4. Determination of the Texture and Sensory Parameters of the Bread

The textural parameters of the breadcrumbs with dimensions of 30 × 30 × 20 mm were determined using a ZWICK Z020/TN2S (ZwickRoell, Ulm, Germany) for a double compression test (TPA–texture profile analysis) to a depth of 50% at a speed of 1 mms^{-1}. The graphs produced by the TPA test were used to compute the hardness, springiness, cohesiveness, and chewiness. In the midst of the bread's central slices, texture measurements were made six times. A panel of 72 untrained consumers (aged 22 to 56) evaluated the bread samples' flavors, aromas, appearances, textures, and general levels of acceptability as part of the sensory analysis. The responses were collected using a 9-point hedonic scale:

1 for extremely dislike, 2 for very dislike, 3 for rather dislike, 4 for moderately dislike, 5 for neither like nor dislike, 6 for moderately like, 7 for somewhat like, 8 for very much like, and 9 for extremely like [40].

2.5. Determination of Antioxidants

2.5.1. Preparation of the Extracts for the Chemical Analyses

To prepare the sample extract, 2 g of grinding sample was mixed with 30 mL of methanol in a 100 mL conical flask. The mixture was placed on a magnetic stirrer and heated to 30 °C. After two hours of stirring, the extract was collected into a separate flask and the raffinate was mixed with another 30 mL of methanol. After two more hours of extraction on a magnetic stirrer, the extract was collected and mixed with the previous portion of extract and centrifuged at 6500 rpm for 15 min.

2.5.2. Determination of the Total Phenolic Content (TPC)

The TPC was examined using the spectrophotometric method described by Singleton et al. [41], with the slight modification introduced by Kobus et al. [42]. Measurements were taken three times.

2.5.3. Determination of the Flavonoid Content (TFC)

The flavonoid content was determined according to the methodology described by Aryal [43], with the slight modification introduced by Kobus et al. [42]. The flavonoid content was measured in triplicate.

2.5.4. Determination of the Antioxidant Activity (DPPH and FRAP)

The antioxidant activity of the tested solutions was determined using the DPPH and FRAP reagents. It was determined according to the methodology described by Blois [44] and that by Benzie and Strain [45], with the slight modification introduced by Kobus et al. [46,47]. Measurements of the antioxidant activity were made in three repetitions.

2.6. Statistical Analysis

Statistica software version 12.0 (Statsoft, Tulsa, OK, USA) was used for the statistical analysis. The mean values were compared using analysis of variance (ANOVA) and Tukey's test ($p < 0.05$). Furthermore, Dunnett's test ($p < 0.05$) was also used to compare the control group to the other experimental groups.

3. Results and Discussion

3.1. Basic Physical Properties of the Gluten-Free Bread with the Addition of Hemp Inflorescence

The results indicated that gluten-free bread with added hemp inflorescence had an increased specific volume (Figure 1a). Bread containing 1% and 4% hemp inflorescence had higher volumes than the control gluten-free bread. The increase in specific volume was approximately 24% in the case of the bread with 1% HI. A higher share of HI (2–5%) caused volume changes of 13% to 20%. There were no significant differences between the volumes of the breads with the addition of 2%, 3%, 4%, and 5% hemp inflorescence. As shown in Figure 1b, a clear increase in pH value was observed as the percentage of hemp inflorescence addition increased from 1% to 5%. The highest pH value was 5.55 for the 5% hemp addition, while the lowest value was 5.3 for the 1% hemp inflorescence addition. pH changes in the control sample were observed only in the case of 4% and 5% HI addition, and they were in the range of 2–3.7%.

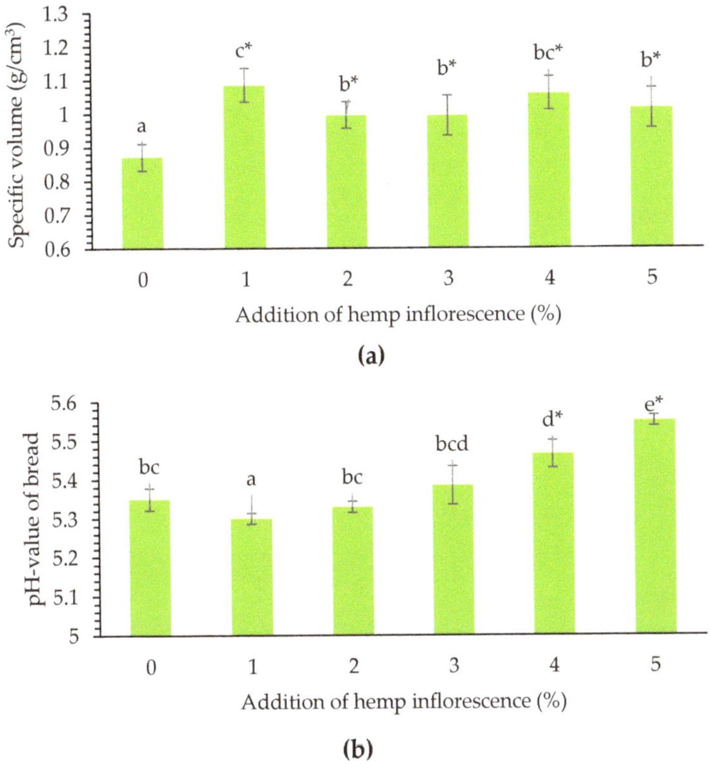

Figure 1. Basic physical properties of gluten-free bread with the addition of hemp inflorescence: (**a**) specific volumes of the breads and (**b**) pH values of the breadcrumbs. The mean values in the same figure marked with different letters were significantly ($p < 0.05$) different, as determined by the Tukey test. The means followed by an asterisk (*) differed from the control sample, as determined by the Dunnett's test ($p < 0.05$).

Thus far, no research has been conducted on the addition of hemp inflorescence to bread; however, other authors have used hemp flour [21–23,48] and discovered that the addition of hemp flour increased the specific volume of bread. Bread with 15% extruded hemp–rice flour had a higher specific volume, lower hardness, and larger pores in the crumb structure according to Wang et al. [49]. Pojic et al. [50] discovered that regardless of the level of substitution, hemp flour affected the water absorption and development time of the dough, and thus, it affected the bread volume, color, and structural and textural properties of the crumb. Hrušková and Švec [48] discovered that the addition of hemp flour reduced the specific volume of wheat bread while increasing the crumb pore density. According to these authors, hemp proteins are primarily composed of low-molecular-weight globulin, and this may have contributed to the volume reduction in the wheat bread.

The addition of HI reduced the crumb lightness significantly, from 75.17 for the control bread to 51.53 for the bread with the highest (5%) share of hemp inflorescence (Table 1). Each amount of HI additive caused a significant change in the lightness of the bread. With a 1% HI addition, the brightness decreased by approximately 13%, and with a 5% HI addition, the brightness decreased by as much as 32%. There were no significant differences in the lightness levels of the breads with 1% and 2% hemp inflorescence additions, and there was a decrease, and then no significant differences, in the lightness levels of the breads with 3% and 4% hemp inflorescence additions. As with the L* index values, the a*, b*, and C* values increased as the amount of HI added to the bread increased. The a-index was significantly different for each HI supplement compared to the control. An addition of 5% increased the value of a* by more than 200%. There were no statistically significant differences between the a* samples containing 1%, 2%, and 4% HI. The control b* value increased by 108% for the 5% hemp bread sample. Figure 2 shows the external appearances and cross-sections of the breads with different hemp inflorescence contents.

Table 1. The colors of the gluten-free breadcrumbs with the addition of hemp inflorescence.

Addition of Hemp Inflorescence (%)	Crumb Color Values					
	L*-Value	a*-Value	b*-Value	C*-Value	h°-Value	ΔE
0	75.17 ± 0.63 d	−0.02 ± 0.03 a	13.59 ± 0.20 a	13.59 ± 0.20 a	90.10 ± 0.12 b	
1	65.83 ± 0.26 c,*	0.79 ± 0.09 b,*	19.39 ± 0.39 b,*	19.41 ± 0.39 b,*	87.68 ± 0.24 a,*	11.0
2	64.98 ± 0.64 c,*	0.86 ± 0.06 b,*	21.34 ± 0.33 c,*	21.36 ± 0.33 c,*	87.69 ± 0.09 a,*	12.8
3	57.78 ± 0.64 b,*	1.64 ± 0.06 c,*	24.95 ± 0.33 d,*	25.01 ± 0.33 d,*	86.24 ± 0.09 a,*	20.8
4	56.35 ± 0.86 b,*	1.79 ± 0.10 c,*	26.54 ± 0.26 e,*	26.60 ± 0.26 e,*	86.15 ± 0.17 a,*	22.9
5	51.53 ± 0.61 a,*	2.14 ± 0.10 d,*	28.22 ± 0.42 f,*	28.31 ± 0.42 f,*	85.66 ± 0.17 a,*	27.9

The a–f values in the same row marked with different letters were significantly different ($\alpha = 0.05$). The means followed by an asterisk (*) differed from the control sample, as determined by the Dunnett's test ($p < 0.05$).

Figure 2. External appearances (**a**) and cross-sections (**b**) of the breads with different hemp inflorescence contents.

The addition of HI to bread reduced the h_o value. It was equal to 90.10 for the control sample and 85.66 for the sample containing 5% HI.

The absolute color difference (ΔE) criterion was also used in the results' elaboration. The ΔE for all the tested samples with HI additions exceeded five, indicating a significant color deviation. All the color parameters changed significantly owing to a significant difference in the color of the hemp inflorescence addition from the rice flour that was used as the main raw material in the recipe. The dried hemp inflorescence had a green-brown color, which influenced the color of the bread significantly. Significant changes and darkening of the bread color were observed in other studies where hemp seed cake and hemp seed grit were used as bread additives (with only 1% of the addition) [26].

When cookies with hemp flour were tested, similar results were obtained [51]. The L* value decreased significantly when the ratio of raw and roasted hemp flour was increased from 0% to 20%.

The crusts and crumb colors of the breads made with extruded hemp–rice flour were significantly darker than the control bread [49]. According to other authors [24], the addition of a hemp product (particularly 15% and 25% hemp flour) as the second auxiliary ingredient of bread results in an interesting color of the crumb due to the content of natural sugars and organic acids.

3.2. Texture and Sensory Analysis of Gluten-Free Bread with the Addition of Hemp Inflorescence

The addition of HI to gluten-free bread slightly affected the hardness of the crumb. A significant reduction in hardness was observed for only the 1–2% of hemp inflorescence (Table 2). Bread hardness is frequently related to the distribution of pores in the bread. The porosity of the crumb increased and the hardness decreased with a smaller addition, i.e., 1–2%; with a larger addition, the bread was more compact and had a hardness similar to the control bread. Overall, the addition of hemp inflorescence may have increased the amount of simple sugars, allowing the yeast to produce more carbon dioxide by creating more pores. A more compact and hard structure was formed with a higher addition of 3–5% hemp inflorescence, which was difficult to overcome by the forming gas bubbles. Compounds found in hemp inflorescence, such as fiber, could bind to rice starch and cause a decrease in hardness once more. The gluten-free bread springiness ranged from 0.74 (for 1% HI addition) to 0.84 (for 4% HI addition). A significant difference from the control sample was only observed for the 4–5% HI addition. In this case, the breadcrumbs' springiness increased by approximately 10%. Bread samples with 1–4% HI added had lower cohesiveness than the control sample; only the addition of 5% HI increased cohesiveness. The addition of 3–4% HI to the control sample increased the chewiness by approximately 7–8%.

Table 2. The textures of the gluten-free breadcrumbs with hemp inflorescence additions.

Addition of Hemp Inflorescence (%)	Hardness (N)	Springiness (−)	Cohesiveness (−)	Chewiness (N)
0	15.17 ± 0.08 ^{b}	0.75 ± 0.01 ^{a}	0.34 ± 0.01 ^{a,b}	3.89 ± 0.17 ^{b}
1	13.11 ± 0.61 ^{a,*}	0.74 ± 0.02 ^{a}	0.33 ± 0.02 ^{a}	3.18 ± 0.22 ^{a,*}
2	13.84 ± 0.89 ^{a,*}	0.77 ± 0.02 ^{a}	0.32 ± 0.01 ^{a}	3.45 ± 0.07 ^{a,b}
3	14.82 ± 0.51 ^{a,b}	0.79 ± 0.02 ^{a,b}	0.33 ± 0.01 ^{a}	3.85 ± 0.16 ^{b}
4	15.76 ± 0.74 ^{b}	0.84 ± 0.04 ^{b,*}	0.32 ± 0.01 ^{a}	4.23 ± 0.18 ^{c,*}
5	14.59 ± 0.52 ^{b}	0.82 ± 0.02 ^{b,*}	0.35 ± 0.01 ^{b}	4.17 ± 0.10 ^{c,*}

The a–c values in the same row marked with different letters were significantly different ($p < 0.05$), as determined by the Tukey test. The means followed by an asterisk (*) differed from the control sample as determined by the Dunnett's test ($p < 0.05$). Abbreviation: SD—standard deviation.

The addition of hemp inflorescence to bread has not been investigated in previous studies. However, when hemp flour was added to wheat bread in the range of 5–20%, a decrease in bread elasticity was observed [25]. The addition of hemp and hemp seed cakes in a proportion greater than 1% caused an increase in hardness and a decrease in elasticity [26].

According to Ertaş and Aslan's research [51], the addition of raw and roasted hemp flour significantly reduced the hardness and softness of wheat biscuits. Other authors [24] have claimed that using a hemp product (especially 15% and 25% hemp flour) as the second auxiliary ingredient of bread improves the structure of the crumb and increases the bread's durability, as well as provides an interesting color of the crumb due to the content of natural sugars and organic acids. Furthermore, the addition of 15% hemp products had the greatest effect on the dough's characteristics, improving crumb quality (porosity and elasticity).

The additions of hemp at 1% and 2% did not significantly alter the taste of the bread, and consumers rated them as the best (Figure 3). The addition of hemp ranging from 3% to 5% significantly reduced the flavor of the gluten-free bread. The bread with 5% hemp added had the worst flavor, leaving a bitter aftertaste when swallowed. The addition of 1% hemp to gluten-free bread improved the taste and smell slightly.

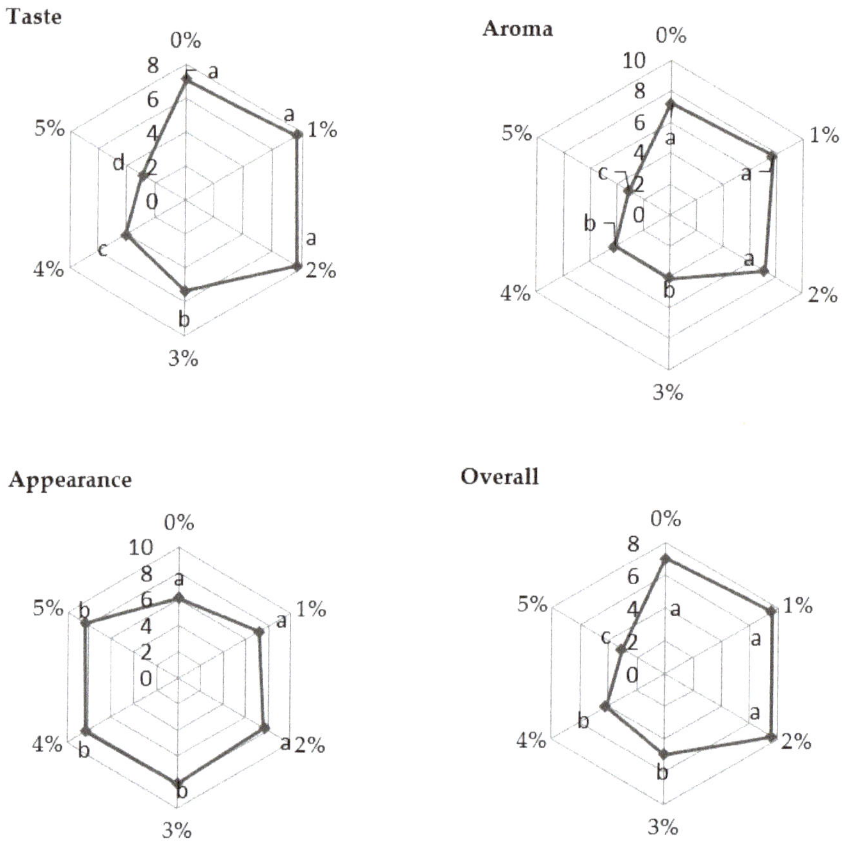

Figure 3. Sensory evaluations of the gluten-free breads with the additions of hemp inflorescence. The mean values in the same figure marked with different letters were significantly different ($\alpha = 0.05$).

The addition of 2% to 5% hemp resulted in a significant deterioration of the bread's aroma. All the hemp-infused bread samples had a better appearance compared to the control sample according to the consumers. In general, the consumers thought that the breads with 1% and 2% hemp were the best; the lower hardness and chewiness of these breads may have influenced the ratings. A lower overall consumer rating was obtained for the bread with the additions of 3%, 4%, and 5% hemp.

Other studies [23] have shown that sensory evaluations of hemp–wheat breads are poor due to the savory flavor and fatty aftertaste. According to Wang et al. [49], adding hemp-rice flour to bread reduced its hardness during storage. However, in the sensory evaluation of biscuits containing up to 20% raw hemp flour and up to 15% roasted hemp flour, they were found to be more desirable by consumers in terms of overall acceptability due to their softness [51]. Wheat flour for bread production can be enriched with functional ingredients by adding partially defatted hemp seed flour. The addition of 5% and 10% partially defatted hemp seed flour resulted in superior sensory and physicochemical values, which were comparable to white wheat bread. The quality of bread obtained with 15% and 20% partially defatted hemp seed flour from P4 and P5 mixture samples (15% and 20% partially defatted hemp seed flour) was comparable to wholegrain bread [52]. The addition of 10% hemp flour improved the sensory properties of sourdough bread, and the authors concluded that the addition of up to 20% hemp flour could be used for bakery products to create an improved nutritional profile [48].

3.3. The Bioactive Compound Content of Gluten-Free Bread with Hemp Inflorescence Addition

The amount of polyphenols in the tested samples was affected by the addition of hemp inflorescence (HI). The polyphenol content increased from 36 to 195% with HI addition in the 1% to 4% tested range of addition (Table 3). There was no statistically significant difference in the polyphenol content when 2% and 3% hemp were added. Similarly, the flavonoid content in the bread with hemp additions compared to the control bread increased from 100% for 1% HI addition to 433% for 5% HI addition (Table 3). All the tested bread samples had statistically significant differences in flavonoid contents. There are numerous studies in the literature on the polyphenol content of hemp. However, a majority of them have focused on the contents of these compounds in seeds, oils, and flours [53–57]. The polyphenol content of the leaves, for example, ranges from 0.09 to 0.56 [54], that of the inflorescences ranges from 3.02 to 9.35 mg QE·g^{-1} [58], while for the seeds it is 51.6 mg GAE·g^{-1} [53,55,56]. According to Izzo's research [14], the inflorescence contains significantly more polyphenols than other parts of the plant (10.5–48.9 mg GAE·g^{-1}) depending on the variety. The polyphenol content in flour ranges between 0.74 and 1.71 mg GAE·g^{-1} [57].

Table 3. The polyphenols, flavonoids, and antioxidant activities of the gluten-free breads with hemp inflorescences added.

Addition of Hemp Inflorescence (%)	TPC (mg GAE·g^{-1} d.m.)	TFC (mg QE·g^{-1} d.m.)	DPPH (μM TE·g^{-1} d.m.)	FRAP (μM TE·g^{-1} d.m.)
0	0.22 ± 0.0 ^{a}	0.03 ± 0.0 ^{a}	1.12 ± 0.08 ^{a}	1.25 ± 0.06 ^{a}
1	0.30 ± 0.01 ^{b,*}	0.06 ± 0.0 ^{b,*}	1.66 ± 0.12 ^{b,*}	1.25 ± 0.18 ^{a}
2	0.41 ± 0.01 ^{c,*}	0.08 ± 0.0 ^{c,*}	2.08 ± 0.16 ^{c,*}	1.60 ± 0.08 ^{b,*}
3	0.44 ± 0.01 ^{c,*}	0.10 ± 0.0 ^{d,*}	2.38 ± 0.06 ^{d,*}	2.09 ± 0.16 ^{c,*}
4	0.58 ± 0.01 ^{d,*}	0.13 ± 0.01 ^{e,*}	2.92 ± 0.02 ^{e,*}	2.27 ± 0.06 ^{c,*}
5	0.65 ± 0.00 ^{e,*}	0.16 ± 0.01 ^{f,*}	3.23 ± 0.06 ^{f,*}	3.00 ± 0.06 ^{d,*}

The a–f values in the same row marked with different letters were significantly different ($\alpha = 0.05$). The means followed by an asterisk (*) differed from the control sample, as determined by the Dunnett's test ($p < 0.05$).

Due to the high polyphenol content of hemp flour, its addition to various types of bakery products helps to strengthen the antioxidant properties of these products. In comparison to the control samples, the addition of raw or roasted hemp flour to wheat flour cookies increased the total phenol content (0.559–1.634 mg GAE·g^{-1}) and antioxidant activity (34.41–40.88%). The TPC values increased with the addition of raw and roasted hemp flour, with the greatest increase in TPC observed for the biscuit samples containing 20% roasted hemp flour (4.03-fold increase in TPC compared to the control sample) [51].

The antioxidant activity in the tested samples, as measured by DPPH and FRAP, increased as the hemp addition increased. The DPPH ranged from 1.66 μM TE·g^{-1} d.m. for 1% hemp addition to 3.23 μM TE·g^{-1} d.m. for 5% hemp addition. Compared to the control sample, with 1% of the additive, there was a 48% increase in the DPPH value, and with 5%, the increase was 188%.

With the addition of 1% and 5% hemp, the antioxidant activity, expressed as FRAP, ranged from 1.25 to 3.00 μM TE·g^{-1} d.m, and so for the highest addition, the value increased by 140%. The antioxidant activity measured by the FRAP test in the bread without the addition of hemp was 1.25 μM TE·g^{-1} d.m (Table 3). There was no statistically significant difference in the antioxidant activity of FRAP between the control bread and the bread with the addition of 1% hemp, as well as between the samples with the additions of 3% and 4% hemp.

4. Conclusions

This study found that hemp inflorescence can be a valuable addition to gluten-free bread made with rice flour. The additive, which was used at concentrations of 1–4%, caused significant changes in the physical properties of the bread. After using 1% of the additive, the volume of the bread increased significantly and the color of the breadcrumb changed significantly. The changes in crumb texture were less clear. Based on the organoleptic evaluation, gluten-free bread with a maximum of 2% HI addition is the most recommended; however, with a higher proportion of hemp inflorescence, the bread has a bitter aftertaste. The addition of hemp inflorescence in the acceptable amount of 2% increased the biological activity of the bread significantly. With this amount of addition, the polyphenol content increased by 86%, the flavonoid content increased by 167%, and the antioxidant activity measured with DPPH increased by 86%, while the activity measured with FRAP increased by 28%.

Author Contributions: Conceptualization, R.R. and A.P.; methodology, R.R. and Z.K.; software, R.R.; validation, R.R. and A.P.; formal analysis, R.R., A.P. and A.B.; investigation, A.P., A.B. and R.R.; resources, A.B.; writing—original draft preparation, A.P., A.B. and R.R; writing—review and editing, A.P., A.B. and R.R; visualization, A.P.; supervision, R.R.; project administration, A.P.; funding acquisition, R.R. All authors have read and agreed to the published version of the manuscript.

Funding: This research received no external funding.

Institutional Review Board Statement: Not applicable.

Informed Consent Statement: Informed consent was obtained from all subjects involved in the study.

Data Availability Statement: Not applicable.

Conflicts of Interest: The authors declare no conflict of interest.

Appendix A

Table A1. Model of experimental parameters.

Ingredients	0	1%	2%	3%	4%	5%
Rice flour (g)	300	297	294	291	288	285
Hemp inflorescence (g)	0	3	6	9	12	15
Water (mL)	360	370	380	390	400	410
Guar gum (g)	6	6	6	6	6	6
Dry instant yeast (g)	4	4	4	4	4	4
Salt (g)	6	6	6	6	6	6

References

1. Rupasinghe, H.P.V.; Davis, A.; Kumar, S.K.; Murray, B.; Zheljazkov, V.D. Industrial Hemp (*Cannabis sativa* subsp. sativa) as an Emerging Source for Value-Added Functional Food Ingredients and Nutraceuticals. *Molecules* **2020**, *25*, 4078. [CrossRef] [PubMed]
2. Spano, M.; Di Matteo, G.; Ingallina, C.; Botta, B.; Quaglio, D.; Ghirga, F.; Balducci, S.; Cammarone, S.; Campiglia, E.; Giusti, A.M.; et al. A Multimethodological Characterization of *Cannabis sativa* L. Inflorescences from Seven Dioecious Cultivars Grown in Italy: The Effect of Different Harvesting Stages. *Molecules* **2021**, *26*, 2912. [CrossRef]
3. Iftikhar, A.; Zafar, U.; Ahmed, W.; Shabbir, M.A.; Sameen, A.; Sahar, A.; Bhat, Z.F.; Kowalczewski, P.Ł.; Jarzębski, M.; Aadil, R.M. Applications of *Cannabis Sativa* L. in Food and Its Therapeutic Potential: From a Prohibited Drug to a Nutritional Supplement. *Molecules* **2021**, *26*, 7699. [CrossRef] [PubMed]
4. Ranalli, P.M.; Di Candilo, G.; Mandolino, G.; Grassi, A.; Carboni. Hemp for sustainable agricultural systems. *Agro-Food Ind. Hi-Tech.* **1999**, *2*, 33–3899.
5. Nissen, L.; Casciano, F.; Babini, E.; Gianotti, A. *Chapter 10—Industrial Hemp Foods and Beverages and Product Properties*; Industrial Hemp; Academic Press: Cambridge, MA, USA, 2022; pp. 219–246.
6. Callaway, J.C. Hempseed as a nutritional resource: An overview. *Euphytica* **1999**, *140*, 65–72. [CrossRef]
7. Ranalli, P.; Venturi, G. Hemp as a raw material for industrial applications. *Euphytica* **2004**, *140*, 1–5. [CrossRef]
8. Crini, G.; Lichtfouse, E.; Chanet, G.; Morin-Crini, N. Applications of hemp in textiles, paper industry, insulation and building materials, horticulture, animal nutrition, food and beverages, nutraceuticals, cosmetics and hygiene, medicine, agrochemistry, energy production and environment: A review. *Environ Chem. Lett.* **2020**, *18*, 1451–1476. [CrossRef]
9. Lewis-Bakker, M.M.; Yang, Y.; Vyawahare, R.; Kotra, L.P. Extractions of Medical Cannabis Cultivars and the Role of Decarboxylation in Optimal Receptor Responses. *Cannabis Cannabinoid Res.* **2019**, *4*, 183–194. [CrossRef]

10. Fiorini, D.; Molle, A.; Nabissi, M.; Santini, G.; Benelli, G.; Maggi, F. Valorizing industrial hemp (*Cannabis sativa* L.) by-products: Cannabidiol enrichment in the inflorescence essential oil optimizing sample pre-treatment prior to distillation. *Ind. Crops Prod.* **2019**, *128*, 581–589. [CrossRef]
11. Jin, D.; Dai, K.; Xie, Z.; Chen, J. Secondary Metabolites Profiled in Cannabis Inflorescences, Leaves, Stem Barks, and Roots for Medicinal Purposes. *Sci. Rep.* **2020**, *10*, 3309. [CrossRef]
12. Spano, M.; Di Matteo, G.; Ingallina, C.; Sobolev, A.P.; Giusti, A.M.; Vinci, G.; Cammarone, S.; Tortora, C.; Lamelza, L.; Prencipe, S.A.; et al. Industrial Hemp (*Cannabis sativa* L.) Inflorescences as Novel Food: The Effect of Different Agronomical Practices on Chemical Profile. *Foods* **2022**, *11*, 3658. [CrossRef]
13. Nagy, D.U.; Cianfaglione, K.; Maggi, F.; Sut, S.; Dall'Acqua, S. Chemical Characterization of Leaves, Male and Female Flowers from Spontaneous Cannabis (*Cannabis sativa* L.) Growing in Hungary. *Chem Biodivers* **2019**, *9*, 90.
14. Izzo, L.; Castaldo, L.; Narváez, A.; Graziani, G.; Gaspari, A.; Rodríguez-Carrasco, Y.; Ritieni, A. Analysis of Phenolic Compounds in Commercial *Cannabis sativa* L. Inflorescences Using UHPLC-Q-Orbitrap HRMS. *Molecules* **2020**, *25*, 631. [CrossRef] [PubMed]
15. Giupponi, L.; Leoni, V.; Pavlovic, R.; Giorgi, A. Influence of Altitude on Phytochemical Composition of Hemp Inflorescence: A Metabolomic Approach. *Molecules* **2020**, *25*, 1381. [CrossRef] [PubMed]
16. Hanuš, L.O.; Meyer, S.M.; Muñoz, E.; Taglialatela-Scafati, O.; Appendino, G. Phytocannabinoids: A unified critical inventory. *Nat. Prod. Rep.* **2016**, *33*, 1357–1392. [CrossRef]
17. Da Porto, C.; Decorti, D.; Natolino, A. Separation of aroma compounds from industrial hemp inflorescences (*Cannabis sativa* L.) by supercritical CO2 extraction and on-line fractionation. *Ind. Crops Prod.* **2014**, *58*, 99–103. [CrossRef]
18. Bertoli, A.; Tozzi, S.; Pistelli, L.; Angelini, L.G. Fibre hemp inflorescences: From crop-residues to essential oil production. *Ind. Crops Prod.* **2010**, *32*, 329–337. [CrossRef]
19. Bakro, F.; Jedryczka, M.; Wielgusz, K.; Sgorbini, B.; Inchingolo, R.; Cardenia, V. Simultaneous determination of terpenes and cannabidiol in hemp (*Cannabis sativa* L.) by fast gas chromatography with flame ionization detection. *J. Sep. Sci.* **2020**, *43*, 2817–2826. [CrossRef]
20. Mikulec, A.; Kowalski, S.; Sabat, R.; Skoczylas, Ł.; Tabaszewska, M.; Wywrocka-Gurgul, A. Hemp flour as a valuable component for enriching physicochemical and antioxidant properties of wheat bread. *LWT* **2019**, *102*, 64–172. [CrossRef]
21. Švec, I.; Hrušková, M. Properties and nutritional value of wheat bread enriched by hemp products. *Potravin. Slovak J. Food Sci.* **2015**, *9*, 304–308. [CrossRef]
22. Švec, I.; Hrušková, M.; Jurinová, I. Technological and nutritional aspect of different hemp types addition: Comparison of flour and whole meal effect. *Croat. J. Food Sci. Technol.* **2015**, *7*, 68–75. [CrossRef]
23. Hrušková, M.; Švec, I.; Jurinová, I. Composite Flours-Characteristics of Wheat/Hemp and Wheat/Teff Models. *Food Sci. Nutr.* **2012**, *3*, 1484–1490. [CrossRef]
24. Bădărău, C.L.; Apostol, L.; Mihăilă, L. Effects of Hemp Flour, Seeds and Oil Additions on Bread Quality. *Int. J. Eng. Res. Appl.* **2018**, *8*, 73–78.
25. Rusu, I.E.; Marc (Vlaic), R.A.; Mureşan, C.C.; Mureşan, A.E.; Mureşan, V.; Pop, C.R.; Chiş, M.S.; Man, S.M.; Filip, M.R.; Onica, B.-M.; et al. Hemp (*Cannabis sativa* L.) Flour-Based Wheat Bread as Fortified Bakery Product. *Plants* **2021**, *10*, 1558. [CrossRef] [PubMed]
26. Wiedemair, V.; Gruber, K.; Knöpfle, N.; Bach, K.E. Technological Changes in Wheat-Based Breads Enriched with Hemp Seed Press Cakes and Hemp Seed Grit. *Molecules* **2022**, *27*, 1840. [CrossRef] [PubMed]
27. Jagelaviciute, J.; Cizeikiene, D. The influence of non-traditional sourdough made with quinoa, hemp and chia flour on the characteristics of gluten-free maize/rice bread. *LWT* **2021**, *137*, 110457. [CrossRef]
28. Hayward, L.; McSweeney, M.B. Acceptability of bread made with hemp (*Cannabis sativa* subsp. sativa) flour evaluated fresh and following a partial bake method. *J. Food Sci* **2020**, *85*, 2915–2922. [CrossRef]
29. Korus, J.; Witczak, M.; Ziobro, R.; Juszczak, L. Hemp (*Cannabis sativa* subsp. sativa) flour and protein preparation as natural nutrients and structure forming agents in starch based gluten-free bread. *LWT* **2017**, *84*, 143–150. [CrossRef]
30. Bugarín, R.; Gómez, M. Can Citrus Fiber Improve the Quality of Gluten-Free Breads? *Foods* **2023**, *12*, 1357. [CrossRef]
31. Aguiar, E.V.; Santos, F.G.; Krupa-Kozak, U.; Capriles, V.D. Nutritional facts regarding commercially available gluten-free bread worldwide: Recent advances and future challenges. *Crit. Rev. Food Sci. Nutr.* **2023**, *63*, 693–705. [CrossRef] [PubMed]
32. Łopusiewicz, Ł.; Kowalczewski, P.Ł.; Baranowska, H.M.; Masewicz, Ł.; Amarowicz, R.; Krupa-Kozak, U. Effect of Flaxseed Oil Cake Extract on the Microbial Quality, Texture and Shelf Life of Gluten-Free Bread. *Foods* **2023**, *12*, 595. [CrossRef] [PubMed]
33. Rybicka, I.; Doba, K.; Bińczak, O. Improving the sensory and nutritional value of gluten-free bread. *Int. J. Food Sci. Technol.* **2019**, *54*, 2661–2667. [CrossRef]
34. Cantero, L.; Salmerón, J.; Miranda, J.; Larretxi, I.; Fernández-Gil, M.d.P.; Bustamante, M.Á.; Matias, S.; Navarro, V.; Simón, E.; Martínez, O. Performance of Apple Pomace for Gluten-Free Bread Manufacture: Effect on Physicochemical Characteristics and Nutritional Value. *Appl. Sci.* **2022**, *12*, 5934. [CrossRef]
35. Różańska, M.B.; Kokolus, P.; Królak, J.; Jankowska, P.; Osoś, A.; Romanowska, M.; Szala, Ł.; Kowalczewski, P.Ł.; Lewandowicz, J.; Masewicz, Ł.; et al. Insight into the Gluten-Free Dough and Bread Properties Obtained from Extruded Rice Flour: Physicochemical, Mechanical, and Molecular Studies. *Appl. Sci.* **2023**, *13*, 4033. [CrossRef]
36. Różyło, R.; Rudy, S.; Krzykowski, A.; Dziki, D. Novel Application of Freeze-Dried Amaranth Sourdough in Gluten-Free Bread Production. *J. Food Process Eng.* **2015**, *38*, 135–143. [CrossRef]

37. Ziemichód, A.; Wójcik, M.; Różyło, R. Seeds of Plantago psyllium and Plantago ovata: Mineral composition, grinding, and use for gluten-free bread as substitutes for hydrocolloids. *J. Food Process Eng.* **2019**, *42*, e12931. [CrossRef]
38. Różyło, R.; Piekut, J.; Wójcik, M.; Kozłowicz, K.; Smolewska, M.; Krajewska, M.; Szmigielski, M.; Bourekoua, H. Black cumin pressing waste material as a functional additive for starch bread. *Materials* **2021**, *14*, 4560. [CrossRef]
39. Różyło, R.; Wójcik, M.; Dziki, D.; Biernacka, B.; Cacak-Pietrzak, G.; Gawłowski, S.; Zdybel, A. Freeze-dried elderberry and chokeberry as natural colorants for gluten-free wafer sheets. *Int. Agroph.* **2019**, *33*, 217–225. [CrossRef]
40. Lim, H.S.; Park, S.H.; Ghafoor, K.; Hwang, S.Y.; Park, J. Quality and antioxidant properties of bread containing turmeric (*Curcuma longa* L.) cultivated in South Korea. *Food Chem.* **2011**, *124*, 1577–1582. [CrossRef]
41. Singleton, V.L.; Orthofer, R.; Lamuela-Raventós, R.M. Analysis of total phenols and other oxidation substrates and antioxidants by means of folin-ciocalteu reagent. *Methods Enzymol.* **1999**, *299*, 152–178.
42. Kobus, Z.; Krzywicka, M.; Pecyna, A.; Buczaj, A. Process efficiency and energy consumption during the ultrasound-assisted extraction of bioactive substances from hawthorn berries. *Energies* **2021**, *14*, 7638. [CrossRef]
43. Aryal, S.; Baniya, M.K.; Danekhu, K.; Kunwar, P.; Gurung, R.; Koirala, N. Total Phenolic Content, Flavonoid Content and Antioxidant Potential of Wild Vegetables from Western Nepal. *Plants* **2019**, *8*, 96. [CrossRef] [PubMed]
44. Blois, M.S. Antioxidant determinations by the use of a stable free radical. *Nature* **1958**, *181*, 1199–1200. [CrossRef]
45. Benzie, I.; Strain, J. The ferric reducing ability of plasma (FRAP) as a measure of "Antioxidant Power: The FRAP Assay". *Anal. Biochem.* **1996**, *239*, 70–76. [CrossRef]
46. Kobus, Z.; Buczaj, A.; Pecyna, A.; Kapica, J.; Findura, P.; Kocira, S. Application of Response Surface Method in Pulsed Ultrasound-Assisted Extraction of Complex Plant Materials—A Case Study on *Cannabis sativa* L. *Appl. Sci.* **2023**, *13*, 760. [CrossRef]
47. Kobus, Z.; Krzywicka, M.; Starek-Wójcicka, A.; Sagan, A. Effect of the duty cycle of the ultrasonic processor on the efficiency of extraction of phenolic compounds from Sorbus intermedia. *Sci. Rep.* **2022**, *12*, 8311. [CrossRef]
48. Hrušková, M.; Švec, I. Hemp products for food and medicine using. In Proceedings of the 8th International Congress FLOUR-BREAD 2015—10th Croatian Congress of Cereal Technologists, Opatija, Croatia, 28–30 October 2015; pp. 106–112.
49. Wang, Y.Y.; Norajit, K.; Ryu, G.H. Influence of Extruded Hemp-Rice Flour Addition on the Physical Properties of Wheat Bread. *Prev. Nutr. Food Sci.* **2011**, *16*, 62–66. [CrossRef]
50. Pojić, M.; Dapčević Hadnađev, T.; Hadnađev, M.; Rakita, S.; Brlek, T. Hemp Seed Cake in Bread Making. *J. Food Qual.* **2015**, *38*, 431–440. [CrossRef]
51. Ertaş, N.; Aslan, M. Antioxidant and physicochemical properties of cookies containing raw and roasted hemp flour. *Acta Sci. Pol. Technol. Aliment.* **2020**, *19*, 177–184.
52. Apostol, L.; Popa, M.; Mustatea, G. *Cannabis sativa* L. partially skimmed flour as source of bio-compounds in the bakery industry. *Rom. Biotechnol. Lett.* **2015**, *20*, 10835–10844.
53. Frassinetti, S.; Moccia, E.; Caltavuturo, L.; Gabriele, M.; Longo, V.; Bellani, L.; Giorgi, G.; Giorgetti, L. Nutraceutical potential of hemp (*Cannabis sativa* L.) seeds and sprouts. *Food Chem.* **2018**, *262*, 56–66. [CrossRef] [PubMed]
54. Mkpenie, V.; Essien, E.; Udoh, I. Effect of extraction conditions on total polyphenol contents, antioxidant and antimicrobial activities of *Cannabis sativa* L. Electron. *J. Environ. Agric. Food Chem.* **2012**, *11*, 300–307.
55. Siano, F.; Moccia, S.; Picariello, G.; Russo, G.L.; Sorrentino, G.; Di Stasio, M.; La Cara, F.; Volpe, M.G. Comparative Study of Chemical, Biochemical Characteristic and ATR-FTIR Analysis of Seeds, Oil and Flour of the Edible Fedora Cultivar Hemp (*Cannabis sativa* L.). *Molecules* **2018**, *24*, 83. [CrossRef]
56. Vonapartis, E.; Aubin, M.-P.; Seguin, P.; Mustafa, A.F.; Charron, J.-B. Seed composition of ten industrial hemp cultivars approved for production in Canada. *J. Food Compos. Anal.* **2015**, *39*, 8–12. [CrossRef]
57. Moccia, S.; Siano, F.; Russo, G.L.; Volpe, M.G.; La Cara, F.; Pacifico, S.; Piccolella, S.; Picariello, G. Antiproliferative and antioxidant effect of polar hemp extracts (*Cannabis sativa* L., Fedora cv.) in human colorectal cell lines. *Int. J. Food Sci. Nutr.* **2020**, *71*, 410–423. [CrossRef] [PubMed]
58. Kobus, Z.; Pecyna, A.; Buczaj, A.; Krzywicka, M.; Przywara, A.; Nadulski, R. Optimization of the Ultrasound-Assisted Extraction of Bioactive Compounds from *Cannabis sativa* L. Leaves and Inflorescences Using Response Surface Methodology. *Appl. Sci.* **2022**, *12*, 6747. [CrossRef]

MDPI
St. Alban-Anlage 66
4052 Basel
Switzerland
www.mdpi.com

Applied Sciences Editorial Office
E-mail: applsci@mdpi.com
www.mdpi.com/journal/applsci

www.ingramcontent.com/pod-product-compliance
Lightning Source LLC
Chambersburg PA
CBHW040759250926
51850CB00007BA/1031
* 9 7 8 3 0 3 6 5 8 5 5 4 3 *